Socio-Cognitive and Affective Computing

Socio-Cognitive and Affective Computing

Special Issue Editors

Antonio Fernández-Caballero
Pascual González
María Teresa López
Elena Navarro

MDPI • Basel • Beijing • Wuhan • Barcelona • Belgrade

MDPI

Special Issue Editors

Antonio Fernández-Caballero
Universidad de Castilla-La Mancha
Spain

Pascual González
Universidad de Castilla-La Mancha
Spain

María Teresa López
Universidad de Castilla-La Mancha
Spain

Elena Navarro
Universidad de Castilla-La Mancha
Spain

Editorial Office
MDPI
St. Alban-Anlage 66
Basel, Switzerland

This is a reprint of articles from the Special Issue published online in the open access journal *Applied Sciences* (ISSN 2076-3417) from 2017 to 2018 (available at: http://www.mdpi.com/journal/applsci/special_issues/Socio_Cognitive_and)

For citation purposes, cite each article independently as indicated on the article page online and as indicated below:

LastName, A.A.; LastName, B.B.; LastName, C.C. Article Title. *Journal Name* **Year**, *Article Number*, Page Range.

ISBN 978-3-03897-198-6 (Pbk)
ISBN 978-3-03897-199-3 (PDF)

Contents

About the Special Issue Editors

Antonio Fernández-Caballero received his M.Sc. Degree in Computer Science from Universidad Politécnica de Madrid, Spain, and he obtained his Ph.D. Degree from the Department of Artificial Intelligence of Universidad Nacional de Educación a Distancia, Spain. He is Full Professor at Universidad de Castilla-La Mancha, Albacete, Spain. He is head of the natural and artificial Interaction Systems (n&aIS) team, belonging to the Laboratory on User Interfaces and Software Engineering (LoUISE) in the Computer Science Research Institute at Albacete (i3A) since 2001. Among his research interests are Computer Vision, Pattern Recognition, Human—Machine Interaction, Affective Computing, Multiagent Systems and Mobile Robots. He has authored over 350 scientific contributions. He is Topic Editor-in-Chief of the International Journal of Advanced Robotic Systems for Vision Systems. He is Associate Editor of Pattern Recognition Letters, and Specialty Chief Editor for Vision Systems Theory, Tools and Applications in Frontiers in ICT, and Frontiers in Robotics and AI. He is Guest Editor of a number of Special Issues of leading international journals.

Pascual González received his M.Sc. Degree and Ph.D. in Computer Science from Technical University of Madrid (Spain) in 1986 and 1999 respectively. Since 1991, he has been lecturer and researcher at the University of Castilla-La Mancha, where he lectures on Software Engineering and Human—Computer Interaction. He is founder and head of the Laboratory on User Interfaces and Software Engineering (LoUISE) research group and he has been principal investigator of several research projects. He has advised 9 Ph.D. students and he is author of more than 120 papers in high-impact international conferences and journals indexed in JCR. He has also been co-editor of several Special Issues in different JCR journals (e.g., Pervasive and Mobile Computing, Cognitive Computation, etc.). His current research topics include Adaptive User Interfaces Design, Affective Computing, Collaborative systems, Post-WIMP systems and Software Engineering applied to tele-rehabilitation interactive systems design.

María Teresa López s an Associate Professor of Computer Science at the University of Castilla-La Mancha (Spain). She received her degree in Physics from the University of Valencia in 1991. She obtained her PHD in Computer Engineering in 2004 from the National University of Distance Education. She is a member of the Natural and Artificial Interaction Systems section of the research group User Interaction Laboratory and Software Engineering (Louise). Her main lines of research are Selective Attention, Multisensory Fusion and Advanced Visual Surveillance. She has authored 50 international scientific contributions since 1995, including 28 articles published in high-impact international journals.

Elena Navarro is an Associate Professor of Computer Science at the University of Castilla-La Mancha (Spain). She received her bachelor's degree and PhD at the University of Castilla-La Mancha, and her master's degree at the University of Murcia (Spain). She is currently an active collaborator of the Louise group of the University of Castilla-La Mancha (Spain). Her current research interests are requirements engineering, software architectures, model-driven development, and aspect-oriented development.

applied
sciences

MDPI

Editorial

Special Issue on Socio-Cognitive and Affective Computing

Antonio Fernández-Caballero [1,2,*], Pascual González [1,2], María T. López [1] and Elena Navarro [1,2]

1   Departamento de Sistemas Informáticos, Universidad de Castilla-La Mancha, 02071 Albacete, Spain;
     Pascual.Gonzalez@uclm.es (P.G.); Maria.LBonal@uclm.es (M.T.L.); Elena.Navarro@uclm.es (E.N.)
2   Centro de Investigación Biomédica en Red de Salud Mental (CIBERSAM), 28029 Madrid, Spain
*   Correspondence: Antonio.Fdez@uclm.es

Received: 2 August 2018; Accepted: 14 August 2018; Published: 15 August 2018

Abstract: Social cognition focuses on how people process, store, and apply information about other people and social situations. [...]

1. Introduction

Social cognition focuses on how people process, store, and apply information about other people and social situations. It focuses on the role that cognitive processes play in our social interactions [1]. On the other hand, the term cognitive computing is generally used to refer to new hardware and/or software that mimics the functioning of the human brain and helps to improve human decision-making [2]. In this sense, it is a type of computing with the goal of discovering more accurate models of how the human brain/mind senses, reasons, and responds to stimuli.

Thus, Socio-Cognitive Computing should be understood as a set of theoretical interdisciplinary frameworks, methodologies, methods and hardware/software tools to model how the human brain mediates social interactions. In addition, Affective Computing is the study and development of systems and devices that can recognize, interpret, process, and simulate human affects, a fundamental aspect of socio-cognitive neuroscience [3]. It is an interdisciplinary field spanning computer science, electrical engineering, psychology, and cognitive science.

Moreover, Physiological Computing is a category of technology in which electrophysiological data recorded directly from human activity are used to interface with a computing device [4]. This technology becomes even more relevant when computing can be integrated pervasively in everyday life environments [5]. Thus, Socio-Cognitive and Affective Computing systems should be able to adapt their behavior according to the Physiological Computing paradigm.

This Special Issue on Socio-Cognitive and Affective Computing aimed at integrating these various albeit complementary fields. Proposals from researchers who use signals from the brain and/or body to infer people's intentions and psychological state in smart computing systems were welcome. Designing this kind of system requires combining knowledge and methods of ubiquitous and pervasive computing, as well as physiological data measurement and processing, with those of socio-cognitive and affective computing. Papers with a special focus on multidisciplinary approaches and multimodality were especially welcome.

2. The Papers

A total of 41 papers were submitted to the Special Issue on Socio-Cognitive and Affective Computing. After peer review, 13 of them were accepted and published.

The first article addresses the issue of mapping between the models of affect representation to enhance both the accuracy and reliability of emotion recognition [6]. The paper proposes a procedure to elaborate new mappings, recommends a set of metrics for evaluation of the mapping accuracy, and delivers new mapping matrices for estimating the dimensions of a Pleasure–Arousal–Dominance model from Ekman's six basic emotions. The results are based on an analysis using three datasets that were constructed based on affect-annotated lexicons. The procedure, as well as the proposed metrics, might be used, not only in evaluation of the mappings between representation models, but also in comparison of emotion recognition and annotation results.

Two additional papers are centered on the user experience and frustration. The first focuses on the learning process of users as an essential cue in the construction of new knowledge when pursuing improvements in user experience [7]. In this paper, the interruption factor is considered in relation to interaction quality due to human–computer interaction (HCI) being seen to affect the learning process. The results obtained from 500 users in an interactive museum in Tijuana, Mexico are presented as a case study. The authors conclude that managing interruptions can enhance the HCI, producing a positive learning process that influences user experience. The second paper investigates how to mitigate user frustration and support student learning through changes in the style in which a computer tutor interacts with a learner [8]. The study examines an approach to mitigate frustration using different etiquette strategies to change the amount of imposition feedback placed on the learner. Changing etiquette strategies from one math problem to the next led to changes in motivation, confidence satisfaction, and performance. The most effective etiquette strategies changed depending on whether the user was frustrated or not.

Next, a series of four papers introduces several devices that enable capturing physiological information from the user. Firstly, an analysis of the effects of the display type (2D vs. 3D) and viewing distance on visual fatigue during a 60-min viewing session based on electroencephalogram relative beta power, and alpha/beta power ratio is provided [9]. In this study, 20 male participants watch four videos. When the viewing distance is increased the visual fatigue is decreased in the case of the 3D display, whereas the fatigue is increased in the case of the 2D display. Moreover, there is approximately the same level of visual fatigue when watching videos in 2D or 3D from a long viewing distance. Another paper introduces the design of a new wearable photo plethysmography sensor and its assessment for mental distress estimation [10]. Temporal, morphological, and frequency markers are computed using time intervals between adjacent normal cardiac cycles to characterize pulse rate variability. The results show that temporal features present a single discriminant power between emotional states of calm and stress. Moreover, a discriminant tree-based model is used to assess the possible underlying relationship among parameters. A third paper describes a study carried out on a group of 35 participants to determine which methods and tools to recognize the emotions of computer users can be used during programming [11]. During the study, data from multiple sensors that are commonly used in methods of emotional recognition are collected. The participants were extensively questioned about the sensors' invasiveness during programming. Moreover, three methods for inducing negative emotions are proposed, and their effectiveness is verified. Lastly, a model that promotes the well-being of the elderly in their homes is presented [12]. The key is that there is no device that provides a more natural interaction than a human body; every one of us sends and receives useful information, which sometimes gets lost. Trends show that the future will be filled with pervasive IoT devices, present in most aspects of human life. We will focus on those aspects that are most important for human well-being and the devices, technologies, and interactions that may be used to collect data directly from users and measure their physiological and emotional responses.

The following paper aims at uncovering hidden communities in social networks by using a parallel programming framework like MapReduce with the objective of improving the efficiency of algorithms [13]. In this paper, nodes are mapped into communities based on the random walk in the network. The proposed approach is compared with some standard existing community detection algorithms for both synthetic and real-world datasets to examine its performance, and it is observed that the proposed algorithm is more efficient than the existing ones.

Finally, five papers have focused their research on machine learning techniques. In the first place, a paper proposes Spark based Single Graph Mining, a Spark-based parallel frequent subgraph mining algorithm in a single large graph [14]. Extensive experiments with four different real-world datasets demonstrate that the proposed algorithm outperforms the existing Graph Mining algorithm by an order of magnitude for all datasets and can work with a lower support threshold. Then, a deep learning architecture, based on long short-term memory networks that model the inter-activity behavior in urban population, is presented [15]. The paper is related to the City4Age H2020 project that is working on the early detection of the risks related to mild cognitive impairment and frailty and on providing meaningful interventions that prevent these risks. The architecture offers a probabilistic model that allows us to predict the user's next actions and to identify anomalous user behaviors. The third article proposes a personalized path decision algorithm that is based on user habits for path planning of self-driving travel [16]. Results show that the algorithm can meet the personalized requirements of the user path selection in the path decision. Fourth, a novel approach to enhance the Feature Engineering and Selection (eFES) optimization process in machine learning is introduced [17]. eFES is built using a unique scheme to regulate error bounds and parallelize the addition and removal of a feature during training. eFES also invents local gain and global gain functions using 3D visualizing techniques to assist the feature grouping function. Results show the promising state of eFES as compared to the traditional feature selection process. Lastly, an inductive transfer learning-based framework (Uncertainty Flow) is put forward to allow knowledge transfer from a single-labeled emotion recognition task to a multi-label affective recognition task to lower the single-label dependency on affective facial analysis [18]. The authors demonstrate that Uncertainty Flow in multi-label facial expression analysis exhibits superiority to conventional multi-label learning algorithms and multi-label compatible neural networks.

Acknowledgments: This issue would not be possible without the contributions of the authors who submitted their valuable papers. We would like to thank all reviewers and the editorial team of *Applied Sciences* for their great work. Lastly, we acknowledge the financial support from the Spanish Ministerio de Ciencia, Innovación y Universidades, Agencia Estatal de Investigación (AEI)/European Regional Development Fund (FEDER, EU) under DPI2016-80894-R and TIN2015-72931-EXP grants, as well as from Centro de Investigación Biomédica en Red de Salud Mental (CIBERSAM) of the Instituto de Salud Carlos III.

Conflicts of Interest: The authors declare no conflict of interest.

References

1. Frith, C.D. Social cognition. *Philos. Trans. R. Soc. Lond. B Biol. Sci.* **2008**, *363*, 2033–2039. [CrossRef] [PubMed]
2. Modha, D.S.; Ananthanarayanan, R.; Esser, S.K.; Ndirango, A.; Sherbondy, A.; Singh, R. Cognitive computing. *Commun. ACM* **2011**, *54*, 62–71. [CrossRef]
3. Picard, R.W. *Affective Computing*; The MIT Press: Cambridge, MA, USA, 2000.
4. Fairclough, S.; Gilleade, K. *Advances in Physiological Computing*; Springer: Dordrecht, The Netherlands, 2014.
5. Fernández-Caballero, A.; González, P.; Navarro, E.; Cook, D.J. Pervasive computing for gerontechnology. *Pervasive Mob. Comput.* **2017**, *34*, 1–2. [CrossRef]
6. Landowska, A. Towards new mappings between emotion representation models. *Appl. Sci.* **2018**, *8*, 274. [CrossRef]
7. Rosales, R.; Castañón-Puga, M.; Lara-Rosano, F.; Evans, R.D.; Osuna-Millan, N.; Flores-Ortiz, M.V. Modelling the interruption on HCI using BDI agents with the fuzzy perceptions approach: An interactive museum case study in Mexico. *Appl. Sci.* **2017**, *7*, 832. [CrossRef]
8. Yang, E.; Dorneich, M.C. Evaluating human-automation etiquette strategies to mitigate user frustration and improve learning in affect-aware tutoring. *Appl. Sci.* **2018**, *8*, 895. [CrossRef]
9. Ramadan, M.Z.; Alhaag, M.H.; Abidi, M.H. Effects of viewing displays from different distances on human visual system. *Appl. Sci.* **2017**, *7*, 1153. [CrossRef]
10. Zangróniz, R.; Martínez-Rodrigo, A.; López, M.T.; Pastor, J.M.; Fernández-Caballero, A. Estimation of mental distress from photoplethysmography. *Appl. Sci.* **2018**, *8*, 69. [CrossRef]
11. Wrobel, M.R. Applicability of emotion recognition and induction methods to study the behavior of programmers. *Appl. Sci.* **2018**, *8*, 323. [CrossRef]
12. Rodrigues, N.; Pereira, A. A user-centred well-being home for the elderly. *Appl. Sci.* **2018**, *8*, 850. [CrossRef]

13. Behera, R.K.; Rath, S.K.; Misra, S.; Damasevicius, R.; Maskeliunas, R. Large scale community detection using a small world model. *Appl. Sci.* **2017**, *7*, 1173. [CrossRef]

14. Qiao, F.; Zhang, X.; Li, O.; Ding, Z.; Jia, S.; Wang, H. A parallel approach for frequent subgraph mining in a single large graph using Spark. *Appl. Sci.* **2018**, *8*, 230. [CrossRef]

15. Almeida, A.; Azkune, G. Predicting human behaviour with recurrent neural networks. *Appl. Sci.* **2018**, *8*, 305. [CrossRef]

16. Chen, P.; Zhang, X.; Chen, X.; Liu, M. Path planning strategy for vehicle navigation based on user habits. *Appl. Sci.* **2018**, *8*, 407. [CrossRef]

17. Uddin, M.F.; Lee, J.; Rizvi, S.; Hamada, S. Proposing enhanced feature engineering and a selection model for machine learning processes. *Appl. Sci.* **2018**, *8*, 646. [CrossRef]

18. Bai, W.; Quan, C.; Luo, Z. Uncertainty flow facilitates zero-shot multi-label learning in affective facial analysis. *Appl. Sci.* **2018**, *8*, 300. [CrossRef]

applied
sciences

MDPI

Article

Towards New Mappings between Emotion Representation Models

Agnieszka Landowska

Department of Software Engineering, Faculty of Electronics, Telecommunications and Informatics,
Gdansk University of Technology, 80-233 Gdansk, Poland; nailie@pg.edu.pl; Tel.: +48-58-347-2989

Received: 24 November 2017; Accepted: 9 February 2018; Published: 12 February 2018

Featured Application: (1) When you need emotions described in one representation model and get results in another form from an affect recognition system; (2) In a late fusion of hypotheses on affect from diverse algorithms (in multimodal emotion recognition); (3) In an evaluation of mappings between emotion representation models.

Abstract: There are several models for representing emotions in affect-aware applications, and available emotion recognition solutions provide results using diverse emotion models. As multimodal fusion is beneficial in terms of both accuracy and reliability of emotion recognition, one of the challenges is mapping between the models of affect representation. This paper addresses this issue by: proposing a procedure to elaborate new mappings, recommending a set of metrics for evaluation of the mapping accuracy, and delivering new mapping matrices for estimating the dimensions of a Pleasure-Arousal-Dominance model from Ekman's six basic emotions. The results are based on an analysis using three datasets that were constructed based on affect-annotated lexicons. The new mappings were obtained with linear regression learning methods. The proposed mappings showed better results on the datasets in comparison with the state-of-the-art matrix. The procedure, as well as the proposed metrics, might be used, not only in evaluation of the mappings between representation models, but also in comparison of emotion recognition and annotation results. Moreover, the datasets are published along with the paper and new mappings might be created and evaluated using the proposed methods. The study results might be interesting for both researchers and developers, who aim to extend their software solutions with affect recognition techniques.

Keywords: affective computing; emotion recognition; emotion representation models; emotion mapping; Ekman's six basic emotions; Pleasure-Arousal-Dominance model

1. Introduction

This paper concerns one of the challenges in automatic multimodal affect recognition, i.e., mapping between emotion representation models. There are numerous emotion recognition algorithms that differ on input information channels, output labels, and representation models and classification methods. The most frequently used emotion recognition techniques that might be considered when designing an emotion monitoring solution include: facial expression analysis, audio (voice) signal analysis in terms of modulation, textual input analysis, physiological signals and behavioral patterns analysis. As literature on emotion recognition methods is very broad and has already been summarized several times, for an extensive bibliography, one may refer to Gunes and Piccardi [1] or Zeng et al. [2].

Hupont et al. claim that multimodal fusion improves robustness and accuracy of human emotion analysis. They observed that current solutions mostly use one input channel only and integration methods are regarded as ad-hoc [3]. Late fusion combines the classification results provided by separate classifiers for every input channel; however, this requires some mapping between emotion representation models used as classifier outputs [3]. Differences in emotion representation models used by emotion recognition solutions are among the key challenges in fusing affect from the input channels.

This paper concentrates on the challenge of mapping between emotion representation models. The purpose of the paper is to propose a method (including a set of metrics) for an evaluation of the mapping accuracy, as well as to propose a new mapping. The main research question addressed in this study is as follows: How to compare results from multiple emotion recognition algorithms, especially when they are provided in different affect representation models?

The studies related to this research fall into two categories: (1) emotion representation models, which are used as an output of affect recognition solutions; and (2) research on mapping algorithms between the models.

(1) There are three major model types of emotional state representation: discrete, dimensional and componential [4]. Discrete models distinguish a set of basic emotions (word labels) and describe each affective state as belonging to a certain emotion from the predefined set. A significant group of emotion recognition algorithms uses emotion representation based on labels only, e.g., distinguishing stress from a no-stress condition [5]. The label-based representation causes serious integration problems, when the label sets are different. Fuzziness of linking concepts with words and a problem of semantic disambiguation are the key issues causing the difficulty. One of the best known and extensively adapted discrete representation model is Ekman's six basic emotions model, which includes joy, anger, disgust, surprise, sadness and fear [6]. Although the model was not initially proposed for emotion recognition, it is the one used most frequently, for example in e-learning affect-aware solutions [7]. Simple affective applications, such as games, frequently use very simple models of two or three labels [8,9]. Furthermore, some more sophisticated solutions, applied in e-learning, incorporating affect recognition, use their own discrete label set [10–12].

Dimensional models represent an emotional state as a point in a multi-dimensional space. The circumplex model of affect, one of the most popular dimensional models, was proposed by Russell [13]. In this model, any emotion might be represented as a point in a space of two continuous dimensions of valence and arousal. The valence (pleasure) dimension differentiates positive from negative emotions, while the dimension of arousal (activation) enables a differentiation between active and passive emotional states. Both dimensions are continuous with neutral affective states represented in the middle of the scale [14]. The model was repeatedly extended with new dimensions, such as dominance and imageability. One of the extended models found some applications in affective computing, for example, the PAD (pleasure-arousal-dominance) model [15,16]. Furthermore, Ekman's six emotions model has been adapted as a dimensional model with each emotion forming one dimension (sometimes a dimension for neutral state is also added). The dimensional models are frequently used by the emotion recognition algorithms and off-the-shelf solutions. Dimensional adaptation of Ekman's six basic emotions is used for solutions based on facial expressions, and Facial Action Coding Scheme (FACS) is the most widely implemented technique [17]. Sentiment analysis of textual inputs (used in opinion mining) mainly explores the valence dimension of the Circumplex/PAD model [18]. Emotion elicitation techniques based on physiology mostly report only on the arousal dimension [19], as positive and negative experiences might cause a similar activation of the nervous system.

Componential models use several factors that constitute or influence the resulting emotional state. The OCC model proposed by Ortony, Clore and Collin defines a hierarchy of 22 emotion types representing all possible states which might be experienced [20]. In contrast to discrete or dimensional models of emotions, the OCC model takes into account the process of generating emotions. However, several papers outline that the 22 emotional categories need to be mapped to a (possibly) lower number of different emotional expressions [21,22].

The analysis of the emotion recognition solutions reveals that there is no one commonly accepted standard model for emotion representation. Dimensional adaptation of Ekman's six basic emotions and the Circumplex/PAD model are the ones widely adopted in emotion recognition solutions. The problem of mapping is multifaceted, including mapping between dimensional and discrete representations, as well as mapping among different dimensional models. There are several solutions to the first issue: using weights, representing discrete labels as points in dimensional spaces and so on. However, the paper concentrates on the latter issue of mapping between dimensional representations. Appreciating label-based and componential models, a further part of this study focuses on mapping between dimensional models only.

(2) There are few studies on mapping between emotion representation models. The papers provide mapping techniques that enable conversions among a personality model, an OCC model and a PAD model of emotion. The mappings have a significant explanatory value of how personality characteristics, moods and emotions are interrelated [16,23]. However, as the personality model is not used in emotion recognition, they are not directly applicable in this study. Exploration of the literature provides one model of mapping between Ekman's five basic emotions and the PAD model [24]. The mapping technique reported in [23,24] provides a linear mapping based on a matrix of coefficients provided in Equation (1).

$$
\begin{aligned}
& PAD[\textit{Anger, Disgust, Fear, Happiness, Sadness}] \\
& = \begin{bmatrix} -0.51 & -0.40 & -0.64 & 0.40 & -0.40 \\ 0.59 & 0.20 & 0.60 & 0.20 & -0.20 \\ 0.25 & 0.10 & -0.43 & 0.15 & -0.50 \end{bmatrix}
\end{aligned}
\tag{1}
$$

The mapping has been used in several further studies [25,26] and remains the most popular one. The existing matrix was not trained on any dataset, instead it was derived from an OCC model. The existing matrix might be considered rather as a theoretical model than a model based on evidence. Nevertheless, this is the only known mapping matrix; therefore, we use it as a reference, as there is no other one to compare our solution to. One might notice, that out of six emotions in Ekman's basic set, the mapping utilizes only five, excluding surprise. An analysis of late fusion studies reveals that two approaches are applied: all emotion recognition algorithms use the common representation model as an output [1], or all representations are mapped into a Circumplex/PAD model of emotions. According to the author's best knowledge, no method nor metrics have been proposed so far for evaluation of the mapping accuracy. No alternative mapping to the one presented in Equation (1) is known to the author.

The study presented in this paper aims at proposing a procedure and metrics to evaluate mapping accuracy, as well as elaborating a new mapping between Ekman's six basic emotions and the Pleasure-Arousal-Dominance Model. The only known mapping, as reported above, is used as a reference in the evaluation procedure. The thesis of the paper was formulated as follows: The proposed metrics set and procedure allows to develop and evaluate mappings between dimensional emotion representation models.

2. Materials and Methods

In this study, mapping techniques among two emotion representation models—Ekman's six basic emotions (dimensional extension) and the PAD model—are explored in detail, using three datasets retrieved from affect-annotated lexicons. In this section, we report: (1) the procedure for obtaining and evaluation of the mapping; (2) the datasets construction; and (3) the metric set used in the evaluation process.

2.1. The Procedure

The procedure of this study was as follows:

(1) Preparation of **the datasets** to train, test and validate the mapping.
(2) Setting up **the metrics** and thresholds (mapping precision).

(3) Estimation using the **state-of-the-art mapping matrix** (for reference).

(4) Training **new models for mapping** with linear regression.

(5) Within-set and cross-set **evaluation** of the proposed new mapping.

The procedure proposed in this study, and especially the metric set and the datasets, might be used in further research for creating more mapping models using classifiers and a machine learning approach. The steps are described in detail in the following paragraphs.

2.2. The Datasets

There are at least two approaches that might be considered for obtaining new mappings between emotion representation models. The first one is to use a set of heuristics based on expert (psychological) knowledge, and that approach was the basis for creation of the state-of-the-art matrix. The second approach is to find a dataset that is annotated with both emotion representation models and to use statistical or machine learning techniques to find a mathematical model for the mapping. The latter approach was chosen for this study. The datasets were obtained by pairing lexicons of affect-annotated words.

The evaluation sets were retrieved from affect-annotated lexicons that use PAD and/or a dimensional adaptation of Ekman's six basic emotions model. The available lexicons might be created by manual, automatic or semi-automatic annotation [27]. As we wanted to use the data retrieved from the lexicons as the "ground truth", only the lexicons with manual annotations were taken into account. The following lexicons were used in this study (historical order):

(1) **The Mehrabian and Russel lexicon.** This lexicon is a relatively old one, as it was developed and published in 1977 by Mehrabian and Russel [28]. It was developed by manual annotation of English words and contains 151 words. The annotators used a PAD model for representation of emotions for the first time and the PAD model has been a reference model ever since then. The annotated sets include mean and standard deviations of evaluations per dimension provided for all participants. Sample words with annotations are presented in Table 1.

Table 1. Sample annotations of words from Russel and Mehrabian lexicon.

Word	Valence Mean (SD)	Arousal Mean (SD)	Dominance Mean (SD)
strong	0.58 (0.24)	0.48 (0.3)	0.62 (0.3)
lonely	−0.66 (0.35)	−0.43 (0.36)	−0.32 (0.3)
happy	0.81 (0.21)	0.51 (0.26)	0.46 (0.38)

There are at least two observations derived from studying the lexicon: firstly, some words are more ambiguous than others; and, secondly, some dimensions are more ambiguous than others (distribution of standard deviations differs).

(2) **The ANEW Lexicon.** The ANEW lexicon was initially developed for English [29], but has many national adaptations. The lexicon contains the most frequently used words (1040). Annotation is based on a PAD model, and mean as well as standard deviations per dimension are provided. Sample words with annotations are presented in Table 2.

Table 2. Sample annotations of words from ANEW lexicon.

Word	Valence Mean (SD)	Arousal Mean (SD)	Dominance Mean (SD)
strong	7.11 (1.48)	5.92 (2.28)	6.92 (2.43)
lonely	2.17 (1.76)	4.51 (2.68)	2.95 (2.12)
happy	8.21 (1.82)	6.49 (2.77)	6.63 (2.43)

Please note the difference in scaling the dimensions for Russel and Mehrabian's and ANEW lexicons. Re-scaling is required for comparisons.

(3) **The Synesketch lexicon**. Synesketch [30] contains 5123 English words annotated manually with overall emotion intensity and Ekman's six basic emotions (anger, joy, surprise, sadness, disgust and fear). The Synesketch lexicon was found partially invalid from the viewpoint of this study—some words had no annotations in any of the dimensions of the basic six emotion set. The all-zeros vector might be interpreted in two ways: a case of neutral annotation or an error. An additional feature provided within the lexicon (overall emotion intensity) was used to differentiate the two cases. If overall emotion intensity was assigned a non-zero value and the other dimensions were not, the vector was considered invalid and excluded from further analysis. Sample words with annotations are presented in Table 3.

Table 3. Sample annotations of words from the Synesketch lexicon.

Word	Emotion Intensity	Happiness	Sadness	Anger	Fear	Disgust	Surprise
strong	0.44	0.72	0.0	0.07	0.07	0.0	0.0
lonely	1.0	0.0	1.0	0.0	0.0	0.0	0.0
happy	1.0	1.0	0.0	0.0	0.0	0.0	0.0

(4) **The NAWL lexicon**. The fourth of the lexicons, NAWL (Nencki Affective Word List), was manually annotated twice by a significant number of annotators using two different representation models, forming a natural set for this study [31,32]. The models used for annotation were: pleasure-arousal-imageability and a subset of Ekman's six, namely: happiness, anger, disgust, sadness and fear. Please note, that for this set the dimensions of dominance and surprise are omitted. The imageability dimension as an additional one was not used in this study. The annotated set includes mean and standard deviations of evaluations per dimension provided for all participants and also for males and females separately. All participants' metrics are used in this study. The number of words in the lexicon is 2902. Sample words with annotations are presented in Table 4.

Table 4. Sample annotations of words from the NAWL (Nencki Affective Word List) lexicon.

Word	Happiness Mean (SD)	Anger Mean (SD)	Sadness Mean (SD)	Fear Mean (SD)	Disgust Mean (SD)	Valence Mean (SD)	Arousal Mean (SD)	Imageability Mean (SD)
strong	4.85 (1.38)	1.96 (1.46)	1.42 (0.90)	2.58 (1.45)	1.38 (0.85)	1.81 (0.98)	2.58 (1.17)	5.92 (1.2)
lonely	1.04 (0.20)	2.58 (1.88)	5.31 (1.57)	4.19 (1.96)	1.92 (1.44)	−1.7 (1.03)	2.48 (1.19)	6.11 (1.01)
happy	5.85 (1.29)	1.58 (1.63)	1.58 (1.53)	1.62 (1.70)	1.31 (1.19)	2.37 (0.88)	3.15 (0.99)	6.41 (0.69)

As the lexicons, and even sometimes the dimensions within one lexicon, were annotated with different scales, re-scaling was performed for comparability of the results for different set and dimensions. A scale of <0.1> was chosen for all dimensions. Re-scaling followed the definitions of mean and standard deviation for addition and multiplication.

The lexicons were automatically paired, which required finding a common subset of words, then the operation of the pair-wise (same word) concatenation of the two annotations was performed. As a result of this pairing procedure, the following datasets were created for this study:

(1) The ANEW-MEHR dataset

The dataset was a result of pairing ANEW and Russel and Mehrabian's lexicons [28,29]. The common subset of words is the same as for the latter (smaller) lexicon (151 words). Please note that, in this dataset, there are two independent annotations in the PAD scale paired and no annotation in Ekman's dimensions. The set was created purposefully for estimating residual error, that is a result of pairing two independent mappings. The metric values, calculated based on the dataset, might be treated as marginal accuracies that might be obtained from the mappings based on pairing lexicons. The detailed specification of the dataset is provided in Appendix A.

(2) The SYNE-ANEW dataset

The second set used in this study was paired based on ANEW [29] and Synesketch [30] lexicons. Only a common subset of words was used as an evaluation set (267 words). The set uses all dimensions of the PAD model (including dominance) and uses a complete set of Ekman's six basic emotions (including surprise). The detailed specification of the dataset is provided in Appendix B.

(3) The NAWL-NAWL dataset

As in NAWL, [31,32], a list of words was annotated twice, the creation of the dataset was (almost) automatic—the only operation to perform was a pair-wise (same word!) concatenation of the two annotations from separate files. The word count equals the size of the lexicon (2902). The size of the lexicon makes it preferable over other datasets: however, one must note that the dataset does not include dimensions of dominance and surprise. The detailed specification of the dataset is provided in Appendix C.

As the sets might be considered complementary rather than competitive, all three are employed in this mapping technique elaboration study. The datasets are available as a supplementary material.

2.3. Metric Set

While evaluating classifiers (and the emotion recognition algorithm is one of these), precision and accuracy are the most popular metrics. Regression models (with continuous outputs) are typically measured by MAE (mean average error), RMSE (root mean squared error) or R^2 metric. All the typically used metrics are invariant to the required precision of the estimate and prone to misinterpretation if the variance of estimate error is high. Mapping between two emotion representation models might be very accurate, if the differences are calculated using precise numbers. However, with all the fuzziness related to emotion concepts, high precision might be misleading. Mathematically, we could find a difference of 0.01 in $(-1, 1)$ scale, but this would not make sense in interpreting the emotion recognition result. Therefore, apart from typical measures of regression model evaluation, a number of metrics was proposed within this study that measures accuracy above a given precision threshold. Moreover, two approaches were adapted, treating dimensions independently and jointly (as 3D space). The definition of the proposed metric set follows. Both the proposed and typical metrics for estimate error are reported within the results section.

2.3.1. Precision in Emotion Recognition

The word annotations, with regard to sentiment, as well the emotional states themselves, are fuzzy and even sometimes ambiguous. One might note that, together with an average annotation of the word, standard deviation is reported. On the other hand, in most of the applications of affect recognition, the decisions are based on simple models, for example, two or three classes of emotions. Therefore, in most of the cases, small mathematical differences in emotion estimates are not interpretable within the application context. In this study, I proposed to use the following precision estimates: (1) absolute distance (mathematical concept equivalent to MAE (mean absolute error)); (2) distance size smaller than 10% of the scale; (3) distance size smaller than 20% of the scale; and (4) distance size smaller than standard deviation of the evaluations (if available). The latter approach uses an additional feature available for the sets, i.e., standard deviation. Please note that the metric thresholds for precision might be adjusted, whenever necessary.

2.3.2. Metrics for the Dimensions Treated Independently

The independently treated dimensions are operationalized with the following metrics (if variables in consecutive equations are assigned the same definition, the definitions are not repeated):

MAE mean absolute distance between the actual and the estimated emotional state per dimension

$$\text{MAE} = \frac{\sum_i abs(x_i - x_i')}{n} \tag{2}$$

where

x_i—the dimension value for ith word (retrieved from the lexicon).
x_i'–the dimension value for ith word (estimated using mapping).
n–number of words within the set.
$X \rightarrow \{P, A, D\}$.

M_{10} relative number of estimates that differ from the actual value by less than 10% of the scale

$$M_{10} = m_{10}/n \tag{3}$$

where

m_{10}—number of estimates that differ from the actual value by less than 10% of the scale.

M_{20} relative number of estimates that differ from the actual value by less than 20% of the scale

$$M_{20} = m_{20}/n \tag{4}$$

where

m_{20}—number of estimates that differ from the actual value less than 20% of the scale.

M_{SD} relative number of estimates that differ from the actual value by less than standard deviation size

$$M_{SD} = m_{SD}/n \tag{5}$$

where

m_{SD}—number of words, for which absolute difference between the estimate and the actual value is smaller than the standard deviation for the word.

2.3.3. Metrics for the Joint Dimensions Accuracy

The joint dimensions accuracy was operationalized with the following metrics:

PAD_{abs} mean absolute distance between the actual and the estimated emotional state in PAD space (calculated using Pythagorean theorem):

$$\text{PAD}_{\text{abs}} = \frac{\sum_i dd_i}{n} \tag{6}$$

where

$$dd_i = \sqrt{(p_i - p_i')^2 + (a_i - a_i')^2 + (d_i - d_i')^2} \tag{7}$$

dd_i—direct distance between estimated and actual emotion calculated for all dimensions using Pythagorean theorem.
p_i, a_i, d_i—the dimension value for ith word (retrieved from the lexicon).
$p_i' \ a_i' \ d_i'$—the dimension value for ith word (estimated using mapping).
n—number of words within the set.

(PAD$_{10}$) relative number of estimates that differ from the actual value by less than 10% of the scale in each dimension.

$$\text{PAD}_{10} = all_{10}/n \tag{8}$$

where

all_{10}—number of words, for estimates, that differ from the actual value by less than 10% of the scale in each dimension.

(PAD$_{20}$) relative number of estimates that differ from the actual value less than 20% of the scale in each dimension.

$$\text{PAD}_{20} = all_{20}/n \tag{9}$$

where

all_{20}—number of estimates that differ from the actual value by less than 20% of the scale in each dimension,

(PAD$_{SD}$) relative number of estimates that differ from the actual value for less than standard deviation size

$$\text{PAD}_{SD} = all_{SD}/n \tag{10}$$

where

all_{SD}—number of words, for which absolute difference between the estimate and the actual value is smaller than the standard deviation for the word.

(DD$_{10}$) relative number of estimates that have direct distance from the estimate smaller than 10% of the scale.

$$\text{DD}_{10} = dd_{10}/n \tag{11}$$

where

dd_{10}—number of words, for which absolute distance between the estimate and the actual value (dd_i) calculated for all dimensions is smaller than 10% of the scale.

(DD$_{20}$) relative number of estimates that have direct distance from the estimate smaller than 10% of the scale.

$$\text{DD}_{20} = dd_{20}/n \tag{12}$$

where

dd_{20}—number of words, for which absolute distance between the estimate and the actual value (dd_i) calculated for all dimensions, is smaller than 20% of the scale.

The above presented set of metrics (four per dimension and six joint-dimensions metrics) are used in the further analysis as accuracy measures.

2.4. Evaluation Calculations

Firstly, the lexicons were re-scaled and paired (details are provided in Section 2.2) forming datasets for further study. Then, the calculations held within the study were performed with Knime analytical tool as described below.

For estimating residual error of pairing lexicons, the ANEW-MEHR dataset was processed with the following steps: (1) calculation of absolute difference for valence, arousal and dominance dimensions per word; (2) calculation of typical metrics (MAE, RMSE, and R^2) for the dataset; (3) calculation of absolute distance in 3D model of emotions per word; (4) comparison with thresholds per word; and (5) calculation of frequency, and threshold-based metrics. The results are reported in Section 3.1.

For reference model mapping evaluation, two datasets, SYNE-ANEW and NAWL-NAWL, were processed with the same procedure: (1) application of the reference mapping to the dataset, adding columns of predicted valence, arousal and dominance; (2) calculation of absolute difference between the actual and predicted values of dimension per word; (3) calculation of typical metrics (MAE, RMSE, and R^2) for the dataset; (4) calculation of absolute distance in 3D model of emotions per word; (5) comparison with thresholds per word; and (6) calculation of frequency and threshold-based metrics. The results are reported in Section 3.2.

For obtaining and evaluating new mapping matrices, two datasets, SYNE-ANEW and NAWL-NAWL, were processed independently using the same procedure: (1) the linear regression model was trained and a mapping matrix was obtained for the set using a ten-fold cross-validation scheme with a random selection of words to separate training and validation subsets and the obtained result sets included both the actual and the estimated value of dimensions per word; (2) calculation of absolute difference between the actual and predicted values of dimension per word; (3) calculation of typical metrics (MAE, RMSE, and R^2) for the dataset; (4) calculation of absolute distance in a 3D model of emotions per word; (5) comparison with thresholds per word; and (6) calculation of frequency- and threshold-based metrics. The proposed matrices are reported in Section 3.3, while the 10-fold cross-validation results are reported in Section 3.4.

Additionally, the cross-set validation was performed for the evaluation of generalizability using the following scheme: (1) application of the new mapping matrix obtained from ANEW_SYNE to the NAWL-NAWL dataset, adding columns of predicted valence, arousal and dominance; (2) calculation of absolute difference between the actual and predicted values of dimension per word; (3) calculatiaon of typical metrics (MAE, RMSE, and R^2) for the dataset; (4) calculation of absolute distance in a 3D model of emotions per word; (5) comparison with thresholds per word; and (6) calculation of frequency- and threshold-based metrics. The results are reported in Section 3.5.

3. Results

Study results are reported for the evaluation sets defined in Section 2.2 and with the metrics operationalized in Section 2.3.

3.1. The Margin Accuracies—Calculations for the ANEW-MEHR Dataset

The ANEW and Russel and Mehrabian's lexicons are both independently created and use the same PAD model for annotation. The differences (averaged over all words) among the annotations derived from the two lexicons are used in this study as an estimate of residual error, that is the result of pairing two independent mappings. The metric values are provided in Table 5.

Table 5. Mapping accuracy metrics for the ANEW-MEHR dataset.

Dimension	Distance-Based Metrics			Threshold-Based Accuracy Metrics				
	MAE/PAD_{ABS}	RMSE	R^2	M_{10}/PAD_{10}	M_{20}/PAD_{20}	M_{SD}/PAD_{SD}	DD_{10}	DD_{20}
P	0.068	0.098	0.860	80.6%	95.5%	94.6%		
A	0.096	0.118	0.556	57.1%	92.0%	97.3%		
D	0.093	0.166	0.577	62.5%	90.2%	99,1%		
PAD	0.172			33.9%	79.5%	91.1%	56.2%	95.5%

The observations derived from pairing the annotations using the same PAD model are the following:

- There is a non-zero residual error for mapping based on pairing independent affect-annotated lexicons.
- The residual error is diverse among the dimensions within the PAD model—it is lowest for valence (P), higher for arousal (A) and the highest for dominance (D). This observation is compliant with

results of dimensions understanding reported in literature (dominance is the least understood dimension resulting in more ambiguous annotation results).

- For a threshold of 10% of the scale, 80.6% of words for valence, 57.1% of words for arousal, and 62.5% for dominance have consistent annotations; for a threshold of 20% of the scale, 95.5% of words for valence, 92% of words for arousal, and 90.2% for dominance have consistent annotations. This observation is compliant with a typical precision-accuracy trade-off.
- Considering the ambiguity of annotations, it is advisable to use the metrics based on the standard deviation for annotated words, as using the same set threshold value for every dimension might cause misinterpretation of accuracy results.
- The traditional metrics (MAE, RMSE and R^2) allow for the interpretation of dimensions independently; however, in practical settings, it would be important to have a mapping that deals with all dimensions within the set threshold; the proposed metrics: PAD_{10}, PAD_{20}, PAD_{SD}, DD_{10} and DD_{20} allow for the interpretation of the mapping accuracies for all dimensions together (treating the PAD model as a typical 3D space).

The residual error obtained from the comparison will be used for reference in an interpretation of the results obtained in the study.

3.2. *The Reference Mapping Accuracies for the SYNE-ANEW and NAWL-NAWL Datasets*

The mapping matrix derived from literature and reported as in Equation (1) was used as a reference model for the model proposed in this study. The results of mapping using the known matrix for the two datasets: SYNE-ANEW and NAWL-NAWL are provided in Tables 6 and 7, consecutively.

Table 6. Accuracy metrics for reference mapping matrix applied on the SYNE-ANEW dataset.

Dimension	Distance-Based Metrics			Threshold-Based Accuracy Metrics				
	MAE/PAD$_{ABS}$	RMSE	R^2	M$_{10}$/PAD$_{10}$	M$_{20}$/PAD$_{20}$	M$_{SD}$/PAD$_{SD}$	DD$_{10}$	DD$_{20}$
P	0.262	0.321	−0.153	25.8%	45.7%	46.8%		
A	0.242	0.293	−1.229	23.9%	48.3%	68.9%		
D	0.252	0.308	−0.778	22.8%	44.6%	56.2%		
PAD	0.487			1.5%	14.6%	25.8%	5.2%	32.2%

Table 7. Accuracy metrics fora reference mapping matrix applied on NAWL-NAWL dataset.

Dimension	Distance-Based Metrics			Threshold-Based Accuracy Metrics				
	MAE/PAD$_{ABS}$	RMSE	R^2	M$_{10}$/PAD$_{10}$	M$_{20}$/PAD$_{20}$	M$_{SD}$/PAD$_{SD}$	DD$_{10}$	DD$_{20}$
P	0.157	0.177	0.327	27.2%	67.9%	59.8%		
A	0.136	0.178	−0.088	49.5%	75.9%	88.9%		
D *	na	na	na	na	na	na		
PA *	0.235			6.7%	46.1%	59.9%	32.0%	80.8%

Note: * dominance dimension is not available for the set.

The absolute measures for the dataset show small differences between the dimensions. A relatively high mean distance is obtained for joined dimensions, which is partially explained by cumulating errors from all dimensions, as the metric PAD_{ABS} is calculated based on the geometrical distance in 3D space.

It seems that setting a proper threshold is crucial and has an enormous influence on the resulting metrics (precision-accuracy trade-off). For a high precision (10% of the scale), accuracy results are around 25% for valence, 24% for arousal, and 23% for dominance. With lower precision requirements, the accuracies increase (46% for valence, 48% for arousal, and 45% for dominance). Setting the accuracy threshold based on standard deviation per word increases accuracies above 55% for the dimensions of arousal and dominance.

The joined dimensions accuracies are lower than for separate dimensions, which could have been expected. If for one dimension mapping is precise, it might not be for the other one and, as a result, the errors cumulate over dimensions. Therefore, we have only 1.5% accuracy for an expected precision of 10% of the scale, 14.6% accuracy for 20% precision and 25.8% for SD-based precision.

For the second dataset, the mapping accuracies are higher for all precision thresholds. This might be a result of the set size and quality (the set is derived from one lexicon purposefully annotated twice); however, one must notice the lack of *Surprise* and *Dominance* dimensions among the annotation labels. Please note, that due to the latter limitation, the results between the two sets are comparable for the single-dimension metrics, but incomparable for joint-dimensions metrics. Accuracies are still dependent on the set precision threshold.

As precision and accuracy are interchangeable, one might go further in proposing new thresholds. Perhaps, the acceptable precision should be set case-by-case, as this might depend on the context of the emotion recognition.

The accuracy results obtained for the known matrix show some room for improvement, especially for higher precision requirements and regarding all dimensions. The results justify undertaking this study towards new mappings.

3.3. The Proposed Mapping Matrices

Two mapping matrices were obtained using linear regression learning with SYNE-ANEW and NAWL-NAWL datasets. The procedure involved a 10-fold cross-validation with linear regression coefficients averaged over repetitions. Cross-validation is a validation technique for assessing how the results of a prediction model would generalize to an independent data set. It is a well-established alternative to the hand-made partitioning of a dataset to training and validation subsets for regression models [33]. One round of cross-validation involves partitioning a sample of data into complementary subsets, performing the analysis on one subset (training set), and validating the analysis on the other subset (validation set). Multiple rounds of partitioning are performed and the validation results are averaged over the rounds to evaluate the model. There are several methods of performing cross-validation [34]. In this procedure, a k-fold technique was used, with 10 folds. In 10-fold cross-validation, the original sample was partitioned randomly into 10 sub-samples of equal size. In a single fold, a single subsample was retained as the validation set, and the remaining nine sub-samples were used as training data. As a result, each of the 10 sub-samples was used exactly once as the validation data. The partitioning was randomized, and the cross-validation procedure was performed using Knime analytical tool.

The cross-validation procedure was repeated 10 times. The matrix coefficients obtained via repetition were similar (with differences <0.03 for all matrix coefficients) and were averaged over the 10 resulting models.

The two resulting matrices are provided as Equations (13) and (14) consecutively. The existing mapping matrix from literature was repeated with a column shuffle in Equation (15) for comparison.

$$\text{PAD}[Happy, \ Sad, \quad Angry, \ Scared, \ Disgusted, \ Surprised, \ 1] =$$
$$= \begin{bmatrix} 0.46 & -0.30 & -0.29 & -0.19 & -0.14 & 0.24 & 0.52 \\ 0.07 & -0.11 & 0.19 & 0.14 & -0.08 & 0.15 & 0.53 \\ 0.19 & -0.18 & -0.02 & -0.10 & -0.02 & 0.08 & 0.50 \end{bmatrix} \tag{13}$$

$$\text{PA}[Happy, \ Sad, \quad Angry, \ Scared, \ Disgusted, \ Surprised, \ 1] =$$
$$= \begin{bmatrix} 0.54 & -0.14 & -0.21 & -0.06 & -0.16 & 0.00 & 0.46 \\ 0.50 & 0.06 & 0.37 & 0.36 & 0.12 & 0.00 & -0.01 \end{bmatrix} \tag{14}$$

$$PAD[Happy, \ Sad, \quad Angry, \quad Scared, \ Disgusted, \ Surprised, \ 1] =$$

$$= \begin{bmatrix} 0.40 & -0.40 & -0.51 & -0.64 & -0.40 & 0.00 & 0.00 \\ 0.20 & -0.20 & 0.59 & 0.60 & 0.20 & 0.00 & 0.00 \\ 0.15 & -0.50 & 0.25 & -0.43 & 0.10 & 0.00 & 0.00 \end{bmatrix} \tag{15}$$

3.4. The Proposed Mapping Accuracies for the SYNE-ANEW and NAWL-NAWL Datasets

For evaluation of the new mapping matrices, a 10-fold cross-validation procedure was performed, as described in Section 3.3 for the two datasets SYNE-ANEW and NAWL-NAWL independently. Cross-validation results for the datasets are presented in Tables 8 and 9.

Table 8. Accuracy metrics for the new mapping for the SYNE-ANEW dataset (cross-validation results).

Dimension	Distance-Based Metrics			Threshold-Based Accuracy Metrics				
	MAE/PAD_{ABS}	RMSE	R^2	M_{10}/PAD_{10}	M_{20}/PAD_{20}	M_{SD}/PAD_{SD}	DD_{10}	DD_{20}
P	0.174	0.297	0.45	33.3%	59.9%	62.5%	-	-
A	0.097	0.122	0.197	58.7%	90.6%	96.8%	-	-
D	0.090	0.113	0.330	61.6%	92.1%	98.3%	-	-
PAD	0.239			17.8%	54.9%	59.9%	32.0%	80.8%

Table 9. Accuracy metrics for the new mapping for the NAWL-NAWL dataset (cross-validation results).

Dimension	Distance-Based Metrics			Threshold-Based Accuracy Metrics				
	MAE/PAD_{ABS}	RMSE	R^2	M_{10}/PAD_{10}	M_{20}/PAD_{20}	M_{SD}/PAD_{SD}	DD_{10}	DD_{20}
P	0.045	0.056	0.918	92.4%	99.9%	99.3%	-	-
A	0.070	0.088	0.574	74.9%	97.6%	99.1%	-	-
D	na	na	na	na	na	na	-	-
PA *	0.090			70.5%	97.6%	98.4%	83.7%	99.8%

Note: * dominance dimension is not available for the set.

The proposed mapping matrix validation results on the SYNE-ANEW dataset are relatively high. For the dimension of valence, over 33% accuracy was obtained for the 10% precision threshold, while, for the 20% and SD-based precision thresholds, the accuracy was around 60%. For the dimension of arousal and dominance, over 58% accuracy was obtained for the 10% precision threshold, while, for the 20% and SD-based precision thresholds, the accuracy exceeded 90%. Single-dimension results are very promising. However, the joint dimension accuracies are lower. Setting a 10% precision threshold resulted in 17.8% accuracy, while the 20% and SD-based thresholds resulted in 54.9% and 59.9% accuracies, respectively. The observed accuracies are higher than for the reference mapping for the same dataset; however, these might still be insufficient for high precision requirements.

The proposed mapping matrix validation results on the NAWL-NAWL dataset are very high. For the dimension of valence, over 90% accuracy was obtained for all precision thresholds. For the dimension of arousal, over 74% accuracy was obtained for the 10% precision threshold, while, for the 20% and SD-based precision thresholds, the accuracy exceeded 97%. Single-dimension results are satisfactory. The joint dimension accuracies are slightly lower, but still exceed 97% for the 20% and SD-based precision thresholds. The observed accuracies are higher than for the reference mapping for the same dataset.

3.5. Cross-Set Accuracies for the Proposed Mapping

Additionally, the cross-set validation was performed for evaluation of generalizability following the procedure described in Section 2.4. The results are presented in Table 10.

Table 10. Accuracy metrics for the mapping obtained from SYNE_ANEW applied on the NAWL-NAWL dataset (cross-set results).

Dimension	Distance-Based Metrics			Threshold-Based Accuracy Metrics				
	MAE/PAD_{ABS}	RMSE	R^2	M_{10}/PAD_{10}	M_{20}/PAD_{20}	M_{SD}/PAD_{SD}	DD_{10}	DD_{20}
P	0.093	0.109	0.744	56.5%	95.8%	87.3%	-	-
A	0.116	0.144	0.282	50.8%	84.4%	92.6%	-	-
D	na	na	na	na	na	na	-	-
PA *	0.162			29.9%	81.1%	81.5%	43.3%	93.1%

Note: * dominance dimension is not available for the set.

The cross-set validation results on the NAWL-NAWL dataset are moderate: higher than reference mapping, but lower than in cross-validation on the same dataset. This result was expected—usually cross-validation procedures provide higher accuracies. For the dimension of valence: over 56% accuracy was obtained for the 10% precision threshold, while, for 20% and SD-based precision thresholds, the accuracy exceeded 87%. For the dimension of arousal: over 50% accuracy was obtained for the 10% precision threshold, while, for 20% and SD-based precision thresholds, the accuracy exceeded 84%. Single-dimension results are lower than in cross-validation, but still satisfactory for lower precision requirements. The joint dimension accuracies are lower. Setting a 10% precision threshold resulted in 30% accuracy, while the 20% and SD-based thresholds resulted in about 81% accuracies. The observed accuracies are higher than for the reference mapping for the same dataset; however, for high precision requirements this might still be insufficient.

4. Summary of Results and Discussion

The mapping technique using the linear transformation based on a coefficient matrix was evaluated in this study. The results may be summarized in the following statements.

Obtaining both accurate and precise mapping results is a challenge. The results confirm standard observations of precision-accuracy conflict. The mapping model provided is better than a reference model, but still might be insufficient for high precision requirements. The summary of results is provided in Table 11.

Table 11. Mapping accuracy metrics for ANEW-MEHR dataset.

Model	Dataset/Analysis	PAD_{ABS}	PAD_{10}	PAD_{20}	PAD_{SD}	DD_{10}	DD_{20}
PAD	ANEW-MEHR simple pairing	0.172	33.9%	79.5%	91.1%	56.2%	95.5%
PAD	SYNE-ANEW reference mapping	0.487	1.5%	14.6%	25.8%	5.2%	32.2%
PAD	SYNE-ANEW proposed mapping cross-validation	0.239	17.8%	54.9%	59.9%	32.0%	80.8%
PA	NAWL-NAWL reference mapping	0.235	6.7%	46.1%	59.9%	32.0%	80.8%
PA	NAWL-NAWL proposed mapping cross-validation	0.090	70.5%	97.6%	98.4%	83.7%	99.8%
PA	Cross-set mapping evaluation on NAWL-NAWL dataset	0.162	29.9%	81.1%	81.5%	43.3%	93.1%

In the SYNE-ANEW dataset, the accuracy results are better for the proposed mapping than for the reference mapping. The same applies to the NAWL-NAWL results. In the latter, the results in cross-validation are even better than for the ANEW-MEHR dataset. Interpretation of cross-set evaluation results is limited because the dataset (and obtained matrices) differed in dimension count (for one of the datasets surprise and dominance was missing).

The proposed mapping matrices might be applied in the comparison and fusion of emotion recognition results; however, the required precision must be taken into account. Acceptable precision should be set based on the emotion recognition goals and application context, and especially on the significance of I type and II type error. Using standard deviation as a threshold for precision is one of the promising directions.

The proposed method and metrics for accuracy proved useful and allowed the mapping technique to be evaluated. Please note, that the proposed metrics are:

- dataset-independent: might be applied to any dataset constructed
- scale-independent: due to setting precision thresholds as percent of the scale or SD-based
- model-dependent: they have been proposed specifically for the pleasure-arousal dominance model; however, they might be adapted to any dimensional emotion representation model
- ambiguity-robust: valid for metrics based on the standard deviation threshold: M_{SD}, PAD_{SD}

The author acknowledges that this study is not free from some limitations. The most important threats to its validity are listed below.

(1) Only three mapping matrices were explored in detail. Perhaps more models might be proposed or retrieved from the datasets available. A comparison of the mapping accuracies obtained from a different mapping model and the same datasets would provide more insight.

(2) The 10%, 20% and SD measures for threshold were chosen arbitrarily. Other thresholds might be considered for other contexts and input channels. The required precision is dependent on the context of application. Although 20% of the scale might seem a broad precision margin, in most cases, it might be sufficient, as frequently only two classes of emotions are analyzed (e.g., a stress and no-stress condition). Our previous study showed, for example, that for word annotations an intra-rater inconsistency of up to 15% of the scale was encountered in 89.1%, 74.3% and 79.4% of the ratings for valence, arousal and dominance, respectively [35]. Some inconsistency in annotations with affect is imminent due to the ambiguous nature of emotion; therefore, in this study we decided to report results for 10% as well as 20% of the scale, followed by the standard-deviation-based metric.

(3) The evaluation is based only on sets retrieved from affect-annotated lexicons. It is expected that the results might be different for sets based on alternative input channels. However, training a mapping model requires a double annotation of some media (the same word/ image/sound/video being assigned values in two emotion representation models). The first challenge in this research was to actually find datasets that are annotated twice, using two different emotion representation models, and that was found true only for the presented lexicons. Basing the mapping on lexicons only might limit the generalizability of the obtained solution, but currently there are no other twice-annotated datasets available.

The new matrices were obtained using the 10-fold cross-validation method for regression model. Cross-validation, although well-established, is also reported as having both advantages and disadvantages. More recent reviews suggest that traditional cross-validation needs to be supplemented with other studies [36]. Therefore, in this study, we report both cross-validation, as well as cross-set evaluation results.

The purpose of the paper was to propose a method and a set of metrics for the evaluation of the mapping accuracy and to propose a new mapping according to the procedure. Despite the limitations, the purpose was achieved, the new mapping was formed, and the thesis *the proposed metric set allows for evaluation of emotion model mapping accuracy* might be accepted.

Implications of the study include the following:

(1) Affective states recognition algorithms provide estimates that might be wrong or imprecise. Additionally, any mapping between the emotion representation models might enlarge uncertainty related to emotion recognition, as there is non-zero accuracy error for all mappings. Therefore, it is always worthwhile considering an emotion recognition solution that provides an estimate in the representation model and that is better fitted into your context of applications.

(2) The new mapping matrices might be applied when you need emotions described with a PA or PAD model and get Ekman's six basic emotions vector form the affect recognition system. The mapping might also be found useful in a late fusion of hypotheses on emotional states obtained from multiple affect recognition algorithms. In this study, two matrices were proposed (Equations (13) and (14)) and it is important to emphasize the differences between them. The first one was

derived from a smaller dataset, and is therefore less accurate, but this applies to a complete list of dimensions. However, if you expect dominance and surprise not to play an important role in your application context, you might consider the latter one, as it was derived from a bigger dataset and therefore achieved higher accuracies for all precision thresholds.

(3)   One of the crucial issues in applying the mapping is setting a precision threshold, which might be context-dependent. A comparison of the metrics used in this study showed that using standard deviation as a threshold might be advisable in some contexts, as it takes into account the ambiguity of the annotated words.

(4)   The proposed procedure and metrics were validated and proven useful in the study of mapping between the emotion representation models. The metrics might also prove useful in a comparison of the two annotations. The datasets developed for this study will be shared. The procedure, metrics and datasets set a framework for future works on mappings between the emotion representation models.

5. Conclusions

The study revealed several interesting observations that are worth exploring further. Future studies will aim at an application of data mining techniques to the proposed sets to retrieve an alternative mapping algorithm. Both linear and nonlinear models would be considered, as well as training diverse classifiers with machine learning methods. Further analysis of distance distribution would be performed, as perhaps asymmetric thresholds might be considered. Another interesting issue on whether the mapping should be input-channel dependent for better accuracies remains open.

The exploration of the mapping models and algorithms will be continued and followed by practical applications. There is an emotion monitor stand constructed at Gdansk University of Technology that uses existing technologies to extend human–systems interaction with emotion recognition and affective intervention. The concept of the stand assumed combining multiple modalities used in emotion recognition to improve the accuracy of affect classification. Integration of the existing technologies, input channels and solutions turned out to be very challenging, and among the reasons for this was the incompatibility and low-accuracy mapping between the emotion recognition models. The mapping might be integrated with the solutions already functioning at the stand.

Supplementary Materials: The following are available online at www.mdpi.com/2076-3417/8/2/274/s1: (1) The ANEW-MEHR dataset that contains a juxtaposition of the annotations with the same, dimensional emotion representation model retrieved from affect-annotated lexicons: Rusel and Mehrabian's and ANEW. (2) The SYNE-ANEW dataset that contains a juxtaposition of the annotations with PAD and Ekman's six-dimensional emotion representation models retrieved from affect-annotated lexicons, Synesketch and ANEW. (3) The NAWL-NAWL dataset that contains juxtaposition of the annotations with pleasure-arousal and a subset of Ekman's six-dimensional emotion representation models retrieved from affect-annotated lexicon NAWL.

Acknowledgments: This study received funding from Polish–Norwegian Financial Mechanism Small Grant Scheme under the contract no Pol-Nor/209260/108/2015, as well as from DS Funds of ETI Faculty, Gdansk University of Technology.

Conflicts of Interest: The author declares no conflict of interest.

Appendix A. Detailed Specification of the ANEW-MEHR Dataset

The ANEW-MEHR dataset was derived from Russel and Mehrabian, and ANEW lexicons of affect-annotated words. The dataset is downloadable as a semicolon-separated .csv file. The dataset is distributed under the terms and conditions of the Creative Commons Attribution (CC BY) license. When referencing the dataset, please cite this publication.

To conform with lexicon permissions in the resulting dataset, the words are coded with numbers. For affective annotations of specific words, please refer to the original lexicons. Please cite Russel and Mehrabian [28] for Mehrabian annotation and Bradley&Lang [29] for ANEW.

Specification of the columns:

	Type	Scale	Description	Source
Word	character	Na	Word code	-
P_mehr	decimal	0–1	Valence dimension in Russel and Mehrabian's publication—mean value	Russel and Mehrabian
A-mehr	decimal	0–1	Arousal dimension in Russel and Mehrabian's publication—mean value	Russel and Mehrabian
D_mehr	decimal	0–1	Dominance dimension in Russel and Mehrabian's publication—mean value	Russel and Mehrabian
P_anew	decimal	0–1	Valence dimension in ANEW lexicon—mean value	ANEW
P_SD	decimal	0–1	Valence dimension in ANEW lexicon—standard deviation	ANEW
A_anew	decimal	0–1	Arousal dimension in ANEW lexicon—mean value	ANEW
A_SD	decimal	0–1	Arousal dimension in ANEW lexicon—standard deviation	ANEW
D_anew	decimal	0–1	Dominance dimension in ANEW lexicon—mean value	ANEW
D_SD	decimal	0–1	Dominance dimension in ANEW lexicon—standard deviation	ANEW

Appendix B. Detailed Specification of the SYNE-ANEW Dataset

The SYNE-ANEW dataset was derived from Synesketch and ANEW lexicons of affect-annotated words. The dataset is downloadable as a semicolon-separated .csv file. The dataset is distributed under the terms and conditions of the Creative Commons Attribution (CC BY) license. When referencing the dataset, please cite this publication.

To conform with lexicon permissions in the resulting dataset, the words are coded with numbers. For affective annotations of specific words, please refer to the original lexicons. Please cite Bradley&Lang [29] for ANEW and Krcadinac et al. [30] for the Synesketch lexicon.

Specification of the columns:

	Type	Scale	Description	Source
Word	character	na	Word code	-
P_anew	decimal	0–1	Valence dimension in ANEW lexicon—mean value	ANEW
P_SD	decimal	0–1	Valence dimension in ANEW lexicon—standard deviation	ANEW
A_anew	decimal	0–1	Arousal dimension in ANEW lexicon—mean value	ANEW
A_SD	decimal	0–1	Arousal dimension in ANEW lexicon—standard deviation	ANEW
D_anew	decimal	0–1	Dominance dimension in ANEW lexicon—mean value	ANEW
D_SD	decimal	0–1	Dominance dimension in ANEW lexicon—standard deviation	ANEW
Hap_syne	decimal	0–1	Happiness dimension of six basic emotions—mean value	Synesketch
Sad_syne	decimal	0–1	Sadness dimension of six basic emotions—mean value	Synesketch
Ang_syne	decimal	0–1	Anger dimension of six basic emotions—mean value	Synesketch
Fea_syne	decimal	0–1	Fear dimension of six basic emotions—mean value	Synesketch
Dis_syne	decimal	0–1	Disgust dimension of six basic emotions—mean value	Synesketch
Sur_syne	decimal	0–1	Surprise dimension of six basic emotions—mean value	Synesketch

Appendix C. Detailed Specification of the NAWL-NAWL Dataset

The NAWL-NAWL dataset was derived from two sentiment annotations of the same lexicon, NAWL. The dataset is downloadable as a semicolon-separated .csv file. The dataset is distributed under the terms and conditions of the Creative Commons Attribution (CC BY) license. When referencing the dataset, please cite this publication.

To conform with lexicon permissions in the resulting dataset, the words are coded with numbers. For affective annotations of specific words, please refer to the original lexicons. Please cite Rieleg et al. and Wierzba et al. [31,32] for NAWL lexicon.

Specification of the columns:

	Type	Scale	Description	Source
Word	character	na	Word code	-
P_nawl	decimal	0–1	Valence dimension in NAWL—mean value	NAWL [31]
P_SD	decimal	0–1	Valence dimension in NAWL—standard deviation	NAWL [31]
A_nawl	decimal	0–1	Arousal dimension in NAWL—mean value	NAWL [31]
A_SD	decimal	0–1	Arousal dimension in NAWL—standard deviation	NAWL [31]
Hap_nawl	decimal	0–1	Happiness dimension of six basic emotions—mean value	NAWL [32]
Sad_nawl	decimal	0–1	Sadness dimension of six basic emotions—mean value	NAWL [32]
Ang_nawl	decimal	0–1	Anger dimension of six basic emotions—mean value	NAWL [32]
Fea_nawl	decimal	0–1	Fear dimension of six basic emotions—mean value	NAWL [32]
Dis_nawl	decimal	0–1	Disgust dimension of six basic emotions—mean value	NAWL [32]

References

1. Gunes, H.; Piccardi, M. Affect recognition from face and body: Early fusion vs. late fusion. In Proceedings of the IEEE International Conference on Systems, Man and Cybernetics, Waikoloa, HI, USA, 12 October 2005; pp. 3437–3443.
2. Zeng, A.; Pantic, M.; Roisman, G.; Huang, T.S. A survey of affect recognition methods: Audio, visual, and spontaneous expressions. *IEEE Trans. Pattern Anal. Mach. Intel.* **2009**, *31*, 39–58. [CrossRef] [PubMed]
3. Hupont, I.; Ballano, S.; Baldassarri, S.; Cerezo, E. Scalable multimodal fusion for continuous affect sensing. In Proceedings of the IEEE Workshop on Affective Computational Intelligence (WACI), Paris, France, 11–15 April 2011.
4. Kołakowska, A.; Landowska, A.; Szwoch, M.; Szwoch, W.; Wrobel, M.R. Modeling emotions for affect-aware applications. In *Information Systems Development and Applications*; Faculty of Management, University of Gdansk: Sopot, Poland, 2015.
5. Baker, R.S.J.D.; D'Mello, S.K.; Rodrigo, M.M.T.; Graesser, A.C. Better to be frustrated than bored: The incidence, persistence, and impact of learners' cognitive—Affective states during interactions with three different computer-based learning environments. *Int. J. Hum. Comput. Stud.* **2010**, *68*, 223–241. [CrossRef]
6. Ekman, P.; Friesen, W.V. Constants across cultures in the face and emotion. *J. Personal. Soc. Psychol.* **1971**, *17*, 124–129. [CrossRef]
7. Landowska, A. Affective learning manifesto-10 years later. In Proceedings of the 13th European Conference of E-Learning, Copenhagen, Denmark, 30–31 October 2014.
8. Obaid, M.; Han, C.; Billinghurst, M. Feed the Fish: An affectaware game. In Proceedings of the 5th Australasian Conference on Interactive Entertainment, Brisbane, Australia, 3–5 December 2008.
9. Szwoch, M. Design elements of affect aware video games. In Proceedings of the International Conference Multimedia, Interaction, Design and Innovation, Warsaw, Poland, 24–25 June 2015.
10. Alexander, S.; Sarrafzadeh, A.; Hill, S. Easy with Eve: A functional affective tutoring system. In Proceedings of the 8th International Conference on Workshop on Motivational and Affective Issues in ITS, Jhongli, Taiwan, 26–30 June 2006.
11. Cabada, R.Z.; Estrada, M.L.B.; Beltr'n V, J.A.; Cibrian R, F.L.; García, C.A.R.; Pérez, Y.H. Fermat: Merging affective tutoring systems with learning social networks. In Proceedings of the IEEE 12th International Conference on Advanced Learning Technologies (ICALT), Rome, Italy, 4–6 July 2012.
12. D'Mello, S.; Jackson, T.; Craig, S.; Morgan, B.; Chip-man, P.; White, H.; Person, N.; Kort, B.; el Kaliouby, R.; Picard, R.; et al. AutoTutor detects and responds to learners affective and cognitive states. In Proceedings of the Workshop on Emotional and Cognitive Issues at the International Conference on Intelligent Tutoring Systems, Montreal, Canada, 23–27 June 2008.
13. Russell, J.A. A circumplex model of affect. *J. Personal. Soc. Psychol.* **1980**, *39*, 1161–1178. [CrossRef]
14. Kołakowska, A.; Landowska, A.; Szwoch, M.; Szwoch, W.; Wróbel, M.R. Emotion recognition and its applications, human-computer systems interaction: Backgrounds and applications 3. *Adv. Intell. Syst. Comput.* **2014**, *300*, 51–62.
15. Mehrabian, A. Pleasure-arousal-dominance: A general framework for describing and measuring individual differences in temperament. *Curr. Psychol.* **1996**, *14*, 261–292. [CrossRef]
16. Mehrabian, A. Analysis of the big-five personality factors in terms of the PAD temperament model. *Aust. J. Psychol.* **1996**, *48*, 86–92. [CrossRef]
17. Ekman, P.; Friesen, W.V. *Facial Action Coding System: A Technique for the Measurement of Facial Movement*; Consulting Psychologists Press: Palo Alto, CA, USA, 1978.
18. Cambria, E.; Schuller, B.; Xia, Y.Q.; Havasi, C. New avenues in opinion mining and sentiment analysis. *IEEE Intell. Syst.* **2013**, *28*, 15–21. [CrossRef]
19. Bach, D.R.; Friston, K.J.; Dolan, R.J. Analytic measures for quantification of arousal from spontaneous skin conductance fluctuations. *Int. J. Psychophysiol.* **2010**, *76*, 52–55. [CrossRef] [PubMed]
20. Ortony, A. *The Cognitive Structure of Emotions*; Cambridge University Press: Cambridge, UK, 1990.
21. Bartneck, C. Integrating the OCC model of emotions in embodied characters. In Proceedings of the Workshop on Virtual Conversational Characters, Melbourne, Australia, 29 November 2002.
22. Steunebrink, B.R.; Dastani, M.; Meyer, J.-J.C. The OCC model revisited. In Proceedings of the 4th Workshop on Emotion and Computing, Paderborn, Germany, 15 September 2009.

23. Yang, G.; Wang, Z.; Wang, G.; Chen, F. *Affective Computing Model Based on Emotional Psychology*; Springer: Berlin/Heidelberg, Germany, 2006.

24. Wei, J.; Wang, Y.; Tu, J.; Zhang, Q. *Affective Transfer Computing Model Based on Attenuation Emotion*; Springer: Berlin, Germany, 2011.

25. Valenza, G.; Lanata, A.; Scilingo, E.P. The role of nonlinear dynamics in affective valence and arousal recognition. *IEEE Trans. Affect. Comput.* **2012**, *3*, 237–249. [CrossRef]

26. Gebhard, P. *ALMA—A Layered Model of Affect*; German Research Center for Artificial Intelligence (DFKI): Saarbrucken, Germany, 2005.

27. Kołakowska, A.; Landowska, A.; Szwoch, M.; Szwoch, W.; Wróbel, M.R. Evaluation criteria for affect-annotated databases. In *Beyond Databases, Architectures and Structure*; Springer: Berlin, Germany, 2015.

28. Russell, J.A.; Mehrabian, A. Evidence for a three-factor theory of emotions. *J. Res. Personal.* **1977**, *11*, 273–294. [CrossRef]

29. Bradley, M.M.; Lang, P.J. *Affective Norms for English Words (ANEW): Instruction Manual and Affective Ratings*; Technical Report C-1; The Center for Research in Psychophysiology, University of Florida: Gainesville, FL, USA, 1977.

30. Krcadinac, U.; Pasquier, P.; Jovanovic, J.; Devedzic, V. Synesketch: An open source library for sentence-based emotion recognition. *Proc. IEEE Trans. Affect. Comput.* **2013**, *4*, 312–325. [CrossRef]

31. Riegel, M.; Wierzba, M.; Wypych, M.; Żurawski, Ł.; Jednoróg, K.; Grabowska, A.; Marchewka, A. Nencki Affective Word List (NAWL): The cultural adaptation of the Berlin Affective Word List–Reloaded (BAWL-R) for Polish. *Behav. Res. Methods* **2015**, *47*, 1222–1236. [CrossRef] [PubMed]

32. Wierzba, M.; Riegel, M.; Wypych, M.; Jednoróg, K.; Turnau, P.; Grabowska, A.; Marchewka, A. Basic emotions in the Nencki Affective Word List (NAWL BE): New method of classifying emotional stimuli. *PLoS ONE* **2015**, *10*. [CrossRef] [PubMed]

33. Picard, R.; Cook, D. Cross-validation of regression models. *J. Am. Stat. Assoc.* **1984**, *79*, 575–583. [CrossRef]

34. Arlot, S.; Alain, C. A survey of cross-validation procedures for model selection. *Stat. Surv.* **2010**, *4*, 40–79. [CrossRef]

35. Landowska, A. Web questionnaire as construction method of affect-annotated lexicon-risks reduction strategy. In Proceedings of the International Conference on Affective Computing and Intelligent Interaction, Xi'an, China, 21–24 September 2015.

36. Varma, S.; Simon, R. Bias in error estimation when using cross-validation for model selection. *BMC Bioinform.* **2006**, *7*. [CrossRef] [PubMed]

applied
sciences

MDPI

Article

Modelling the Interruption on HCI Using BDI Agents with the Fuzzy Perceptions Approach: An Interactive Museum Case Study in Mexico

Ricardo Rosales [1,*], Manuel Castañón-Puga [2], Felipe Lara-Rosano [3], Richard David Evans [4], Nora Osuna-Millan [1] and Maria Virginia Flores-Ortiz [1]

[1] Accounting and Administration School, Autonomous University of Baja California, Tijuana 22390, Mexico; nora.osuna@uabc.edu.mx (N.O.-M.); vflores@uabc.edu.mx (M.V.F.-O.)
[2] Chemistry and Engineering School, Autonomous University of Baja California, Tijuana 22390, Mexico; puga@uabc.edu.mx
[3] Complexity Science Center, National Autonomous University of Mexico, Mexico 04510, Mexico; flararosano@gmail.com
[4] Business Information Management and Operations Department, University of Westminster, London NW1 5LS, UK; R.Evans@westminster.ac.uk
[*] Correspondence: ricardorosales@uabc.edu.mx; Tel.: +52-664-294-5884

Received: 27 July 2017; Accepted: 9 August 2017; Published: 13 August 2017

Abstract: Technological advancements have revolutionized the proliferation and availability of information to users, which has created more complex and intensive interactions between users and systems. The learning process of users is essential in the construction of new knowledge when pursuing improvements in user experience. In this paper, the interruption factor is considered in relation to interaction quality due to human–computer interaction (HCI) being seen to affect the learning process. We present the results obtained from 500 users in an interactive museum in Tijuana, Mexico as a case study. We model the HCI of an interactive exhibition using belief–desire–intention (BDI) agents; we adapted the BDI architecture using the Type-2 fuzzy inference system to add perceptual human-like capabilities to agents, in order to describe the interaction and interruption factor on user experience. The resulting model allows us to describe content adaptation through the creation of a personalized interaction environment. We conclude that managing interruptions can enhance the HCI, producing a positive learning process that influences user experience. A better interaction may be achieved if we offer the right kind of content, taking the interruptions experienced into consideration.

Keywords: human–computer interaction; ambient intelligence; interruption factor; belief–desire–intention agents; perceptual computing; Type-2 fuzzy inference system

---

1. Introduction

We are currently involved in an information revolution where technology facilitates tasks and the daily activities of people, making them more productive; however, it is important to evaluate if the evolution of technology is truly helping people. In some cases, technology is not necessarily appropriately adopted; for example, if something that is new or novel is being misused, impacting misinformation and ignorance, it could mean the opposite [1]. In order to argue how technology can help in the interaction among educators and learners during the learning process, it is desirable that educators act as facilitators in the learning process, meaning that they are more interested in the learners learning than in the delivery of teaching.

1.1. The Learning Process

The learning process is a relatively permanent change in an individual's behaviour, affecting their knowledge, attitudes, and skills, and can occur at any place or time consciously or subconsciously. In nearly all cases, the educator and learner are essential in the delivery of the process. They both build and play a leading role in teaching and learning. The learners gain more autonomy as they progress in the educational system, improving their interactions with the help of technology. It is important to also consider the educator as an identifying model; the learner is often influenced by the educator's different ways of thinking, speaking, and acting. The success of the learning process depends on effective collaboration between educators and learners [2].

1.2. Interruption Factor on the Learning Process

It is important to consider the interruption factor in the learning process. Interruptions are manifestations in a real-world context where multiple tasks are often taking place in parallel with the educator–learner interaction. The concept of interruption is relatively complex, and researching human interruption can be difficult. Interruption is a problem commonly encountered in educational institutions because educators and learners have cognitive limitations that restrict their abilities during the interaction, thereby resulting in temporary or full suspension of the learning process. These human limitations for handling interruptions may produce critical mistakes and affect the outcome of the learning interaction [3].

1.3. Interruption Factor in Human–Computer Interaction (HCI)

The interruption factor is not only present in the educator–learner interaction; interruption also emerges in HCI because humans have cognitive limitations affecting their performance, making them vulnerable to errors and delays in the interaction. In some situations, even very brief interruptions can have detrimental effects on human interaction. On the other hand, the computational expandability of interactive content may introduce greater applications that proactively push information towards the human; however, this information is not always appropriate, and incorrect information may increase a computer's potential to disrupt situations inappropriately [4]. The fact that the interruption factor may be continuous raises further questions: Can we keep the attention and interest of educators and learners during interactions using technology? How does the use of technology help to create solutions that support these interruptions?

1.4. HCI Model Representation

Interruption issues during the educator–learner interaction are immediate; this research proposes a model of interaction phenomena based on the learning process. The proposed model is a representation of HCI, where the learner is the human side and the educator is the computer.

In order to simulate the user and computer in the model, we use belief–desire–intention (BDI) agents: the user is the learner BDI agent (LA), and the educator is the educator BDI agent (EA); the LA simulates the learner's performance influenced by various factors (interaction level and distance), including emotional state, topic interest, and intentions. The EA monitors the learner's performance using situational perceptions to deliver adequate and appropriate content types, avoiding interruptions and keeping the interest of the learner in the context of the learning process interaction.

Our proposed model fits and supports the cognitive limitations of the learner; the learner will be able to perform various activities successfully at the time of the interaction.

1.5. Degree of Engagement

One objective of the EA is to keep a degree of engagement between educator and learner. A learner's participation or engagement levels vary in the interaction due to the degree of engagement; this is determined to be relative to the attentiveness or the interaction of a participant with the focus of attention. Some interruption factors contribute to an increase or decrease in the degree of engagement during the interaction; these include disturbances, noise, lack of feedback, prejudices, and level of interest in the content topic. Our model can represent the response in maintaining the degree of engagement [5].

1.6. Model Validation

In order to validate the model, we analysed the interactions of 500 users at an interactive museum in Tijuana, Mexico; the museum user was considered the learner and the exhibition as the educator. Data were gathered by observing user–exhibition interactions.

In order to achieve the above, we developed reactive modelling to maintain the frequent interaction in the learning process environment; we gave the learner a personalised environment (contents, services, information) to interact with his or her own interaction context. This research is also aided by different paradigms, such as multi-agent modelling to represent the involved elements in agents (learner and educator). We consider the behaviour of the learner's beliefs, desires, and intentions using the BDI paradigm. We use HCI to simulate the learner and educator interaction and Type-2 fuzzy logic to develop fuzzy perception in the educator BDI agent.

2. Related Work

Fuzzy logic has been seen to play an important role in situational perceptions research; prior research has been conducted on the reduction of interruption factors on HCI ([6], which introduces a fuzzy perception model for BDI agents to support the simulation of decision-making processes in environments with imperfect information; [7] introduced a graded BDI (g-BDI) agent development framework; g-BDI enabled the creation of agents as multi-context systems that provide the reason for three fundamental and graded mental attitudes (i.e., beliefs, desires, and intentions); [8] constructed an accurate model of the prevailing situation in order to make effective decisions about future courses of action within constraints using classical reasoning, augmented with a fuzzy component, generating beliefs in a fuzzy context of energy-awareness; [9] integrated the concepts of the BDI agent architecture into spatial issues; as a result, a novel spatial agent model was designed and implemented to analyse urban land use planning, the result of which showed the effects of spatial agents' behaviour, such as intention, commitment, and interaction on their decisions, always considering the uncertainty presented.

2.1. Intelligent Environments

The remarkable and rapid progress of research and technology allows us today to consider environments where the individual or group moves as a single entity able to understand the specific characteristics of a person or group to suit their needs and then respond intelligently to requests or appropriately respond in a natural and intuitive way. Of course, this environment must recognise the security and privacy needs of the individuals with respect to encountered situations. In the research field of intelligent environments (IEs), many scenarios can be provided as an example, such as a traveller who arrives at the airport in a foreign city where an IE program has been established; with this example system, the person is identified and verified immediately by the immigration authorities; they are given guidance on the rental of a car and a recommended route to their hotel. When the traveller arrives at the hotel, the room intelligently matches her/his personality with room features, including temperature, music, lighting, etc. IE may be seen as the coming together of different research areas [10], and it is seen to have some principal characteristics: ubiquity, context-awareness, natural interaction, and intelligence.

Perceiving the Learning Environment

A person's perspective can be described as the intention of a user to understand the situation that another user perceives [11]. As this definition highlights, the precise mechanisms of perspective are not highly specified, although research in cognitive psychology has suggested a shifting of attention to features of the environment that have less emphasis on an egocentric viewpoint [12,13].

The learning environment can be considered a place where learning is fostered and supported. For the learning environment to be fuzzy, it cannot be fully predefined. Sometimes, the users are involved in choosing learning activities and controlling the pace and direction of delivery, and accordingly, uncertainty and lack of control come into play. This place is the guide in the condition of continuing tentativeness and guardedness, but despite much care and attention, the system will often be chaotic to users. The complex nature of the learning environment interaction requires careful planning and design to avoid issues arising. An effective learning environment needs to be complemented by additional power resources of other users and the surrounding environment and culture [14].

2.2. Computational Intelligence

Computational intelligence (CI) is a well-established paradigm with current systems having defined computer characteristics that perform different tasks; sometimes, these tasks can be complicated and performed only by conventional actions. CI involves adaptive mechanisms to perceive and learn the intelligent behaviours presented in complex and chaotic environments; it also possesses attributes of abstraction, discovery, and association [15].

CI is a fast-moving, multidisciplinary field; it covers disciplines such as algorithms, data structures, neuro-computing, and artificial intelligence [16]. Nowadays, CI has attracted more attention over traditional artificial intelligence due to its tolerance of imprecise information, partial truth, and uncertainty [17]. Artificial intelligence is inefficient when solving problems with large input sizes (e.g., in data mining), whereas CI can support these.

Fuzzy Inference Systems

Computing using inference based on fuzzy logic is a popular method of computing, with many applications in areas such as control classification, expert systems, robotics, and pattern recognition adopting this method. The fuzzy inference system (FIS) is known by many names, such as the fuzzy expert system, the fuzzy model, fuzzy logic associative memory, and the fuzzy controller (also known as fuzzy rule-based systems). The FIS represents the major unit of a logic system. It can formulate adequate rules , based on which the decision is made. Figure 1 depicts the FIS structure.

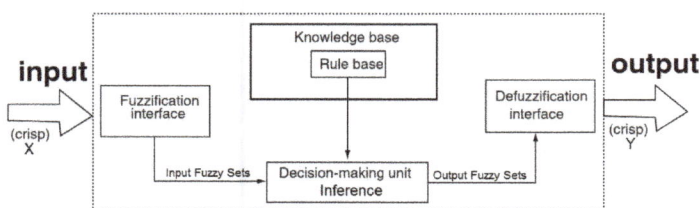

Figure 1. Fuzzy inference system.

An important element of the FIS is fuzzy sets, whereby knowledge is partitioned into particular situations; the fuzzy set is defined by its vague and ambiguous features, and limits may also be specified ambiguously. Crisp sets are those without ambiguity in their membership functions. The fuzzy set theory can deal powerfully with the presented ambiguity.

The work in [18,19] proposed the concept of Type-1 and Type-2 fuzzy sets. Type-1 fuzzy sets are described by membership functions that are aligned with numeric values between [0, 1], whereas Type-2 fuzzy sets are described by membership functions that are themselves fuzzy, with linguistic and subjective values. Type-1 fuzzy sets are certain. Type-2 fuzzy sets are useful in scenarios where it is complicated to determine exact membership functions due to the uncertainty present; accordingly, they can be used to handle rule uncertainties and measure uncertainties. For these reasons, we chose to work with Type-2 FIS, because it allowed for the handling of arising uncertainties; additionally, the case study reported is a dynamic context where uncertainty exists all of the time and at every moment.

2.3. Agent-Based Modelling

Agent-based modelling (ABM) is increasingly applied to numerous empirical situations [20]. Its methodological advantage lies explicitly in its ability to simulate human decision-making processes while considering high degrees of heterogeneity [21,22]. ABM addresses the primary drawback of current simulation programs, which are limited in uncertainty prediction and provide a dynamic response to those uncertainties [23]. ABM takes into account the behaviours that emerge from the interactions of numerous autonomous agents [24]. It is capable of addressing the uncertainty of real-world actions using fuzzy logic techniques, rough sets, Bayesian networks, etc. [25].

An agent based in ABM can think and act like humans operating under autonomous control and perceive its environment autonomously, adapting to changes in order to achieve certain goals [26]. In making behavioural decisions, ABM outperforms simple reactive "if–then" rules by allowing agents to learn and change behaviours in response to their experiences [27]. Even at the simplest level, an ABM consists of agents and the relationships between them; there could be valuable findings of the system as a whole [28].

BDI Agent Architecture

Nowadays, new cooperative strategies for multi-agent systems and the combination of high-level compressed state representations and hybrid reward functions produce the best results in terms of task completion rates and learning efficiency. In the BDI paradigm, agents' states are represented through three types of components: beliefs, desires, and intentions. From the viewpoint of sociology and psychology, it has been an important direction using the BDI model to study agent modelling [29].

The BDI model is an abstraction of human deliberation based on rational actions theory in the human cognitive process [30]. Intention is subsequently planned and executed. A deliberation process selects the optimum goal from a set of possible options that meet a specific desire.

Goal-oriented approaches were advanced, as well as the requirements for the modelling method to assess likely user actions [31]; this incorporates a notion of the awareness of soft human behaviours into the system design and has been adopted by the goal-oriented requirements agents [32].

The BDI model has become almost a norm in the field of multi-agent systems (MASs) [33].

3. Interactive Museum Case Study

In order to validate the proposed model, we analyse and observe, identifying the involved elements during the learning process among educator and learner. The case study was carried out by modelling scenes on interactive environments that may represent a magnificent place for modelling interaction. In these environments, we find a variety of interactive exhibitions from which different situations emerge due to the presence of groups of people; otherwise, we find interruption factors causing incomplete interactions, reducing performance and increasing error rates, affecting user attention, as well as the emergence and variety of scenarios that can provide feedback to the research.

Due to its facilities and dynamism during daily activities, the Interactive Museum of "El Trompo", located in Tijuana, Mexico, was chosen as a suitable place for our study. This is because it is an interactive educational museum dedicated to youths. Its primary goal is to provide a place to interact and play while learning.

3.1. Methodology

3.1.1. Room Selection

In order to analyse user–exhibition interaction in a better way, we studied (data sheets) the museum rooms, their themes, objectives, goals, methods of interaction, and their logistic location physically and theoretically; additionally, we observed the methods of interaction found in every room in order to select a suitable room that allowed us to analyse the behaviour, actions, performance, interruption factors, interaction distance, and interaction levels of users, as well as the interactive content type, information, and/or services of exhibitions provided. We also considered whether the content was adequate for users, suitable in relation to the kind of interactions of users, and adequate in maintaining the attention of the user. We further examined whether the content was harmful in causing interruptions or useful in avoiding interruptions. Additionally, we analysed the objective to determine whether it was adequate at encouraging a good interaction for the users and the media interface of the exhibition modules to determine whether they were adequate to have a good interaction.

3.1.2. Exhibition Module Selection

After analysis of the different exhibition modules, an interesting interactive module was chosen with features that allowed us to obtain the majority of the parameters to analyse in the research. The name of the exhibition module chosen was "Move Domain". The educational experience involved users interacting and playing with one of four objects (car, plane, bike, or balloon), which were displayed simultaneously on four separate screens, demonstrating the four different methods of moving in the simulated virtual world. Users were able to have the experience of using all four transportation means; they were able to interact in the virtual world and see how other users travel and interact around the virtual world. The exhibition's objective was to allow users to develop hand-eye coordination skills and spatial orientation using technology. The content was based on eye coordination and interaction with electronic games, with the exhibition's message being "I can learn about virtual reality through playing". The suggested number of users allowed at the same time was four.

3.1.3. Exhibition Module Interface

The module interface consisted of four sub-modules attached to connectors. Each module included a cover stand for the 32 inch screen, software that simulated the virtual world, and a cabinet to protect the computer.

The exhibition module was provided with a joystick to handle the plane, a steering wheel and pedals to drive the car, handlebars to ride the bike, and a rope to fly the balloon.

This interactive exhibition module is one of the most visited in the museum, and allowed us to obtain important data for analysis, processing, and validation of the model. Figure 2 depicts the analysed module.

Figure 2. The analysed interactive exhibition module. The figure shows the analysed module; this is composed of four sub-modules with different user interfaces.

3.2. Study Subjects

As subjects for the study, users were randomly selected from those children and adults who participated in supervised tours as a part of a permanent program of collaboration between local schools and the museum. The institution has the necessary agreements in place with the schools to conduct noninvasive interactive module evaluations to improve their design.

We evaluated user interaction–interruption behaviour by performing ethnographic research (notes style) to observe in a noninvasive way. Personal data were not required; therefore, information was produced directly in the museum room through real-time observations in line with institution committee recommendations to guarantee the anonymity of users.

Evaluation Interaction Parameters

We analysed and studied parameters such as interaction level (Which influencing factors increase or decrease the level of interaction? What is the quality of interaction? What is the interaction time? Which are the interaction abandonment factors?), presence (Do the users have a constant presence? Do the users have intermittent presence?), interactivity (Do the users have interactivity directly or indirectly with the exhibition? Do the users have shared interactivity with the exhibition?), control (Do the users have full control over the exhibition?), feedback (Do the users receive some feedback about the content?), creativity (Do the users change the way they interact according to their creativity?), productivity (Do the users propose something that changes the interaction?), communication (Do the users have communication directly from the exhibition?), adaptation (Do the users adapt their actions according to the interactive content type delivered by the exhibition?), and distance (is the users' distance adequate to interact with the exhibition? Is the distance a factor in interrupting or improving the interaction?).

All data collected were analysed in order to develop the FIS with all possible factors involved in order to obtain an adequate content interactive type. Figure 3 depicts in detail the average results of the parameters of the 500 users analysed.

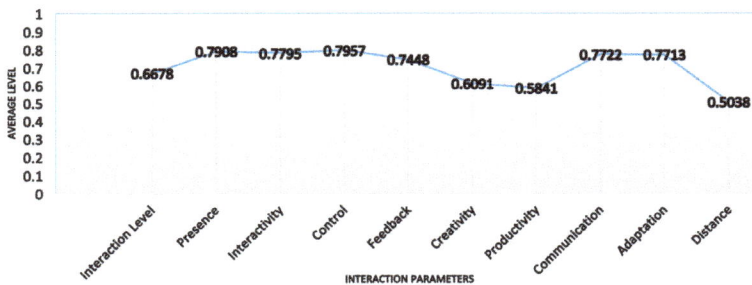

Figure 3. Average of interaction parameters. The figure shows the interaction parameters necessary to develop the adequate fuzzy inference system (FIS) in order to obtain the suitable interactive content type.

4. Modelling Interaction on HCI

To support user interactions, HCI is operating as a background process, using invisible sensing computational entities to interact with users. In our research, these entities are simulated by the learner BDI agent and the educator BDI agent. The entities' collaboration permits the HCI to deliver a customised interactive content type to users in a noninvasive manner which is context-aware.

The relationships between users (museum users) and computers (exhibition) need to be systematically modelled and represented to be ready for the emergent context; for this reason, we represent user–exhibition relationships. In addition, if we add tools such as JT2FIS [34] to facilitate these interactions, it should be possible to deal with the uncertainty presented by the inputs (interaction level, distance), task conflicts, and interruptions that occur when a user is involved in multiple interactions simultaneously.

Interactions—both personal and business related—are an essential part of our daily lives; these interactions are responsible for our well-being and productive work or business environment. The proliferation of mobile devices and pervasive computing technologies has radically changed the ways in which people communicate and the interaction between them. These devices support human interactions as they are equipped with software that helps users in coordinating, competing, and collaborating with each other [35].

In our research, we represent HCI simulated on a museum modelled with embedded agents (LA, EA) that allow the user–exhibition interaction to be supported. Our modelling proposal provides dynamic support for interactions, and it is aware not only of the user's physical context, but also of the social context (i.e., when a user interacts with another user). Our model consists of contextual attributes, such as the location of the user and what they are doing during the interactions. The social relationships are among users and other actors (teacher, museum guide, classmates, etc.) that influence an individual's action and performance during the interaction.

The idea for modelling the interaction on HCI in this research is important because it helps to improve the interaction level experience, offering an adequate interactive content type based on interaction level and distance to avoid possible interruption factors. Intensive research has been carried out in context modelling and context-awareness systems, focusing on physical contexts. However, these studies are not particularly explicit in addressing or modelling user–exhibition interactions and coordination functionality to improve the interaction quality experience.

Moreover, modelling the interaction on HCI provides opportunities to improve the interactions among the user and computer engaging in spontaneous activities as in our case study. Our HCI modelling selects and delivers adequate content and information to support the interactions of a large number of diverse users. Our proposed model supports spontaneous interactions and interruptions, coordinating available content during the course of the interaction.

Related work on supporting the interactions of users with available content in interactive environments has tended to focus on recognising activities based on the identification of meanings attached to a place. However, they do not account for the new requirement of spontaneity. A way to realize spontaneity is the approach of task-oriented computing [36].

4.1. Representing HCI in a Museum

Currently, museums are defined as non-profit institutions that serve to acquire, conserve, research, communicate, exhibit, and study educational material for the purpose of learning and customer enjoyment. Given the significant influence of museums on society, researchers have paid them much attention; for instance, Reference [37] studied the dynamic interaction between perceived quality and emotion as determinants of visitor satisfaction.

In our research, we modelled interactive museum elements; the museum user was represented by LA and the exhibition by the EA. In order to know the interactive content type offering adequate interactive information avoiding interruption, we used inputs like interaction level [38] and distance measuring [39]; the distance is randomly simulated, based on the measurement of the effective reading distance of passive tags embedded in the environment.

In the measurement of the distance and the level of interaction, some numerical results were obtained. These could be the inputs to our FIS. This practice measuring entails the following reflection: Does a metric allow us to properly measure the distance and interaction level of the user? For this reason, our model is composed of linguistic input variables with their respective membership functions to the interaction level (intlevel0, intlevel1, intlevel2, intlevel3, intlevel4, intlevel5) and distance (far, medium, near); linguistic variables can be considered a measure of the belief of the user (LA) and exhibition (EA).

Within the existing model, the implementation of the BDI paradigm was chosen because the propositions of beliefs represent the level of interaction and distance where the BDI agent assumes the user has a specific evaluated interaction level and distance. Intentions are a subset of desires that should encourage and assist the user in the process of interaction through interactive activities. In order to approach the relevance to integrate fuzzy logic modelling to formalise a BDI agent's beliefs, recognising that transition ranges exist on interaction levels and distance, the use of fuzzy logic is deemed appropriate for exhibition (EA) reasoning, in order to deliver the adequate content type.

This research analyses data obtained from BDI agents (EA and LA) simulating the interaction among the user and exhibition. The educator BDI agent confronts its desires with its beliefs using fuzzy logic to infer relevant information on the interaction level and the distance of users' performance. This information is obtained using fuzzy perceptions.

4.1.1. Modelling Museum Elements

In our research, we define HCI as interactive museum elements composed of two principal actors: user (LA) and exhibition (EA). Through their performance (distance and interaction level), the user (LA) is evaluated by fuzzy perceptions of the exhibition (EA), obtaining (using the proposed FIS) the adequate interactive content type, ensuring the user's interest is maintained and avoiding the interruption factor. These agents have direct communication, constantly requesting and receiving information. The exhibition (EA) contains all information of the interactive exhibition it can collect through sensing using fuzzy perceptions; it can sense the performance data (interaction level and distance) of the users (LA). Interaction level and distance operate as feedback data to be processed by reasoning of FIS to offer the adequate interactive content type to help and increase the interaction quality experience of the user (LA), maintaining attention for the exhibition and avoiding the interruption factor.

The user (LA) and exhibition (EA) involved in the HCI receive and request all changes occurring during the interaction; consequently, the agents are then ready for emergent changes. This is of great importance, as it feedback information due to their awareness of the context at all times. Figure 4 represents the two agents involved in HCI.

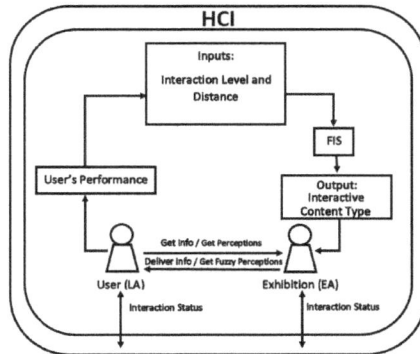

Figure 4. Agents involved in human–computer interaction (HCI). The figure shows the principal agents' interactions among each other and their context. LA, learner agent; EA, education agent.

The definition of the involved agents is based on the characteristics of the HCI. This means that the resources that provide services and information are based on user performance (distance, interaction level). The HCI serves as a mediator to update the agents with the status and changes that occur during the interaction. The EA is a representation of a BDI agent with mental states. The beliefs and desires of the agent involve definitions, estimating that the user has a certain distance and the level of interaction during a given period; this estimation can deal with uncertainty, determined by a set of membership functions.

The EA can perceive its interaction context with perceivers (fuzzy perceptions) and act accordingly with effectors (intentions). The exhibition (EA) perceivers must perceive and act to satisfy the user's requirements. Perception is an internal action where the agent captures information from its interaction context, formulating its beliefs from perceptions combined with prior knowledge of the interaction context (i.e., the EA has an interactivity perceiver that can measure the interaction level of a user to some degree of error; due to its prior knowledge, it creates the belief if the user has a high or low interaction level).

On the other hand, acting is the performance of an external action where the agent generates changes that modify the interaction context, offering adequate interactive content or customised services to the user. In this research, we developed powerful agents that covered the prior characteristics, implementing BDI paradigms such as such Jason [40], based on the AgentSpeak (L) introduced by Rao and Georgeff [41]. Figure 5 illustrates the process of the rationalization of the exhibition (educator BDI agent).

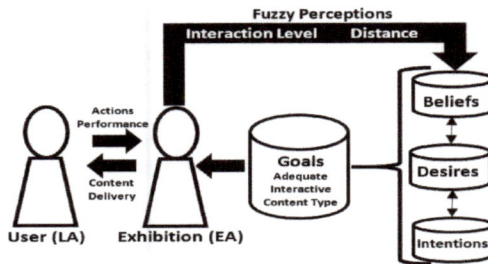

Figure 5. Exhibition (educator BDI agent) rationalization. The figure illustrates the rationalization process elements: the beliefs, desire, and intentions which are key to obtain the goals—in this case, to deliver the adequate interactive content.

The exhibition's belief base has several defined plans. Each plan is specified by different membership functions of linguistic variables that are received from the interaction, level, and distance in the process of the interaction. These plans have implications through beliefs: beliefs over desires and desires over intentions; consequently, this assists in determining the most appropriate interactive content type for the user, avoiding interruptions because it is based on the user's performance.

4.1.2. Development of Museum Elements

In our research, HCI is represented in an interactive museum. The involved elements—user (LA) and exhibition (EA)—are developed and simulated with a Jason-based programming language on AgentSpeak (L) ([40–43]).

The insertion of the proposed interaction context consists of hybrid BDI–fuzzy elements, binding the BDI and fuzzy logic paradigm. The idea here is to offer adequate interactive content types, avoiding interruptions that provoke incomplete interactions in the HCI.

The developed agents sometimes conduct simple tasks, performing only occasional database queries, sending data and information processed to responsible agents. Our reactive and involved agents—LA and EA—are in constant readiness to obtain information from the interaction context.

The communication among the agents is composed of four elements: receiver agent, sender or issuer agent, type of action, and message content. Messages can be sent and received to instances of LA and EA. The type of action defines which message will be sent (i.e., reporting the status of the HCI or informing about a user's performance).

The message content has different types of information values, including interaction level (value among 0–5 depending on the user interactivity), user–exhibition distance (value among near, medium, and high), interactive content type delivered (audio, graphic, text, and video; considering that audio content requires low interaction, graphics content requires medium interaction, text content requires high interaction, and video content requires extremely high interaction). Figure 6 depicts the communication process among agents.

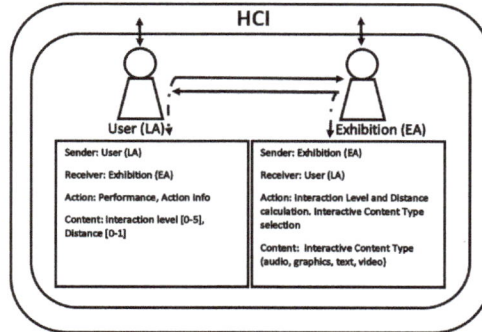

Figure 6. Communication process among agents. The figure shows the agent's communication process and is composed of four elements: sender, receiver, action, and content.

4.1.3. Fuzzy Perceptions in the Museum Elements

The logic in using BDI and the fuzzy logic paradigm in the museum elements is to help handle uncertain information in order to present adequate and appropriate interactive content. A belief is prepared in accordance with environmental inputs (interaction level and distance), with the respective membership functions. These variables act as inputs to the FIS, which define the output (interactive content type). The result of these beliefs is written by a fuzzy value and, in this case, given a linguistic value. The update process is dynamic and is altered according to the user's performance. The membership functions are modelled considering an initial belief based on fuzzy perceptions and ensuring an accurate result for assessing the user interaction.

The implementation of the FIS is solely for the purpose of effective utilization; this requires the use of programs that directly apply fuzzy logic functions. Some utility programs have specific modules to facilitate the accomplishment of this task and provide the necessary tools to conduct effective fuzzification; these include the utility JT2FIS [34], which is a Java class library for an interval Type-2 fuzzy inference system, used to build intelligent object-oriented applications. It also provides an effective fuzzification method and tools. The JT2FIS utility is used in this research.

The HCI inputs are the input variables that can be perceived by the EA. They are called the performance data of the user's interaction. On identifying the input variables (interaction level and distance) and the output variable (interactive content type) for the FIS, these are associated with a set of membership functions.

These functions comprise linguistic variables for the input "interaction" (intlevel0, intlevel1, intlevel2, intlevel3, intlevel4, intlevel5), for the input "distance" (far, medium, and near), and for the output "interactive content type" (audio, graphics, text, and video). Gaussian functions were used in all inputs and outputs because this type of membership function has a soft non-abrupt decay.

The FIS was implemented using the Mamdani method with the minimal implication operator and the defuzzification method of the centroid.

We defined 18-inference IF–THEN rules covering all linguistic variables; the rules are composed by the operator associated with the minimum method. Aggregation rules are made by the maximum method. The proposed FIS is flexible and permits the addition or deletion of rules; this can be seen as an advantage, as it can be adapted to different contexts or, if different variables exist, can be increased.

The FIS was also configured considering the performance of the 500 users. It is one of the most important elements of the proposed model, as it represents the inference that can be considered as the knowledge base. Table 1 shows the base rules (the knowledge base representation used in this research).

Table 1. Inference fuzzy rules of the FIS.

No	Inference Fuzzy Rules
1	If (IntL is IntL5) and (Dist is Near) then (IntConType is video)
2	If (IntL is IntL5) and (Dist is Med) then (IntConType is video)
3	If (IntL is IntL5) and (Dist is Far) then (IntConType is text)
4	If (IntL is IntL4) and (Dist is Near) then (IntConType is video)
5	If (IntL is IntL4) and (Dist is Med) then (IntConType is video)
6	If (IntL is IntL4) and (Dist is Far) then (IntConType is text)
7	If (IntL is IntL3) and (Dist is Near) then (IntConType is video)
8	If (IntL is IntL3) and (Dist is Med) then (IntConType is text)
9	If (IntL is IntL3) and (Dist is Far) then (IntConType is grfs)
10	If (IntL is IntL2) and (Dist is Near) then (IntConType is text)
11	If (IntL is IntL2) and (Dist is Med then (IntConType is text)
12	If (IntL is IntL2) and (Dist is Far) then (IntConType is audio)
13	If (IntL is IntL1) and (Dist is Near) then (IntConType is grfs)
14	If (IntL is IntL1) and (Dist is Med) then (IntConType is grfs)
15	If (IntL is IntL1) and (Dist is Far) then (IntConType is audio)
16	If (IntL is IntL0) and (Dist is Near) then (IntConType is grfs)
17	If (IntL is IntL0) and (Dist is Med) then (IntConType is audio)
18	If (IntL is IntL0) and (Dist is Far) then (IntConType is audio)

4.1.4. Fuzzy Perceptions Process

An important process that impacts all museum elements is how the HCI—aided by exhibition (EA)—obtains fuzzy perceptions from the user's performance. This process begins with the perceiver (χ) observation of changes in the surrounding environment; this change can be represented as a set of indicators (ζ), with every indicator being described by set membership function values (π). The indicators (ζ) can be perceived by perceivers χ, which can sense these values and consider the inputs (interaction level and distance) of the FIS. Through its inference, the FIS generates the output (interactive content type). The fuzzy value result will be considered a belief atom $\kappa(\tau)$, where $\kappa(\tau) \, \varepsilon \, \Omega$ is a belief set. Figure 7 depicts the fuzzy perception process of the exhibition EA.

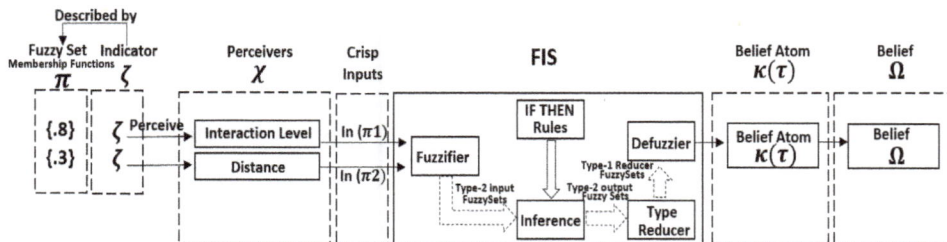

Figure 7. Fuzzy perception process. The figure shows the fuzzy perception process. This process is the key to creating the fuzzy perceptions of the EA in order to deliver the adequate interactive content type, considering the user's performance (interaction level and distance).

4.1.5. Illustrative Example

The following example illustrates how the entities in this case (museum visitor, represented by LA) can change their performance (interaction level and distance) during a museum tour day and how the exhibition can perceive the performance changes using fuzzy perceptions.

Let us consider that a visitor takes a museum tour. At the beginning of the tour, the visitor starts the interaction on the first exhibition modules; at this point, the visitor's performance is interaction level is "5" and distance is "near" the exhibition (let us suppose the visitor has much energy and interest in the exhibition).

Therefore, the EA, using fuzzy perceptions, implies its beliefs, then based on its beliefs, the resultant belief is a video (fuzzy value) because it perceives high interaction (0.9) and near distance (0.9). Later, let us suppose that the visitor is in the middle of their museum tour (the visitor decreases their performance and interest). Now, the interaction level is "2.5", and the distance is "medium" with respect to the exhibition. Therefore, the EA, using fuzzy perceptions, implies its beliefs, then based on its beliefs, the resultant belief is graphics (fuzzy value), because it perceives medium interaction (0.4) and medium distance (0.4). Finally, let us suppose that the visitor is at the final part of the museum tour (the visitor's performance and interest continues to decrease during their visit). Now, the interaction level is "0.5", and the distance is "far" from the exhibition. Therefore, the EA, using fuzzy perceptions, implies its beliefs, then based on its beliefs, the resultant belief is audio (fuzzy value) because it perceives low interaction (0.1) and distance far (0.1).

Table 2 depicts the exhibition (domain) agent's mind state, elements, interaction level, and distance presented.

Table 2. Mind state values of exhibition EA.

Interaction Level	Distance	Element	Description	Source
5	Near	Belief	video(fuzzy value(.9),(.9))	percept
2.5	Medium	Belief	graphics(fuzzy value(.4),(.4))	percept
0.5	Far	Belief	audio(fuzzy value(.1),(.1))	percept
5	Near	Events	+!select(video)	self
2.5	Medium	Events	+!select(graphics)	self
0.5	Far	Events	+!select(audio)	self
5	Near	Intentions	+!deliver(video)	self
2.5	Medium	Intentions	+!deliver(graphics)	self
0.5	Far	Intentions	+!deliver(audio)	self

5. Results

In this section, the results obtained from the sample of 500 users visiting the interactive museum "El Trompo" in Tijuana, Mexico are presented and analysed. The users were evaluated and processed using a custom fuzzy c-means method of data mining named Data Mined Type-2 (DMT2F) [38].

5.1. FIS Configuration

Once all data were mined with DMT2F and 18 inference rules were added to build the knowledge base, we obtained as a result of the FIS configuration parameters that the FIS was configured with two inputs (interaction level and distance). These inputs were composed with the exact parameters considering the performance of the 500 users. These inputs are essential for feedback to the FIS; also, the FIS was configured with one output (interactive content type). This output is important because it delivers the adequate interactive content type in order to avoid interruption during the interaction.

Additionally, DMT2F considers the uncertainty of each user. This uncertainty is assigned automatically by the evaluation method of the FIS Type-2. The implementation of DMT2F is adequate in the interaction context because it is in context where uncertainty is contemplated and present at all times. During this process, an FIS with higher accuracy was obtained in the realized configuration, as we considered and undermined all possible input variables; consequently, an accurate interactive content type result was obtained during real-time interaction, improving the user's interest, attention, and interaction, and avoiding interruptions. Table 3 depicts the configuration parameters of the data mined Type-2 FIS.

Table 3. Inputs, outputs data mined Type-2 FIS configuration.

Type	Member Function	Params (Values)
Input1 (IntLevel)	GaussUncertainty MeanMemberFunction	IntLevel0 = [0.486 1.709 2.196] IntLevel1 = [0.373 2.387 2.761] IntLevel2 = [0.300 3.021 3.321] IntLevel3 = [0.284 3.607 3.891] IntLevel4 = [0.286 3.676 3.963] IntLevel5 = [0.376 4.132 4.508]
Input2 (Distance)	GaussUncertainty MeanMemberFunction	Far = [0.196 0.312 0.509] Medium = [0.200 0.372 0.573] Near = [0.217 0.492 0.71]
Output (IntConType)	GaussUncertainty MeanMemberFunction	IntCnTypAud = [0.090 0.463 0.554] IntCnTypGph = [0.066 0.597 0.663] IntCnTypTxt = [0.070 0.618 0.689] IntCnTypVid = [0.088 0.792 0.880]

5.2. Interactive Content Type Results

After using the DMT2F in our FIS, an evaluation was conducted to obtain adequate interactive content type results for the sample of 500 users. Table 4 depicts these analysed users.

Table 4. Results of the interactive content type using the data mined Type-2 fuzzy method (DMT2F).

Subject	Interaction Level	Distance	Interactive Content Type Using DMT2F
1	1.6738	0.4168	0.5553 (audio)
2	2.0087	0.3130	0.0.5451 (audio)
3	3.0073	0.6225	0.5921 (graphics)
4	4.1161	0.8769	0.7665 (video)
5	4.3935	0.3055	0.7370 (video)
6	3.8101	0.3505	0.7166 (text)
...	...	...	...
500	4.3870	0.6969	0.7679 (video)

Interactive Content Type Percentage

In this research, we obtained noteworthy results about the percentage of interactive content types delivered. The sample results obtained were: 20% of users received content type "video"; 21% of users received content type "audio"; 32% of users received content type "text"; and 27% of users received content type "graphics". This means that the interactive content type "text" is the most adequate to interact with in this kind of interactive environment, helping to avoid possible interruptions originating from inadequate content. Figure 8 depicts these results.

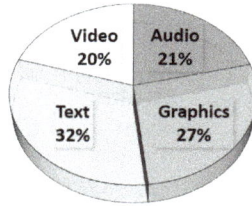

Figure 8. Interactive content type percentage. The figure shows the interactive content type percentage of the 500 users analysed.

6. Discussion

Nowadays, technology is being increasingly used in museums around the world. However, in Tijuana, Mexico, not many museums are adopting new technologies, with many simply offering the same content type in all of their exhibitions. On the other hand, we identify some museums, such as our case study "El Trompo" Interactive Museum in Mexico, which has adopted interactive exhibitions through the use of technology.

6.1. Technology Use

The use of technology has been seen to help attract more visitors to museums, with visitors demanding adequate and appropriate interactive content types that permit and encourage their interest throughout the museum, avoiding interruptions that impact interaction. One way to achieve this is through the proposed model, which offers museum visitors experiences that directly impact their knowledge and interest. Our model evaluates and perceives user uncertainty in their interaction (level and distance), aided by fuzzy perceptions in EA. After this perception, the output interactive content type is identified, offering suitable services and information; having an impact and improving interest avoids interruptions or abandonment of the museum's exhibitions.

6.2. Applying Fuzzy Logic

Our agents—particularly the domain (EA) with mental states—can perceive interruptions at the moment of the first interaction through the recognition of the user's performance (interaction level and distance). The use of fuzzy logic to reconstruct the beliefs on BDI EA can deal with uncertainty in a model built in part from mental states. This integration allows the analysis of uncertainty in the user's performance process, resulting in a suitable profile of each user's performance.

6.3. Interruption Factor

The user's low performance is a consequence of interruption factors during the interaction, and in some cases, it takes time to recover from or could represent a complete interruption of interaction. In order to identify how to avoid interruptions, some questions arise: How can we recover from irregular interruptions, or how we can avoid interruptions? The response is through the proposed research where we analysed factors and offered adequate interactive content types, which can avoid these interruptions. If we deliver the adequate and appropriate content, we can help users recover from suspended interaction.

Interruption Factor Recovery

When a user is interrupted, it is necessary to recall their progress before the interruption occurred, but if we are able to recall this progress with the adequate content, the interruption's recovery is much faster; otherwise, the interruption's recovery can be slower. We can compare the proposed research with other recovery techniques [44,45], analysing ways to resume after interruption; through comparison, we can see that our research improves and strengthens the interaction in order to increase attention in the exhibition. Moreover, we analysed if negative effects of the interruption are avoided if the user's performance increases having a complete and uninterrupted interaction. Comparing our research with other research [46,47] that used avoidance of interruption techniques, none were seen to use fuzzy perceptions and did not consider users' uncertainty, which increases the error degree to avoid interruption.

6.4. Performance of Users

The user evaluation during HCI is completed to determine the adequate interactive content type based on performance, the gathering of relevant effects of interruptions, and the identification of what happens when users are interrupted by a particular interruption factor. This could include interruptions with content related to the exhibition, where the performance could be maintained, reduced, or minimised, but not representing the interaction. Otherwise, for interruptions that are not content related, the performance may degrade until null performance, representing an interaction abortion. If we deliver adequate content using our proposed model, it is important to resume interaction after interruptions. Understanding such adaptations will improve interaction level experience.

6.5. Noninvasive Evaluation

The sample of users was examined during their natural interaction experience. The proposed evaluation method is noninvasive because it operates in background mode. The users can interact naturally, obtaining real information in a natural way, influencing the quality of results.

7. Conclusions

During the learning process, we can identify many interruption factors. The proposed model helps the learner experience a quality means of learning, which is provided by the educator with all of the adequate interactive content types, improving performance and creating an effective learning process.

We have explored the idea that HCI enhancement—with adequate content types that avoid interruptions—could intensify the positive influence on learning and improve the relationship with other users; this can be beneficial to elevate the interactive level experience, allowing a direct repercussion in the quality of HCI.

Through experimental research, we model and represent with agents the learner (LA)–educator (EA) interaction; both agents are provided with the BDI approach, which permits the creation of beliefs, desires, and intentions, allowing mental states for reasoning. The EA uses the fuzzy logic approach to have a fuzzy perception of the user's performance, improving interaction and avoiding interruptions.

We have demonstrated that if adequate and appropriate content types are offered, the interaction can be completed inclusively with some interruptions. We analysed if the content of the interruption is similar to the interaction context, identifying that not all interruptions can be deemed negative, with some interruptions helping users to complete their interaction. Otherwise, if the interruption content is completely different, it can increase the interruption level until the user's interaction is abandoned.

This research can act as an alternative for HCI researchers to approach successful interaction; it can represent an option to understand how to minimize or avoid the interruption factors during interactions.

Acknowledgments: We would like to thank The National Council for Science and Technology of Mexico, the Autonomous University of Baja California and the Interactive Museum 'El Trompo' for the support granted during this research.

Author Contributions: All of the authors conceived of and designed the study. Manuel Castañón-Puga and Felipe Lara-Rosano provided methodological advice. Ricardo Rosales, Nora Osuna-Millan and Maria Virginia

Flores-Ortiz performed the field survey and analysed the data. Ricardo Rosales and Richard David Evans wrote the paper. All authors read and approved the manuscript.

Conflicts of Interest: The authors declare no conflict of interest.

References

1. Morley, D.; Parker, C. *Understanding Computers: Today and Tomorrow*; Cengage Learning: Boston, MA, USA, 2013.
2. Kolb, D.A. *Experiential Learning: Experience as the Source of Learning and Development*; Prentice-Hall: Upper Saddle River, NJ, USA, 1984.
3. Maureen, A.C.; Marsh, R. Interest level improves learning but does not moderate the effects of interruptions: An experiment using simultaneous multitasking. *Learn. Individ. Differ.* **2014**, *30*, 112–117.
4. McFarlane, D.C.; Latorella, K. Modifying the Classroom Environment to Increase Engagement and Decrease Disruption with Students Who Are Deaf or Hard of Hearing. *J. Deaf Stud. Deaf Educ.* **2012**, *17*, 518–533.
5. Guardino, C.; Antia, S. The Scope and Importance of Human Interruption in Human–Computer Interaction Design. *Hum.–Comput. Interact.* **2002**, *17*, 1–61.
6. Farias, G.; Dimuro, G.; Costa, A. BDI agents with fuzzy perception for simulating decision making in environments with imperfect information. In Proceedings of the Multi-Agent Logics, Languages, and Organisations Federated Workshops, Lyon, France, 30 August–2 September 2010; Volume 627, 23–41.
7. Casali, A.; Godo, L.; Sierra, C. A graded BDI agent model to represent and reason about preferences. *Artif. Intell.* **2011**, *175*, 1468–1478.
8. Song, S.; O'Hare, G.; O'Grady, M. Fuzzy decision making through energy-aware and utility agents within wireless sensor networks. *Artif. Intell. Rev.* **2007**, *27*, 165–187.
9. Behzadi, S.; Alesheikh, A.; Sierra, C. Introducing a novel model of belief-desire-intention agent for urban land use planning. *Eng. Appl. Artif. Intell.* **2013**, *26*, 2028–2044.
10. Weiser, M. The computer for the 21st century. In *Human Computer Interaction: Toward the Year 2000*, 2nd ed.; Morgan Kaufmann Publishers: Burlington, MA, USA, 1995.
11. Tomasello, M.; Kruger, A.; Ratner, H. Cultural learning. *Behav. Brain Sci.* **1993**, *16*, 495–552.
12. Frischen, A.; Loach, D.; Tipper, S. Seeing the world through another person's eyes: Simulating selective attention via action observation. *Cognition* **2009**, *111*, 212–218.
13. Tversky, B.; Hard, B. Embodied and disembodied cognition: Spatial perspective-taking. *Cognition* **2009**, *110*, 124–129.
14. Gayle, W. *Constructivist Learning Environments: Case Studies in Instructional Design*; Educational Technology Publication Inc.: Washington, DC, USA, 1996.
15. Hassaniena, A.; Al-Shammari, E.; Ghali, N. Computational intelligence techniques in bioinformatics. *Comput. Biol. Chem.* **2013**, *47*, 37–47.
16. Konar, A. *Computational Intelligence Principles, Techniques and Applications*; Springer: Berlin/Heidelberg, Germany, 2005.
17. Andina, D.; Pham, D. *Computational Intelligence for Engineering and Manufacturing*; Springer: Berlin/Heidelberg, Germany, 2007.
18. Zadeh, L. Fuzzy sets. *Inf. Control* **1965**, *8*, 338–353.
19. Zadeh, L. The concept of a linguistic variable and its application to approximate reasoning. *Inf. Sci.* **1975**, *8*, 199–249.
20. Smajgl, A. Challenging beliefs through multi-level participatory modelling in Indonesia. *Environ. Model. Softw.* **2010**, *25*, 1470–1476.
21. Gilbert, N. *Agent-Based Models*; SAGE Publications: Thousand Oaks, CA, USA, 2008.
22. Parker, D.; Manson, M.; Janssen, M.; Hoffmann, M.J.; Deadman, P. Multiagent system models for the simulation of land-use and land-cover change: A review. *Ann. Assoc. Am. Geogr.* **2003**, *93*, 314–337.
23. Lee, Y.; Malkawi, A. Simulating human behavior: An agent-based modeling approach. In Proceedings of the 13th IBPSA Conference, Chambery, France, 26–28 August 2013.
24. Sokolowski, J.; Banks, C. *Principles of Modeling and Simulation: A Multidisciplinary Approach*; John Wiley and Sons: Hoboken, NJ, USA, 2009.

25. Ramos, A.; Augusto, J.; Shapiro, D. Ambient intelligence-the next step for artificial intelligence. *IEEE Intell. Syst.* **2008**, *23*, 15–18.
26. Russell, S.; Norvig, P. *Artificial Intelligence; A Modern Approach*, 3rd ed.; Pearson Education International: London, UK, 2009.
27. Macal, C.; North, M. Tutorial on agent-based modelling and simulation. *J. Simul.* **2010**, *4*, 151–162.
28. Bonabeau, E. Agent-based modeling: methods and techniques for simulating human systems. *Proc. Nat. Acad. Sci. USA* **2002**, *99*, 7280–7287.
29. Dongning, L; Yong, T. Intelligent agents belief temporal substructure logic model. *Comput. Appl.* **2010**, *27*, 2448–2451.
30. Bratman, M. *Intentions, Plans, and Practical Reason*, 1st ed.; Center for the Study of Language and Information: Stanford, CA, USA, 1999.
31. Yu, E. *Modelling Strategic Relationships for Process Reengineering, Social Modeling for Requirements Engineering*; MIT Press: Cambridge, MA, USA, 2011.
32. Jian, Y.; Li, T.; Liu, L.; Yu, E. Goal-oriented requirements modelling for running systems. In Proceedings of the First International Workshop on Requirements@Runtime, Sydney, Australia, 28–28 September 2010; Volume 1, pp. 1–8.
33. Wooldridge, M.; Jennings, N. Intelligent Agents: Theory and Practice. *Knowl. Eng. Rev.* **1995**, *10*, 115–152.
34. Castañón-Puga, M.; Castro, J.R.; Flores-Parra, J.M.; Gaxiola-Pacheco, C.G.; Martínez-Méndez, L.-G.; Palafox-Maestre, L.E. JT2FIS A Java Type-2 Fuzzy Inference Systems Class Library for Building Object-Oriented Intelligent Applications. *Adv. Soft Comput. Appl. Lecture Notes Comput. Sci.* **2013**, *8266*, 204–215.
35. Castelfranchi, C. Modelling social action for AI agents. *Artif. Intell.* **1998**, *103*, 157–182.
36. Wang, Z.; Garlan, D. *Task-Driven Computing*; Carnegie Mellon University: Pittsburgh, PA, USA, 2000.
37. De Rojas, C.; Camarero, C. Visitors' experience, mood and satisfaction in a heritage context: Evidence from an interpretation center. *Tour. Manag.* **2008**, *29*, 525–537.
38. Rosales, R.; Flores, D.; Palafox, L. Representation of interaction levels using data mining and Type-2 fuzzy inference system. *Front. Comput. Sci.* **2017**, under review.
39. Luh, Y.-P.; Liu, Y.-C. Measurement of Effective Reading Distance of UHF RFID Passive Tags. *Mod. Mech. Eng.* **2013**, *13*, 115–120.
40. Bordini, R.; Bazzan, A.; Jannone, R.; Bassao, D.M.; Vicari, R.M.; Lesser, V.R. Agentspeak: Efficient intention selection in BDI agents via decision-theoretic task scheduling. In Proceedings of the first international joint conference on autonomous agents and multi-agent systems, Bologna, Italy, 15–19 July 2002; pp. 1294–1302.
41. Rao, A.; Georgeff, M. AgentSpeak(L): BDI agents speak out in a logical computable language. In Proceedings of the Seventh Workshop on Modelling Autonomous Agents in a Multi-Agent World, Einhoven, The Netherlands, 22–25 January 1996; Volume 1038, pp. 42–55.
42. Moreira, A.; Vieira, R.; Bordini, R. Extending the operational semantics of a BDI agent-oriented programming language for introducing speech-act based communication. In Proceedings of the International Workshop on Declarative Agent Languages and Technologies, New York, NY, USA, 19 July 2004; pp. 135–154.
43. Bordini, R.; Hubner, J. Jason: A Java-based AgentSpeak interpreter used with saci for multiagent distribution over the net. In Proceedings of the 6th International Workshop CLIMA VI, London, UK, 27–29 June 2005.
44. Altmann, E.; Trafton, J. Memory for goals: an activation based model. *Cognit. Sci.* **2002**, *26*, 39–83.
45. Brumby, D.; Cox, A.; Back, J.; Gould, S.J. Recovering from an interruption: Investigating speed-accuracy tradeoffs in task resumption strategy. *J. Exp. Psychol.: Appl.* **2013**, *19*, 95–107.
46. Mark, G.; Voida, S.; Cardello, A. A pace not dictated by electrons: an empirical study of work without email. In Proceedings of the Conference on Human Factors in Computing Systems, Austin, TX, USA, 5–10 May 2012; pp. 555–564.
47. Relihan, E.; O'Brien, V.; O'Hara, S.; Silke, B. The impact of a set of interventions to reduce interruptions and distractions to nurses during medication administration. *Qual. Saf. Health Care* **2010**, *19*, 1–6.

applied sciences

MDPI

Article

Evaluating Human–Automation Etiquette Strategies to Mitigate User Frustration and Improve Learning in Affect-Aware Tutoring

Euijung Yang and Michael C. Dorneich *

Industrial and Manufacturing Systems Engineering, Iowa State University, 3004 Black Engineering, Ames, IA 50011, USA; eui.j.yang@gmail.com
* Correspondence: dorneich@iastate.edu; Tel.: +1-515-294-8018

Received: 3 March 2018; Accepted: 18 May 2018; Published: 30 May 2018

Abstract: Human–automation etiquette applies human–human etiquette conventions to human–computer interaction (HCI). The research described in this paper investigates how to mitigate user frustration and support student learning through changes in the style in which a computer tutor interacts with a learner. Frustration can significantly impact the quality of learning in tutoring. This study examined an approach to mitigate frustration through the use of different etiquette strategies to change the amount of imposition feedback placed on the learner. An experiment was conducted to explore how varying the interaction style of system feedback impacted aspects of the learning process. System feedback was varied through different etiquette strategies. Participants solved mathematics problems under different frustration conditions with feedback given in different etiquette styles. Changing etiquette strategies from one math problem to the next led to changes in motivation, confidence satisfaction, and performance. The most effective etiquette strategies changed depending on if the user was frustrated or not. This work aims to provide mechanisms to support the promotion of individualized learning in the context of high level math instruction by basing affect-aware adaptive tutoring system design on varying etiquette strategies.

Keywords: intelligent tutoring systems; affect-aware tutoring; human–computer interaction; etiquette strategy; user frustration

1. Introduction

This work investigates the intersection of human–human etiquette strategies, student frustration, and the interaction style between a learner and an intelligent tutoring system. Preliminary results of this work were presented in [1]. Human emotion can drive the direction of conversation and plays a key role in communication [2]. Both positive emotions (e.g., happiness and fulfilment) and negative emotions (e.g., boredom and frustration) are significant components in communication, especially in learning [3–5]. Negative emotions, notably frustration, have significant consequences such as lower task productivity [6–9], longer decision-making time [10,11], and lower learning efficiency [12].

1.1. Intelligent Tutoring Systems

Student learning is supported by a human tutor's ability to respond to questions, analyze answers, and provide customized feedback. In much the same way, computer-based intelligent tutoring systems (ITSs) enable learning by providing customized feedback to users through instructional content and teaching strategies [13,14]. ITS research aims to apply the best practices of human tutors while attempting to develop new methods for ITS teaching and learning [15–17]. However, in contrast to human tutors, ITSs have limited ability to adjust their interaction behavior based on the emotional state of the student to appropriately meet the needs of the student [18]. Affect-aware systems (also called affective systems) include emotion as a factor, and typically adjust the task difficulty level of problems and provide adaptive feedback to consider user emotions [8,19].

An ITS is a form of adaptive system and is a computer-based system designed to be responsive to the current contact by changing its behavior without explicit human control. Adaptive systems can adjust their behavior by tracking the condition of the users [20], and have four categories: (1) adapting the allocation of functions between the human and the automation system; (2) adapting the information displayed to the user; (3) changing the user's task priority by directing their attention, and (4) changing the interaction style between the human and the system. The interplay of human factor considerations when changing the interaction style has been one reason that this approach has been less utilized than the others. For instance, while humans use various interaction styles when they face certain situations, adjusting the way computers deliver information violates the human factors principle of consistency in the context of human–computer interaction (HCI) [20]. However, a consistent feedback style may not always be the best in every situation. Furthermore, given the complexity and subtly of the interplay between frustration and HCI, mitigating frustration in human–computer interaction through system changes has been less studied [8].

1.2. Lessons Learned from Human–Human Communication and Learning

People interact differently in human–human interaction when they perceive the emotional states of others [21]. For example, special communication skills are used by physicians to deliver bad news when they detect their patients' negative emotions [22]. A human tutor may change his or her speaking style to enhance a student's motivation or mitigate frustration by considering other factors besides performance in order to maximize student learning. In education, various factors influence effective student learning. Keller [23] proposed four steps for encouraging and sustaining students' motivation in the learning processes: attention, relevance, confidence, and satisfaction (ACRS). The ARCS model has been used to improve learning effectiveness in distance learning [24], employee education [25], and manufacturing training [26]. Higher levels of motivation, confidence, perceived satisfaction, and overall performance lead to higher rates of engagement in a combination of classroom and online learning [27]. Feedback can be used to not only enhance performance, but also to enhance precursors to performance such as motivation, confidence, and satisfaction [23].

1.3. Ettiquette Strategies

Communication in human–human interaction can serve as a basis to investigate the utility of changing the interaction style of an ITS. Social behaviors in human interactions are governed by expectations between the speaker and hearer based on conventional norms. Conventional requirements for social behavior are codified in etiquette. When people share the same model of etiquette, they expect the same level of social behaviors from each other. Interactions between people with inappropriate etiquette may be unproductive, confusing, or even potentially dangerous [28]. Etiquette includes three independent factors: social power, social distance, and imposition. It is possible for people to have expectations when interacting with computers.

Etiquette strategies between humans were developed to redress the affronts posed by face-threatening acts (FTAs) [29,30]. FTAs are an act by the speaker that opposes the desires of the hearer, damaging their face. Positive face is characterized by the desire to be liked and admired. Ignoring someone threatens positive face. Negative face is the desire to be unimpeded in one's action, where the speakers does not impose on the hearer [29].

Etiquette has independent factors including three social variables: social power (i.e., ability of one person to impose their will on another), social distance (e.g., level of familiarity), and imposition (i.e., degree of threat of an FTA). The social power and social distance are decided by the relationship between speakers and hearers. It may take a long time to change the aspects of social power and social distance between two entities, if they can be changed at all. However, the level of imposition can be determined by using different interaction styles since it refers to the amount of demand or burden [29,31]. Consequently, the concept of different etiquette strategies is based on the idea that it is easier to adjust the imposition from speaker to hearer to mitigate FTAs [29].

Cooperation to maintain each other's face is facilitated by etiquette strategies. Four types of etiquette strategies have been identified [29]. A *bald* strategy does not consider the level of imposition on the hearer from the speaker. "Pass the salt" is a direct request that does not attempt to minimize the threat to the hearer's face. *Positive politeness* minimizes the imposition and social distance between speaker and hearer by giving compliments or making assertions of familiarity and solidarity. "That is a nice coat, where did you get it?" prefaces the request for information by paying a compliment. *Negative politeness* assumes that the speaker is in some way imposing on the hearer. "I don't want to bother you but..." or "I was wondering if..." attempt to be respectful, but the speaker knows that there is some level of imposition in the request. *Off-record* utterances by the speaker makes requests on the hearer only indirectly, use general language that requires the hearer to infer the true meaning. For example, a speaker could say "Wow, it sure is getting cold in here." This requires the hearer to infer that the speaker is really asking for the temperature to be raised [29].

The effectiveness of different interaction styles with etiquette was examined to see how these strategies could potentially enhance or inhibit effective tutoring [32]. Human tutors were able to select from one of three different etiquette strategies as they saw fit: bald, positive politeness, or negative politeness when they communicated with their students. Etiquette strategies were used by human tutors in tutoring conversations, both positively and negatively. Observations from conversation examples showed that positive politeness were used to encourage the students when they struggled to solve problems. However, the tutors' responses about the problem answer (e.g., "No, that is wrong") may have led to negative impressions for students even though it was not part of the intentional feedback based on etiquette strategies. The study suggested that human tutors use different interaction strategies to tailor tutoring even though there were violations of the rules of conversations.

1.4. Application of Ettiquette Strategies in Tutoring

The concept of etiquette and politeness has been applied to automation [33]. Miller et al. [34] developed computational models of communication focused on politeness and etiquette, and established roles of social interactions such as managing power, familiarity relationship, urgency, and indebtedness. Etiquette was used to make natural and polite interactions between humans and computer systems [35–37].

Various systems for training and tutoring have explored the concept of etiquette. A virtual manufacturing plant factory training system was developed to teach employees, based on two levels of politeness: direct and indirect (polite). Results showed that indirect interaction leads to higher student motivation [38]. The virtual factory training system demonstrated beneficial effects of two etiquette strategies (positive and negative politeness) on learning efficiency [39]. In a similar manner, a language and culture learning system explicitly delivered language contents and taught social norms by using face-to-face interactions with etiquette and anthropomorphism [40]. A disease and hospital information system were developed to convey information politely [41]. The participants' ratings of politeness and appropriateness were higher in bald, positive politeness, and negative politeness conditions, but lower in off-record condition because it required subtlety and consideration of context to be properly comprehended.

1.5. Ability of an ITS to Adapt Interaction Style

To summarize the discussion above, research suggests that feedback can be used to address performance, motivation, confidence, and satisfaction [23]. Furthermore, observation of human tutors reveals that they change their etiquette strategies to support student learners [32]. Finally, using etiquette strategies in human-computer interaction may be a viable strategy to adapt the interaction style of an ITS. Taken together, this leads to the hypothesis H1 that asks whether changes in etiquette strategies by an ITS can lead to different outcomes in performance, motivation, confidence, and satisfaction.

Hypothesis H1. *Changing etiquette strategies in tutoring leads to differences in performance, motivation, confidence, and satisfaction.*

1.6. Frustration in Human–Computer Interaction

Emotion can influence the quality of the interaction between a human and a computer. In general, human operators accept machines as a team member, and therefore expect appropriate reactions from machines [42]. Even though computer systems provide benefits in productivity, frustration is one of the most common experiences in HCI [43]. Frustration is an emotional state where achieving a goal is blocked by obstacles [44]. Aggression is one of the consequences of frustration, which is a complex emotion related to disappointment and anger [45]. Frustration has been shown to reduce the quality of ongoing performance by eliciting responses that interfered with the completion of a given task [6]. In an experiment conducted on children, frustration significantly reduced perceptual-motor performance, especially in boys [7].

Despite ever increasing technological capabilities, frustration remains a recurring problem for users of computer-based systems. Therefore, frustration continues to be of significant interest in HCI. Frustration has been shown to be both frequent and damaging to productivity. On average, users waste 42–43% of their time due to frustration when using computers [46].

Previous work found that task performance is influenced by the level of frustration. For example, a higher level of frustration led to a lower performance score on a digit–symbol substitution test [47]. Likewise, operators' task performance was diminished when they were frustrated by system delays in a robot vehicle teleoperating task [48]. Frustration led to lower user satisfaction, lower motivation, and drove the users to seek alternative systems [46,49].

1.7. The Impact of Frustration on Learning with an ITS

In learning, higher frustration caused slower response times [50] and delayed content acquisition [51]. Frustration also reduced the motivation of students [52] and led to a lack of confidence of students in computer science [53]. Studies have explored how to account for user frustration in the development of effective tutoring systems. Different heuristic strategies have been used to mitigate user frustration, including mirroring student actions to show empathy; adjusting the authority level of the tutoring system to reduce pressure; and changing the voice, motion, and gestures of the avatar in the tutoring system to provide encouragement for the students [18]. The intelligent tutor's strategies effectively supported the students by encouraging them to continue their tasks although they were frustrated [18,54]. These studies showed that frustration is a topic worth exploring for reasons other than its relation to productivity. Affect-aware computer system would benefit from a more human-like ability to sense and respond to frustration [55].

The concept of automation etiquette applies human–human etiquette conventions to HCI [33]. If a system can incorporate an understanding of the user affective state into its reasoning, the interaction between the user and the computer system could be made more sophisticated. Computers could appropriately modify their behavior with users to further joint performance. For instance, in tutoring, human tutors are finely attuned to their students' emotional states. If computers could be more attuned, they may be able to provide appropriate responses in stressful situations where human emotion is impacting the ability to function. Initial studies explored the effects of various interaction styles and etiquette strategies to potentially enhance human–human tutoring [32], increase the situation awareness of users in HCI [28], and lead to higher reliability of the system from the user's perspective [35]. In combination with advances of tutoring, human-computer interfaces that incorporate more empathy and affect could enable ITSs to more authentically embody the richness of human social interactions [18,19].

1.8. Application of Ettiquette Strategies to Mitigate Frustration

To summarize the discussion above, human tutors are attuned to their student's emotional states. Human tutors have been shown to change their etiquette strategies to enhance outcomes [32]. Frustration can decrease performance [51], motivation [52], satisfaction [46,49], and confidence [53]. While some heuristics have been used to mitigate frustration [18], it is an open question whether the frustration level has an effect on which etiquette strategy most effectively impacts an outcome. Hypothesis H2 is therefore presented to test if the most effective etiquette strategy changes for different levels of frustration of the learner.

Hypothesis H2. *When users are frustrated, the most effective etiquette strategies are different from when they are not frustrated.*

1.9. Impact

Understanding the effects of different etiquette strategies on users' performances, motivation, confidence, and satisfaction can contribute to the design of an effective HCI system to enhance the quality of interactions between users and systems. Such a system could support a student emotionally as well as cognitively. An experiment was conducted to investigate the effects of etiquette strategies in tutoring while the participants solved mathematics problems under different levels of frustration. This work aims to provide mechanisms to support the promotion of individualized learning in the context of high level math instruction. The goal was to develop an understanding of how different etiquette strategies can have differential effects not only performance, but also on the learning precursors of motivation, confidence, and satisfaction. In the same way human tutors adapt their feedback to learners when they become frustrated, an adaptive ITS system could change its communication style.

2. Materials and Methods

The objective of this study was to explore the ability of etiquette strategies to mitigate user frustration and improve task performance, motivation, confidence, and satisfaction in tutoring.

2.1. Participants

A total of 40 university students (23 males, 17 females) averaging 21.1 years old (range: 18–29). They averaged 5.7 h (range: 1–15) of computer-use daily. Participants last attended mathematics class an average of 1.35 years ago (range: 1–3). All subjects gave their informed consent for inclusion before they participated in the study. The study was conducted in accordance with the Declaration of Helsinki, and the protocol was approved by the Ethics Committee of Iowa State University (15–142).

2.2. Task

Participants solved mathematics problems in algebra, geometry, trigonometry, calculus, statistics, and probability. The Graduate Record Examination (GRE) practice book provided the problem. The GRE is an exam used for admissions into graduate school. Twenty trials with one math problem each were provided (see Figure 1). All problems had a historical GRE correct rate of 30–40%. The same level of task difficulty ensured that participants would require feedback frequently in order to solve the problem. Problems were displayed on a computer monitor. Participants were provided pencils and scratch paper.

> Two trains leave a station at 12:00 pm going in perpendicular directions. If the first train travels 60 mph and the second train travels at 80 mph. At what time is the distance between them exactly 33 and 1/3 miles?

> Jeff received a 10% increase in his salary in each of the last 3 years. If his present salary is $26,620, what was his starting salary?

Figure 1. Example problems.

2.3. Independent Variables

The independent variables were *frustration* (levels: high, low) and *etiquette strategy* (levels: bald, positive-politeness, negative-politeness, off-record, no-feedback). Frustration was induced by imposing a time constraint and by changing the label of the level of task difficulty on the problems, even though all problems had the same level of difficulty. Frustration comes from unfulfilled expectations [56]. All problems were of a similar difficulty level (30–40% GRE correct rate). However, half of the twenty problems were labeled as 'easy' problems, the other half were labeled "hard". Thus, if a problem is labeled as easy but is actually hard, participants will get frustrated because their experience with the problem is different from their expectation. Recognizing the difference between expected and actual difficulty has been shown to cause frustration [57,58]. A pilot test determined the level of difficulty such that the problems produced a measurable level of frustration (when mislabeled "easy") but not so hard that subjects gave up. Additionally, a time constraint was also employed to manipulate frustration [59]. The time constraint was the average of their last five practice problems. Beeps at 1 min, 30 s and 10 s remaining reminded the participant of the time constraint. The manipulations were designed to elicit frustration to a level that did not cause the user to give up on the task.

The independent variable of etiquette strategies had five levels: the four different etiquette strategies and a baseline condition of no feedback. Table 1 shows the same feedback being presented in each etiquette strategy.

Table 1. Example sentences of etiquette strategies.

Etiquette Strategies	Definition	Example Sentences
Bald	Direct without consideration to level of imposition.	Use appropriate formula.
Positive-politeness	Minimize imposition via statements of friendship, solidarity, and compliments.	Why don't you try other formulas? Let's check them together!
Negative-politeness	Respectful but assumes some level of imposition.	If it's alright with you, could you please check other formulas as well?
Off-record	Indirect feedback.	Various formulas are provided.

2.4. Dependent Variables

2.4.1. Etiquette Strategies Preference

One possible use for etiquette strategies would be to match the appropriate etiquette strategy to the preference of the learner, much as it has been argued that the presentation of information should be matched to a student's learning style. Past research has documented that learners will express a preference of how information should be presented to them [60]. The participants were asked before the experiment to rate their preferences for the four etiquette strategies. Participants were asked to read the definitions and examples of four etiquette strategies and complete their preference rating (10-point Likert scale). This baseline data was used to compute the correlation between their preference and trial results.

2.4.2. Independent Variable Manipulation Verification (Frustration)

The NASA task load index (TLX) frustration subscale [61] scores served as a subjective measure of frustration. To verify the independent variable manipulation, participant responses were compared between low and high frustration in the no feedback condition.

2.4.3. Task Performance

Both an objective and subjective measure of performance were used. A rubric was used to objectively grade their score (see Table 2). The TLX performance subscale scores provided a subjective measure of performance.

Table 2. Scoring rubric.

Score	Answer
1	Correct, variables and equations demonstrated
0.75	Correct equation with calculation mistakes
0.50	Correct approach but wrong or no equations
0.25	Participant defined variables or drew shapes but incorrect approach
0	Blank

2.4.4. Motivation, Confidence, and Satisfaction

After each trial, participants were asked to rate motivation, confidence, and satisfaction on a 10-point Likert scale.

2.4.5. Feedback Appropriateness and Effectiveness

After each trial, participants were asked to rate feedback appropriateness and feedback effectiveness using a Likert scale from 0–10.

2.4.6. Mental and Temporal Workload

NASA TLX is a subjective assessment tool that rates perceived workload in multiple dimensions. The participants' mental demand and temporal demand were measured through NASA TLX subscales after each trial.

2.5. Experimental Design

This was a within-subject, 2 (frustration: low, high) × 5 (etiquette strategy: bald, positive–politeness, negative-politeness, off-record, no-feedback) experimental design. A within-subject design was used to block the effect of individual differences such as math skill level. Each combination of independent variables condition was tested twice for a total of 20 trials. Condition order was counterbalanced using Latin squares to account for learning effects.

2.6. Procedure

After the consent process, briefing, and demographic survey, participants reviewed and practiced problems until they felt comfortable. The time constraint for high frustration trials was the average completion time of the last five practice trials. Between trials, participants completed a post-trial survey and the NASA TLX. Opinions and tactics were gathered in a post-experiment survey. Finally, a debriefing explained the true goal of the study, as participants were initially told that the study purpose was to test their mathematics problem-solving ability.

2.7. Data Analysis

The Shapiro–Wilk test was used to check the normality of data. Bartlett's test was used to test the homogeneity of variance. Measured data were analyzed with ANOVA tests. Post-hoc analysis used Tukey's honest difference test (HSD) in order to distinguish pairwise means that are significantly different from each other. Tukey results are presented as a series of letters for each group. If two groups do not share a letter, then they are significantly different from each other. The results are reported as significant for alpha < 0.05, and marginally significant for alpha < 0.10 [62]. Cohen's d was calculated to check effect size [63]. The Cohen's d results are reported as small effect for 0.20 < d <0.50, medium effect for 0.50 < d < 0.80, and large effects for d > 0.80. Spearman's rank order correlation coefficient was computed to test the association between two ranked variables: participants' baseline rating of etiquette strategies versus each dependent variable.

3. Results

3.1. Interaction Style Preferences

Before starting the trials, participants' had significantly different preferences of etiquette strategies, $F(3,117) = 12.6$, $p < 0.001$. Figure 2a indicates the baseline ratings of each etiquette strategy. Significant pairwise differences between strategies are indicated in the figure when the two groups do not share a letter, based on Tukey's HSD. For example, bald and positive-politeness were not significantly different from each other (and therefore are both labelled as A in Figure 2a), and likewise negative-politeness and off-record were not different from each other (labelled as B in Figure 2a). However, every group labelled A was significantly different ($p < 0.05$) from every group labelled B. The following pairs of groups were found to be significantly different: bald and negative-politeness; bald and off-record; positive-politeness and negative-politeness; and positive-politeness and off-record.

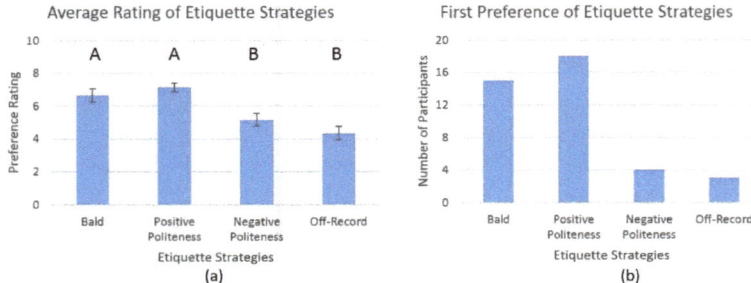

Figure 2. (**a**) Average and standard error of strategies preference ($n = 40$); (**b**) Count of preferred strategy.

From the participant rating data, it was possible to determine each participant's first preference for a strategy by identifying their highest rank among four strategies they rated. Figure 2b illustrates the distribution of participant's first preference of etiquette strategies. The baseline etiquette strategy ratings were not correlated to any of the dependent variables measured after each trial (math problem).

3.2. Independent Variable Manipulation Verification (Frustration)

The TLX frustration subscale was significantly higher for high frustration than low frustration, $F(1,39) = 48.5$, $p < 0.001$, d = 0.72 (see Figure 3). The figure indicates significant pairwise differences between groups when they do not share a letter. This verifies the manipulation of frustration through problem labelling and time constraints. Anecdotal participant's comments in the high frustration conditions included: "I do not have enough time to solve problems," "Is it really an easy problem?" "I am so frustrated," "There is no hope."

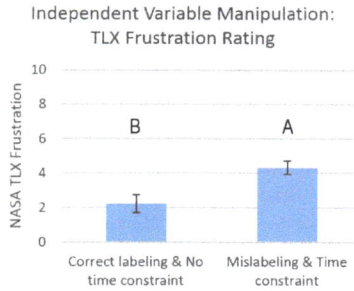

Figure 3. Mean and standard error of frustration ($n = 40$).

3.3. Task Performance

The participants correctly solved significantly more problems in low frustration than high frustration, $F(1,39) = 127.4$, $p < 0.001$, d = 0.81. The main effect of etiquette strategies on task performance (score) was significant, $F(4,156) = 2.77$, $p = 0.029$. Figure 4a indicates significant ($p < 0.05$) pairwise differences between groups when they do not share a letter, based on Tukey's HSD.

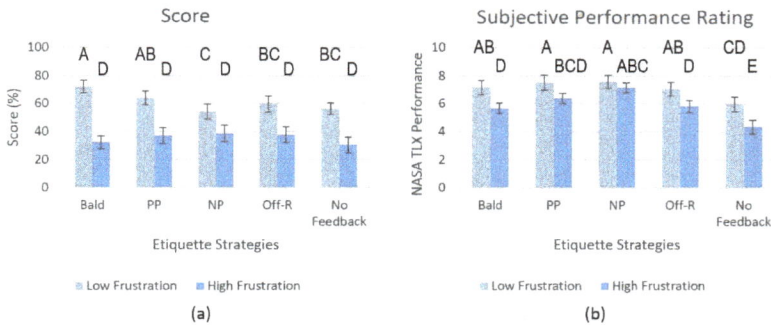

Figure 4. Mean and standard error of (**a**) problem score and (**b**) NASA task load index (TLX) performance ($n = 40$).

In the low frustration condition, bald was significantly different from negative-politeness, off-record, and no-feedback; positive-politeness was significantly different from negative-politeness, off-record, and no-feedback. Every strategy in the high frustration condition was significantly different from every strategy in the low frustration condition. The interaction was significant, $F(4,156) = 3.28$, $p = 0.013$.

The participants rated their own subjective performance significantly lower in high frustration than low frustration, $F(1,39) = 30.2$, $p < 0.001$, d = −0.41. The main effect of etiquette strategies on subjective rating of performance was significant, $F(4,156) = 11.6$, $p < 0.001$. The interaction was not significant, $F(4,156) = 1.01$, $p = 0.41$. Figure 4b indicates significant ($p < 0.05$) pairwise differences when two groups do not share a letter, based on Tukey's HSD. In the low frustration condition, no-feedback was significantly different from all four etiquette strategies. In the high frustration condition, bald and off-record were significantly different from negative-politeness; no-feedback was significantly different from all four etiquette strategies. Across frustration conditions, high/bald, high/off-record, and high/no-feedback were all significantly different than all four etiquette strategies in low frustration; high/positive-politeness was significantly different from low/positive-politeness, low/negative-politeness.

3.4. Motivation

The main effect of frustration on motivation was not significant, $F(1,39) = 0.11$, $p = 0.75$. The main effect of etiquette strategies on motivation was significant, $F(4,156) = 5.45$, $p < 0.001$. The interaction was not significant, $F(4,156) = 0.96$, $p = 0.43$. Figure 5a indicates significant ($p < 0.05$) pairwise differences between groups when they do not share a letter, based on Tukey's HSD. In the low frustration condition, off-record was significantly different from bald, positive-politeness, and negative-politeness; no-feedback was significantly different from all four etiquette strategies. In the high frustration condition, positive-politeness was significantly different from bald, negative-politeness, and no-feedback. Across frustration conditions, low/positive-politeness was significantly different from bald, negative-politeness, off-record, and no-feedback in the high frustration condition; low/no-feedback was significantly different from positive-politeness and off-record in the high frustration condition.

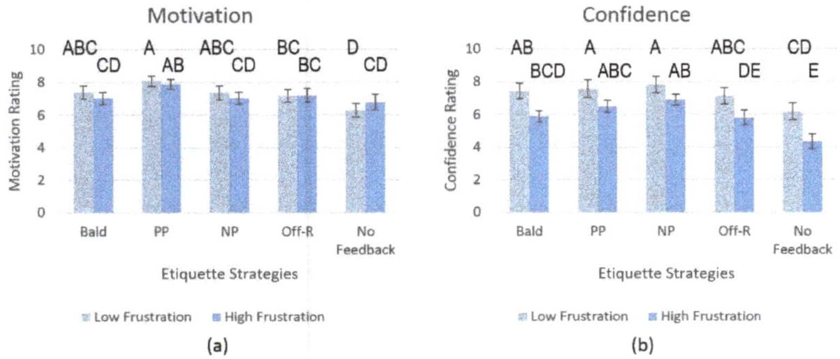

Figure 5. Mean and standard error of (**a**) motivation and (**b**) confidence ($n = 40$).

3.5. Confidence

Participants had significantly more confidence about tasks in low frustration than high frustration, $F(1,39) = 12.8$, $p < 0.001$, d = 0.47. The main effect of etiquette strategies on confidence was significant, $F(4,156) = 9.66$, $p < 0.001$. The interaction was not significant, $F(4,156) = 0.71$, $p = 0.59$. Figure 5b indicates significant ($p < 0.05$) pairwise differences between groups when they do not share a letter, based on Tukey's HSD. In the low frustration condition, no-feedback was significantly different from all four etiquette strategies. In the high frustration condition, both off-record and no-feedback were significantly different from bald, positive-politeness, and negative-politeness. Across frustration conditions, high/bald was significantly different to positive-politeness, negative-politeness, off-record, and no-feedback in the low frustration condition; high/bald is different to low/positive-politeness and low/negative-politeness; high/negative-politeness is significantly different from low/no-feedback; high/off-record is significantly different to all four etiquette strategies in low frustration; high/no-feedback is significantly different to all four etiquette strategies and no-feedback in the high frustration condition.

3.6. Satisfaction

Participants were significantly more satisfied with overall feedback in low frustration than high frustration, $F(1,39) = 7.32$, $p = 0.010$, d = 0.22. The main effect of etiquette strategies on satisfaction with feedback was significant, $F(4,156) = 9.43$, $p < 0.001$. The interaction was not significant, $F(4,156) = 0.56$, $p = 0.69$. Figure 6a indicates significant ($p < 0.05$) pairwise differences between groups when they do not share a letter, based on Tukey's HSD. In the low frustration condition, positive-politeness was significantly different from off-record; and no-feedback was significantly different from all four etiquette strategies. In the high frustration condition, positive-politeness was significantly different from bald, negative-politeness, off-record, and no-feedback; no-feedback was significantly different from bald, positive-politeness, and negative-politeness. Across frustration conditions, high/bald, high/negative-politeness, and high/off-record were all significantly different from low/bald, low/positive-politeness, and low/negative-politeness; high/positive-politeness was significantly different from low/no-feedback; high/no-feedback was significantly different from all four etiquette strategies in the low frustration condition.

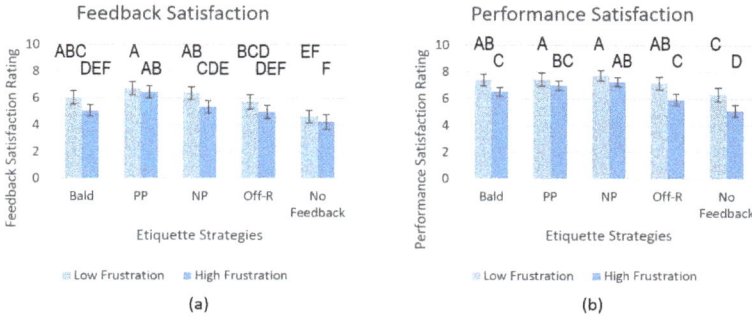

Figure 6. Mean and standard error of satisfaction with (**a**) feedback and (**b**) performance (*n* = 40).

Participants were significantly more satisfied with their own performance in low frustration than high frustration, $F(1,39) = 33.6$, $p < 0.001$, d = 0.31. The main effect of etiquette strategies on satisfaction with performance was significant, $F(4,156) = 10.54$, $p < 0.001$. The interaction was not significant, $F(4,156) = 0.93$, $p = 0.45$. Figure 6b indicates significant ($p < 0.05$) pairwise differences between groups when they do not share a letter, based on Tukey's HSD. In the low frustration condition, no-feedback was significantly different from all four etiquette strategies. In the high frustration condition, negative-politeness was significantly different from bald and off-record; no-feedback was significantly different from all four etiquette strategies. Across frustration conditions, both high/bald and high/off-record were significantly different from all four etiquette strategies in low frustration; high/positive-politeness was significantly different from high/positive-politeness and high/negative-politeness; high/negative-politeness was significantly different from low/no-feedback; high/no-feedback was significantly different from all four etiquette strategies in low frustration.

3.7. Feedback Appropriateness and Effectiveness

The main effect of frustration on the participant's rating of feedback appropriateness was not significant, $F(1,39) = 1.33$, $p = 0.26$. The main effect of etiquette strategies on the participant's rating of feedback appropriateness was significant, $F(4,156) = 12.31$, $p < 0.001$. The interaction was not significant, $F(4,156) = 1.26$, $p = 0.29$. Figure 7a indicates significant ($p < 0.05$) pairwise differences between groups when they do not share a letter, based on Tukey's HSD. In the low frustration condition, bald was significantly different from positive-politeness; no-feedback was significantly different from all four etiquette strategies. In the high frustration condition, positive-politeness was significantly different from bald, negative-politeness, off-record, and no-feedback; negative-politeness was significantly different from no-feedback. Across frustration conditions, high/bald was significantly different from low/positive-politeness and low/negative-politeness, and low/negative-politeness; high/positive-politeness was significantly different from low/bald, low/off-record, and low/no-feedback; high/negative-politeness was significantly different from low/positive-politeness and low/no-feedback; high/off-record was significantly different from low/positive-politeness and low/negative-politeness; high/no-feedback was significantly different from all four etiquette strategies in low frustration.

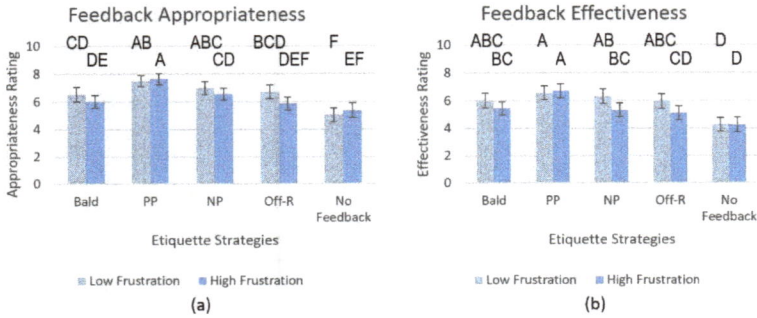

Figure 7. Mean and standard error of feedback (**a**) appropriateness and (**b**) effectiveness (*n* = 40).

Feedback was marginally significantly more effective in low frustration than high frustration, $F(1,39) = 3.06$, $p = 0.088$, d = 0.14. The main effect of etiquette strategies participant's rating of feedback effectiveness was significant ($F(4,156) = 10.31$, $p < 0.001$. The interaction was not significant, $F(4,156) = 1.07$, $p = 0.37$. Figure 7b indicates significant ($p < 0.05$) pairwise differences between groups when they do not share a letter, based on Tukey's HSD. In the low frustration condition, no-feedback was significantly different from all four etiquette strategies. In the high frustration condition, positive-politeness was significantly different from bald, negative-politeness, off-record, and no-feedback; negative-politeness was significantly different from no-feedback. Across frustration conditions, both high/bald and high/negative-politeness were significantly different from low/positive-politeness and low/no-feedback; high/positive-politeness was significantly different from low/no-feedback; high/off-record was significantly different from low/positive-politeness and low/negative-politeness; high/no-feedback was significantly different from all four etiquette strategies in low frustration.

3.8. Mental and Temproal Workload

The main effect of frustration on mental demand was not significant, $F(1,39) = 0.03$, $p = 0.87$. The main effect of etiquette strategies on mental demand was significant, $F(4,156) = 6.69$, $p < 0.001$. The interaction was not significant, $F(4,156) = 0.32$, $p = 0.87$, Figure 8a indicates significant ($p < 0.05$) pairwise differences between groups when they do not share a letter, based on Tukey's HSD. In the low frustration condition, negative-politeness was significantly different from bald and no-feedback. In the high frustration condition, bald was significantly different from negative-politeness and off-record; positive-politeness was significantly different from negative-politeness; negative-politeness was significantly different from no-feedback. Across frustration conditions, high/bald was significantly different from low/negative-politeness; high/negative-politeness was significantly different from low/bald. Low/positive-politeness, low/off-record, and low/no-feedback; high/off-record was significantly different from low/bald and low/no-feedback; high/no-feedback was significantly different from low/negative-politeness.

Feedback was significantly more temporally demanding in high frustration than low frustration, $F(1,39) = 70.3$, $p < 0.001$, d = 1.23. The main effect of etiquette strategies on temporal workload was significant, $F(4,156) = 4.82$, $p = 0.001$. The interaction was significant, $F(4,155) = 2.54$, $p = 0.042$. Figure 8b indicates significant ($p < 0.05$) pairwise differences between groups when they do not share a letter, based on Tukey's HSD. In the low frustration condition, both bald and positive-politeness were significantly different from negative-politeness and off-record; negative-politeness was significantly different from no-feedback. In the high frustration condition, all strategies were not significantly different from each other. Across frustration conditions, all high frustration conditions were significantly different from all etiquette strategies in low frustration.

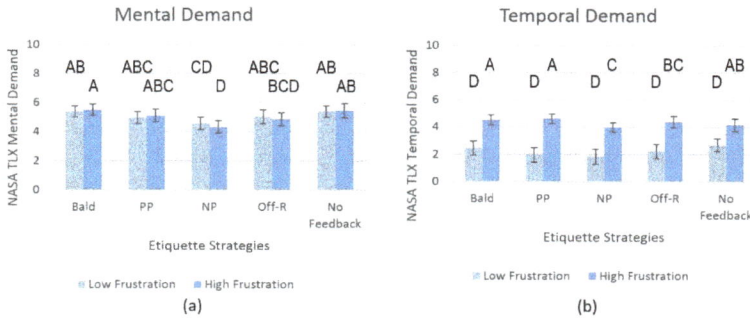

Figure 8. Mean and standard error of TLX (**a**) mental demand and (**b**) temporal demand (*n* = 40).

4. Discussion

Results demonstrated that etiquette strategies significantly influenced motivation, confidence, satisfaction, and performance. However, the null of hypothesis H1 was only partially rejected. Mathematical problem scores in low frustration condition were higher when the bald strategy was provided (as hypothesized in H1), but in the high frustration condition, there were no differences in scores between any etiquette strategies. However, the time constraints in the high frustration condition may have resulted in a ceiling effect, as some participants ran out of time to solve a given problem. When compared to positive politeness, negative politeness lead to higher performance in the high frustration condition.

Positive politeness resulted in higher motivation and satisfaction when compared to the no feedback in the low frustration condition. On the other hand, motivation and satisfaction were not driven by the interaction style of the feedback in the high frustration condition. In the high frustration condition, participants provided feedback with negative politeness had higher confidence in their work when compared to when they were not given any feedback. Moreover, positive politeness led to higher satisfaction with feedback than no feedback in high frustration condition. Thus, positive and negative politeness effectively worked to increase confidence and satisfaction with feedback. These results demonstrated that user's motivation, confidence, satisfaction, and performance vary depending upon the etiquette strategies used in tutoring. Thus, it may be feasible to build an adaptive tutoring system that changed interaction styles in order to make improvements to performance, motivation, confidence, and satisfaction.

The results did not lead to the rejection of hypothesis H2. When participants were frustrated and provided feedback with positive and negative politeness, their self-assessed performance, motivation, confidence, and satisfaction were higher than when they were provided bald, off-record, and no feedback. Thus, the most effective etiquette strategies were different when users are frustrated.

The results provided evidence that people's performance, motivation, confidence, and satisfaction can be affected by a change of etiquette strategy. In addition, there was no correlation between the four dependent variables and participants' baseline etiquette strategy preference ratings, and so no evidence that the best strategy for these participants was fixed and based on their own preferences.

Although frustration is a common and natural emotion people experience while learning, it impacts on learners' self-esteem, distractibility, and ability to follow directions [64]. A tutor's feedback can be a great help to mitigate students' frustration and ultimately reduce the consequences of frustration. The results of this study show that different feedback interaction styles impact different aspects of the learning process. For example, the participants performed better by receiving feedback based on bald and positive politeness under low frustration while they performed better with negative politeness feedback under high frustration. Their satisfaction with performance showed a similar pattern: participants were more satisfied when they received positive politeness feedback under low frustration, but negative politeness feedback under high frustration. These results demonstrated that different etiquette strategies were helpful to improve the participants' performances when they were highly frustrated. It provides the evidence that choosing the proper interaction style can mitigate the influences of frustration. Likewise, the participants' ratings of motivation, satisfaction, and confidence showed a similar tendency. Since motivation, satisfaction, and confidence are directly connected to the students' learning goals, providing appropriate feedback to support these is crucial to enhance effective learning [23]. These results can be applicable for not only a human tutor but also a computer tutor.

5. Conclusions

The results of this work lay the foundation for using etiquette strategies as a method to realize affect-aware ITSs that can support a student emotionally as well as cognitively. Results demonstrated that varying the interaction style of feedback presentation in an ITS has differential effects depending on the emotional state of the learner. Furthermore, results demonstrated that there is not one "best" strategy to simultaneously improve motivation, confidence, satisfaction and performance. Different etiquette strategies influence these factors differently, depending on the learner's current emotional and learning state. Further research is needed to establish the interaction of strategy impacts.

Frustration is one of the most frequently occurring emotions in the use of computers [43] and in learning [18]. In the same way human tutors adapt their feedback to learners when they become frustrated, an adaptive computer system could change its communication style. Based on an understanding of the user's emotional state, the system could adapt its interaction style to mitigate frustration, improve human–computer interaction, and potentially improve task performance. This study provided a basic understanding of the role of different interaction styles of feedback under varying user emotional states and can be used to form the basis of an adaptive tutoring system.

This experiment used only math problems. It is possible that the type of task will greatly influence the best feedback strategy. Further work will be needed to generalize the results of this study. The level of frustration, although moderate on an absolute scale, had a significant effect on the appropriateness and effectiveness ratings of the feedback. Future work will study the effect of higher levels of frustration on motivation, confidence, satisfaction and performance. Future work could also consider personality factors such as learner attributional style, perceived competency, or self, which may influence motivation and hence learning [65].

In human–computer tutoring, most of the real-time adaptation is triggered by poor performance and results in a change to the task difficulty or problem content. However, a good human tutor will be aware of the emotional state of the learner and adapt their interaction style to support aspects of the student's learning that underlie performance such as a student's motivation, confidence, or satisfaction. This work aims to provide mechanisms to support the promotion of individualized learning in the context of high level math instruction. Future work will look at the ability to adapt interaction styles depending on the emotional state of the students as well as the goal of the tutor. The results presented here could be used to derive the logic of etiquette strategies adaptation to form the basis of an adaptive tutoring agent. In on-going work, an adaptive tutoring system was designed to improve the learning factors of motivation, confidence, satisfaction, and performance using a rule set developed based on the current data set [66], to trigger the most appropriate etiquette strategy for a given combination of factors and frustration level.

Author Contributions: E.Y. and M.C.D. conceived and designed the experiments; E.Y. performed the experiments; E.Y. analyzed the data; E.Y. and M.C.D. wrote the paper.

Acknowledgments: For their contributions to the pilot experiment implementation and data analysis, the authors would like to thank Maria Dropps, Mariangely Iglesias-Pena, David Montealegre, and Jordan Zonner. This material is based in part upon work supported by the National Science Foundation under Grant No. 1461160.

Conflicts of Interest: The authors declare no conflict of interest.

References

1. Yang, E.; Dorneich, M.C. Evaluation of Etiquette Strategies to Adapt Feedback in Affect-Aware Tutoring. In Proceedings of the Human Factors and Ergonomics Society Annual Meeting, Washington DC, USA, 19–23 September 2016.
2. Ferdig, R.E.; Mishra, P. Emotional Responses to Computers: Experiences in Unfairness, Anger, and Spite 1. *J. Educ. Multimed. Hypermed.* **2004**, *13*, 143.
3. Kort, B.; Reilly, R.; Picard, R.W. An affective model of interplay between emotions and learning: Reengineering educational pedagogy-building a learning companion. In Proceedings of the IEEE International Conference on Advanced Learning Technologies, Madison, WI, USA, 6–8 August 2001.
4. Fisher, C.D.; Noble, C.S. A within-person examination of correlates of performance and emotions while working. *Hum. Perform.* **2004**, *17*, 145–168. [CrossRef]
5. Woolf, B.; Burleson, W.; Arroyo, I.; Dragon, T.; Picard, R. Emotional intelligence for computer tutors. In Proceedings of the Workshop on Modeling and Scaffolding Affective Experiences to Impact Learning at 13th International Conference on Artificial Intelligence in Education, Los Angeles, CA, USA, 9–13 July 2007.
6. Waterhouse, I.K.; Child, I.L. Frustration and the quality of performance. *J. Personal.* **1953**, *21*, 298–311. [CrossRef]
7. Solkoff, N.; Todd, G.A.; Screven, C.G. Effects of frustration on perceptual-motor performance. *Child Dev.* **1964**, *35*, 569–575. [CrossRef] [PubMed]
8. Klein, J.; Moon, Y.; Picard, R.W. This computer responds to user frustration: Theory, design, and results. *Interact. Comput.* **2002**, *14*, 119–140. [CrossRef]
9. Powers, S.R.; Rauh, C.; Henning, R.A.; Buck, R.W.; West, T.V. The effect of video feedback delay on frustration and emotion communication accuracy. *Comput. Hum. Behav.* **2011**, *27*, 1651–1657. [CrossRef]
10. Toda, M. Emotion and decision making. *Acta Psychol.* **1980**, *45*, 133–155. [CrossRef]
11. Lerner, J.S.; Li, Y.; Valdesolo, P.; Kassam, K.S. Emotion and decision making. *Psychology* **2015**, *66*, 799–823. [CrossRef] [PubMed]
12. Graesser, A.C.; Chipman, P.; Haynes, B.C.; Olney, A. AutoTutor: An intelligent tutoring system with mixed-initiative dialogue. *IEEE Trans. Educ.* **2005**, *48*, 612–618. [CrossRef]
13. Wenger, E. *Artificial Intelligence and Tutoring Systems*; Morgan Kaufmann: Los Altos, CA, USA, 1987.
14. Gilbert, S.B.; Blessing, S.B.; Guo, E. Authoring Effective Embedded Tutors: An Overview of the Extensible Problem Specific Tutor (xPST) System. *Int. J. Artif. Intell. Educ.* **2015**, *25*, 428–454. [CrossRef]
15. Murray, T. An Overview of Intelligent Tutoring System Authoring Tools: Updated analysis of the state of the art. In *Authoring Tools for Advanced Technology Learning Environments*; Springer: Dordrecht, The Netherlands, 2003; pp. 491–544.
16. Broderick, Z. Increasing Parent Engagement in Student Learning Using an Intelligent Tutoring System with Automated Messages. Ph.D. Thesis, Worcester Polytechnic Institute, Worcester, MA, USA, 2011.
17. Koedinger, K.; Tanner, M. Things you should know about intelligent tutoring systems. In Proceedings of the EDUCAUSE Learning Initiative (ELI), Denver, CO, USA, 4–6 February 2013.
18. Woolf, B.; Burleson, W.; Arroyo, I.; Dragon, T.; Cooper, D.; Picard, R. Affect-aware tutors: Recognizing and responding to student affect. *Int. J. Learn. Technol.* **2009**, *4*, 129–164. [CrossRef]
19. Picard, R.W.; Papert, S.; Bender, W.; Blumberg, B.; Breazeal, C.; Cavallo, D.; Machover, T.; Resnick, M.; Deb, R.; Carol, S. Affective learning—A manifesto. *BT Technol. J.* **2004**, *22*, 253–269. [CrossRef]
20. Feigh, K.M.; Dorneich, M.C.; Hayes, C.C. Toward a characterization of adaptive systems a framework for researchers and system designers. *Hum. Factors* **2012**, *54*, 1008–1024. [CrossRef] [PubMed]
21. Ekman, P. Universal facial expressions of emotion. *Calif. Ment. Health Res. Dig.* **1970**, *8*, 151–158.
22. Back, A.L.; Arnold, R.M.; Baile, W.F.; Fryer-Edwards, K.A.; Alexander, S.C.; Barley, G.E.; Gooley, T.A.; Tulsky, J.A. Efficacy of communication skills training for giving bad news and discussing transitions to palliative care. *Arch. Intern. Med.* **2007**, *167*, 453–460. [CrossRef] [PubMed]
23. Keller, J.M. Development and use of the ARCS model of instructional design. *J. Instr. Dev.* **1987**, *10*, 2–10. [CrossRef]
24. Malik, S. Effectiveness of Arcs Model of Motivational Design to Overcome Non Completion Rate of Students in Distance Education. *Turk. Online J. Distance Educ.* **2014**, *15*, 194–200. [CrossRef]

25. Visser, J.; Keller, J.M. The clinical use of motivational messages: An inquiry into the validity of the ARCS model of motivational design. *Instr. Sci.* **1990**, *19*, 467–500. [CrossRef]
26. Shellnut, B.; Knowltion, A.; Savage, T. Applying the ARCS model to the design and development of computer-based modules for manufacturing engineering courses. *Educ. Technol. Res. Dev.* **1999**, *47*, 100–110. [CrossRef]
27. Mohammad, S.; Job, M.A. Confidence-Motivation–Satisfaction-Performance (CMSP) Analysis of Blended Learning System in the Arab Open University Bahrain. *Int. J. Inf. Technol. Bus. Manag.* **2012**, *3*, 23–29.
28. Wu, P.; Miller, C.A.; Funk, H.; Vikili, V. Computational models of etiquette and Culture. In *Human-Computer Etiquette: Cultural Expectations and the Design Implications They Place on Computers and Technology*; Taylor Francis Group: Oxford, UK, 2010.
29. Brown, P.; Levinson, S.C. Universals in language usage: Politeness phenomena. In *Questions and Politeness: Strategies in Social Interaction*; Cambridge University Press: Cambridge, UK, 1978; pp. 56–311.
30. Mills, S. *Gender and Politeness*; Cambridge University Press: Cambridge, UK, 2003; Volume 17.
31. Kasper, G. Linguistic etiquette. In *Intercultural Discourse and Communication: The Essential Readings*; John Wiley & Sons: Hoboken, NJ, USA, 2005; pp. 58–67.
32. Pearson, N.K.; Kreuz, R.J.; Zwaan, R.A.; Graesser, A.C. Pragmatics and pedagogy: Conversational rules and politeness strategies may inhibit effective tutoring. *Cognit. Instr.* **1995**, *13*, 161–188. [CrossRef]
33. Miller, C.A.; Funk, H.B. Associates with etiquette: Meta-communication to make human-automation interaction more natural, productive and polite. In Proceedings of the 8th European Conference on Cognitive Science Approaches to Process Control, Munich, Germany, 24–26 September 2001; pp. 24–26.
34. Miller, C.A.; Wu, P.; Funk, H.B. A computational approach to etiquette: Operationalizing Brown and Levinson's politeness model. *Intell. Syst.* **2008**, *23*, 28–35. [CrossRef]
35. Parasuraman, R.; Miller, C.A. Trust and etiquette in high-criticality automated systems. *Commun. ACM* **2004**, *47*, 51–55. [CrossRef]
36. Hayes, C.C.; Miller, C.A. (Eds.) Should computers be polite? In *Human-Computer Etiquette: Cultural Expectations and the Design Implications They Place on Computers and Technology*; Taylor Francis: Boca Raton, FL, USA, 2010.
37. Dorneich, M.C.; Ververs, P.M.; Mathan, S.; Whitlow, S.; Hayes, C.C. Considering etiquette in the design of an adaptive system. *J. Cognit. Eng. Decis. Mak.* **2012**, *6*, 243–265. [CrossRef]
38. Qu, L.; Wang, N.; Johnson, W.L. Using learner focus of attention to detect learner motivation factors. In *User Modeling 2005*; Springer: Berlin/Heidelberg, Germany, 2005; pp. 70–73.
39. Wang, N.; Johnson, W.L.; Mayer, R.E.; Rizzo, P.; Shaw, E.; Collins, H. The politeness effect: Pedagogical agents and learning outcomes. *Int. J. Hum. Comput. Stud.* **2008**, *66*, 98–112. [CrossRef]
40. Johnson, W.L.; Friedland, L.; Schrider, P.; Valente, A.; Sheridan, S. The Virtual Cultural Awareness Trainer (VCAT): Joint Knowledge Online's (JKO's) solution to the individual operational culture and language training gap. In Proceedings of the International Exhibition and Conference, Minsk, Belarus, 24–27 May 2011.
41. Bickmore, T. Relational agents for chronic disease self management. In *Health Informatics: A Patient-Centered Approach to Diabetes*; MIT Press: Cambridge, MA, USA, 2010; pp. 181–204.
42. Nass, C.; Fogg, B.J.; Moon, Y. Can computers be teammates? *Int. J. Hum. Comput. Stud.* **1996**, *45*, 669–678. [CrossRef]
43. Ceaparu, I.; Lazar, J.; Bessiere, K.; Robinson, J.; Shneiderman, B. Determining causes and severity of end-user frustration. *Int. J. Hum. Comput. Interact.* **2004**, *17*, 333–356. [CrossRef]
44. Lawson, R. *Frustration; the Development of a Scientific Concept*; Macmillan: Basingstoke, UK, 1965.
45. Dollard, J.; Miller, N.E.; Doob, L.W.; Mowrer, O.H.; Sears, R.R. *Frustration and Aggression*; American Psychological Association: Washington, DC, USA, 1939.
46. Lazar, J.; Jones, A.; Hackley, M.; Shneiderman, B. Severity and impact of computer user frustration: A comparison of student and workplace users. *Interact. Comput.* **2006**, *18*, 187–207. [CrossRef]
47. Hokanson, J.E.; Burgess, M. Effects of physiological arousal level, frustration, and task complexity on performance. *J. Abnorm. Soc. Psychol.* **1964**, *68*, 698. [CrossRef]
48. Yang, E.; Dorneich, M.C. The Emotional, Cognitive, Physiological, and Performance Effects of Variable Time Delay in Human-Robot Interaction. *Int. J. Soc. Robot.* **2017**, *9*, 491–508. [CrossRef]

49. Hoxmeier, J.A.; DiCesare, C. System response time and user satisfaction: An experimental study of browser-based applications. In Proceedings of the Americas Conference on Information Systems, Long Beach, CA, USA, 10–13 August 2000.

50. Chen, J.S.; Gross, K.; Amsel, A. Ontogeny of successive negative contrast and its dissociation from other paradoxical reward effects in preweanling rats. *J. Comp. Physiol. Psychol.* **1981**, *95*, 146. [CrossRef]

51. Amsel, A. *Frustration Theory: An Analysis of Dispositional Learning and Memory (No. 11)*; Cambridge University Press: Cambridge, UK, 1992.

52. Weiner, B. An attributional theory of achievement motivation and emotion. *Psychol. Rev.* **1985**, *92*, 548. [CrossRef] [PubMed]

53. Hansen, S.; Eddy, E. Engagement and frustration in programming projects. *ACM SIGCSE Bull.* **2007**, *39*, 271–275. [CrossRef]

54. Arroyo, I.; Ferguson, K.; Johns, J.; Dragon, T.; Meheranian, H.; Fisher, D.; Barto, A.; Mahadevan, S.; Woolf, B.P. Repairing disengagement with non-invasive interventions. In Proceedings of the Artificial Intelligence in Education, Los Angeles, CA, USA, 9–13 July 2007; pp. 195–202.

55. Picard, R.W.; Klein, J. Computers that recognize and respond to user emotion: Theoretical and practical implications. *Interact. Comput.* **2002**, *14*, 141–169. [CrossRef]

56. Berkowitz, L. Frustration-aggression hypothesis: Examination and reformulation. *Psychol. Bull.* **1989**, *106*, 59. [CrossRef] [PubMed]

57. Hone, K. Empathic agents to reduce user frustration: The effects of varying agent characteristics. *Interact. Comput.* **2006**, *18*, 227–245. [CrossRef]

58. Glass, A.; McGuinness, D.L.; Wolverton, M. Toward establishing trust in adaptive agents. In Proceedings of the 13th International Conference on Intelligent User Interfaces, Gran Canaria, Spain, 13–16 January 2008; pp. 227–236.

59. Wahlström, J.; Hagberg, M.; Johnson, P.; Svensson, J.; Rempel, D. Influence of time pressure and verbal provocation on physiological and psychological reactions during work with a computer mouse. *Eur. J. Appl. Physiol.* **2002**, *87*, 257–263. [PubMed]

60. Pashler, H.; McDaniel, M.; Rohrer, D.; Bjork, R. Learning styles: Concepts and evidence. *Psychol. Sci. Public Interest* **2009**, *9*, 105–119. [CrossRef] [PubMed]

61. Hart, S.G.; Staveland, L.E. Development of NASA-TLX (Task Load Index): Results of empirical and theoretical research. *Adv. Psychol.* **1988**, *52*, 139–183.

62. Gelman, A. Commentary: P values and statistical practice. *Epidemiology* **2013**, *24*, 69–72. [CrossRef] [PubMed]

63. Cohen, J. *Statistical Power Analysis for the Behavioral Sciences*; Lawrence Earlbaum Associates: Hillsdale, NJ, USA, 1988; pp. 20–26.

64. Liu, Z.; Pataranutaporn, V.; Ocumpaugh, J.; Baker, R.S. Sequences of Frustration and Confusion, and Learning. In Proceedings of the 6th International Conference on Educational Data Mining, Memphis, TN, USA, 6–9 July 2013; pp. 114–120.

65. Pintrich, P.R.; De Groot, E.V. Motivational and self-regulated learning components of classroom academic performance. *J. Educ. Psychol.* **1990**, *82*, 33. [CrossRef]

66. Yang, E.; Dorneich, M.C. Affect-Aware Adaptive Tutoring Based on Human-Automation Etiquette Strategies. *Hum. Factors* **2018**. [CrossRef] [PubMed]

*applied
sciences*

MDPI

Article

Effects of Viewing Displays from Different Distances on Human Visual System

Mohamed Z. Ramadan [1], Mohammed H. Alhaag [1,*] and Mustufa Haider Abidi [2]

[1] Industrial Engineering Department, King Saud University, Riyadh 11421, Saudi Arabia; mramadan1@ksu.edu.sa

[2] Raytheon Chair for Systems Engineering (RCSE), Advanced Manufacturing Institute, King Saud University, Riyadh 11421, Saudi Arabia; mabidi@ksu.edu.sa

* Correspondence: inengmohamed@yahoo.com; Tel.: +966-559-034-918

Academic Editor: Antonio Fernández-Caballero
Received: 3 October 2017; Accepted: 6 November 2017; Published: 9 November 2017

Abstract: The current stereoscopic 3D displays have several human-factor issues including visual-fatigue symptoms such as eyestrain, headache, fatigue, nausea, and malaise. The viewing time and viewing distance are factors that considerably affect the visual fatigue associated with 3D displays. Hence, this study analyzes the effects of display type (2D vs. 3D) and viewing distance on visual fatigue during a 60-min viewing session based on electroencephalogram (EEG) relative beta power, and alpha/beta power ratio. In this study, twenty male participants watched four videos. The EEGs were recorded at two occipital lobes (O1 and O2) of each participant in the pre-session (3 min), post-session (3 min), and during a 60-min viewing session. The results showed that the decrease in relative beta power of the EEG and the increase in the alpha/beta ratio from the start until the end of the viewing session were significantly higher when watching the 3D display. When the viewing distance was increased from 1.95 m to 3.90 m, the visual fatigue was decreased in the case of the 3D-display, whereas the fatigue was increased in the case of the 2D-display. Moreover, there was approximately the same level of visual fatigue when watching videos in 2D or 3D from a long viewing distance (3.90 m).

Keywords: 3D display; visual fatigue; electroencephalogram (EEG); causative factor

---

1. Introduction

Numerous studies have been conducted to evaluate the visual fatigue caused because of watching 3D displays using electroencephalogram (EEG) [1–16]. The EEG method is selected as it is the most significant and reliable physiological measure for evaluating mental fatigue [17–21]. The EEG signals detect the slight electrical potentials (generally less than 300 μV) produced by the brain as a continuous graphical distribution of spatiotemporal of voltage over time. Using the frequency domain of the EEG signals, particularly frequency bandwidths, help in revealing the functional states of the brain [22,23]. De Waard [22], and Fisch and Spehlmann [23] classified the EEG signal into four characteristic waves based on their frequency, namely delta wave (δ, 0–4 Hz), theta wave (θ, 4–8 Hz), alpha wave (α, 8–13 Hz), and beta wave (β, 13–30 Hz). The delta (δ) wave is usually related to the depth of sleep; moreover, this wave is associated with specific encephalopathic diseases and underlying lesions [13]. The theta (θ) wave is related to drowsiness. The wave is associated with cases observed during hypnogogic states, light sleep, clear dreaming, preconscious state just before falling asleep; and just after waking up. The alpha (α) wave is related to an alert, relaxed state of consciousness, which decreases with increase in visual flow; and with open eyes and extreme sleepiness. The beta (β) wave is often related to various states such as busy, active, or active concentration and worried thinking [24].

Studies have been conducted on visual fatigue using EEG signals, because in-depth information associated with the physiological states is included in the analysis of EEG signals. Evaluating the β-power frequency of the EEG is one of the methods used to measure visual fatigue. Kim and Lee [1], and Li et al. [9] used this method to measure the visual fatigue of participants after viewing three dimensional (3D) television (TV). Kim and Lee [1], and Li et al. [9] showed that the β-power frequencies (frequency band >12 Hz) were negligible in the 2D cases, but not in the 3D cases; the frequencies increased with the increase in the viewing duration. The results showed that watching a 3D movie increased the EEG power compared to the case of watching a 2D movie. In particular, Kim and Lee [1] revealed that α band had less significance compared to the β band concerning comparative viewing of 2D and 3D movies. Tran et al. [25] explained that the four characteristic waves of the EEG signals are closely associated with brain activity, and the level of visual fatigue changed with the change in the energy of the EEG waves. Belyavin and Wright [26] observed increases in the δ and θ wave frequency bands and decreases in the β frequency waveband during fatigue; the results were in good agreement with the study conducted by Subasi [27], who showed that θ frequency waveband increased with fatigue.

In other studies, the power indices of the EEG were used to evaluate visual fatigue. Cheng et al. [17] used EEG power indices (α, θ, θ/α, β/α, and $((\alpha + \theta)/\beta)$) to evaluate the mental fatigue caused during a visual-display terminal task. The EEG power indices were significantly different before and after performing the task, and the amplitude significantly reduced after the task was performed. The results showed that after three hours of watching the visual-display terminal task, the participants appeared mentally fatigued to a significant extent. Jap et al. [28] presented four ratio algorithms to evaluate the fatigue, which were $(\alpha + \theta)/\beta$, α/β, $(\alpha + \theta)/(\alpha + \beta)$, and θ/β. The ratios increased with increase in fatigue. In addition, the algorithms show that the β wave decreased, but the change was more than α wave with respect to the fatigue. Similarly, Eoh et al. [29] showed that $(\alpha + \theta)/\beta$ and α/β significantly increased whereas α and β waves were reduced. Chen et al. [6] used the four algorithms and four frequency wavebands to evaluate the visual fatigue associated with viewing 3D TVs. The results showed that in some regions in the brain, the indicators for participants watching 3D TV varied more significantly than when the participants watched 2D TV, except for the θ rhythm. When the participants were viewing 3D TV, the energy of α and β frequency bands significantly decreased, while the energy of the δ wave band significantly increased. The energy of the θ waveband remained stable.

Park et al. [5] assessed the connection between the emotional states of a teenager and relative β-band power at occipital lobe region while viewing 3D TV using some active and some passive glasses. The results showed that the relative β-band power was relatively higher while using passive (filter) glasses when compared to using active (shutter) glasses. Active 3D glasses interact wirelessly with images on a screen to enhance 3D viewing, whereas passive glasses do not [5]. With passive 3D glasses, the TV screen is coated so that light from alternate scan lines is polarized differently. The TV then interlaces two images on the screen, one for each eye.

Chen et al. [7] showed that the power spectral entropy is strongly correlated to visual fatigue. The results showed that the power spectral entropy reduced significantly after viewing 3D TV for a long time, thus indicating the decrease in the attention level of the participants. Hsu and Wang [10] investigated whether the EEG power indices (α, β, θ/α, β/α, and $(\alpha + \theta)/\beta$) were useful measures of visual fatigue when playing TV video games. The results revealed that the power indices of the EEG were valid indicators to measure visual fatigue. Hsu and Wang [10] found that β/α, β, and α EEG power were good choices to measure visual fatigue, but β/α was the best. The decreases in β/α and β power indices and increase in α power index were related to visual fatigue.

The current stereoscopic 3D displays have several human-factor issues, including visual-fatigue symptoms such as eyestrain, headache, fatigue, nausea, and malaise. The viewing distance and viewing time are considered significant contributing factors to the visual fatigue associated with 3D displays [30–33]. The viewing distance is considered an essential environmental condition and is described in the guidelines for video display terminals (VTDs) [30]. Park and Mun [34] showed that discomfort due to negative and positive conflicts in binocular disparity depends strongly on viewing distance. In addition, prolonged viewing of a 3D display increases the degree of visual fatigue [31]. Matthews [35] studied the long-term effects of visual fatigue from watching chromatic displays. The American Optometric Association found that prolonged use of visual systems could result in inefficient visual processing functions, which they considered visual fatigue [36].

Previous studies have focused on spectral variations in the various EEG frequency bands [5–10,17,25–27]. It has often been hypothesized that through changes in attention or onset of visual fatigue, the low-frequency waves in the EEG such as δ, θ, and α will increase, while the higher frequency waves such as β will decrease. In previous studies on the physiological responses for watching display images, responses were recorded in only pre- and post-watching periods and were then compared in order to obtain reliable methods for measuring visual fatigue. Hence, the physiological responses that occurred during watching TV displays were neglected in literature. Hence, this study investigated the effects of the display type, viewing distance, and viewing time on the visual fatigue based on the EEG relative β power of the occipital lobe (O1), α/β power of EEG, and total subjective visual discomfort score.

2. Materials and Methods

2.1. Participants

Twenty male university students with normal vision acuity with no medical history participated in this study as volunteers. The average age was 27.7 years, and the standard deviation was 2.53 years. The participants gave their informed consent before participating in the experiment, which was approved by the Human Participants Review Sub-committee of the Institutional Review Board (IRB) of King Saud University, College of Medicine, and King Khalid University Hospital (project identification code E-14-1182).

2.2. Experimental Design

In this study, four independent variables were selected: two within-subject variables and two between-subject variables. Thus, a mixed design (i.e., A × B (C × D × S)) was used to represent the experiment. The two independent between-subject variables were (A) viewing distance (3H vs. 6H, where H is the height of the screen) and (B) the display type (2D vs. 3D). Thus, it was a 2 × 2 × 8 × 4 design. Each of the (A) and (B) = 2 × 2 = 4 distance–display conditions contained S = 5 participants, and each participant watched all four movies for one hour. Two types of displays were chosen because they have been used in previous studies [6,8,11,37,38], thus allowing comparison. Moreover, the viewing distance is selected as three times and six times of the display height based on the recommendation made by the International Telecommunication Union (ITU) standards [39]. The two independent variables based on the repeated measures or within the subject variable were (C) viewing time (pre-test, T10, T20, T30, T40, T50, T60, and post-test) and (D) movies (Jurassic World, Avengers, San Andreas, and Godzilla). Park et al. [5,40] selected the viewing duration to be 45 min; however, Chen et al. [41] used 10 min, and Park et al. [38] used 1 h as the viewing period. Hence, in this study, it was decided to perform the experiment for 60 min. The EEG signal was continuously recorded starting at pre-test (3 min prior to the experiment), T10 (first 10-min period), T20 (second 10-min period), T30 (third 10-min period), T40 (fourth 10-min period), T50 (fifth 10-min period), T60 (sixth and final 10-min period), and post-test (3 min period after the experiment). The dependent variables included the relative β power and α/β power of the EEG, and total subjective visual discomfort score. Pairwise comparisons were performed to investigate the source of any significant effect considering the time factor.

2.3. Experimental Setup

The equipment used in this study included a commercial 50-in LG 3D smart TV (50LF650T) (LG Electronics Inc., Seoul, South Korea). An eight-channel Biomonitor ME6000 (Mega Electronics Ltd., Kuopio, Finland) was used with four channels to record the EEG signals. A Snellen chart, GPM Vernier calipers (DKSH, Zürich, Switzerland), and measuring tape were used to test visual acuity, inter pupillary distance, and head dimensions, respectively. A visual discomfort questionnaire, and neurophysiological and cardiovascular assessment were used to obtain general information history. An emotive EEG headset was used to place the EEG electrodes. The software system used to record signals was Mega Win 3.0.1 (Mega Electronics Ltd.). Other materials include 70%-isopropyl-alcohol swab, cotton squares, band aid, Ag/AgCl disk electrodes for EEG, and high-viscosity electrolyte gel for active electrodes.

2.4. Experimental Procedures

First, the experimental procedure was started by setting up the environment and experimental variables. The LG 3D smart TV (50LF650T) with passive row interlaced technology was used to display the 2D and 3D videos. Table 1 lists the main specifications of the LG 3D smart TV.

Table 1. Specifications of 3D LG Smart Television.

Variables	Specifications	Variables	Specifications
Size	50 in	Diagonal	126 cm
Resolution	1920 × 1080 pixels	Width	112.7 cm
Height	65 cm	Refresh rate	100 Hz
Aspect ratio	16:9	Picture mode	Vivid dynamic
Backlight	100 nt	Contrast	100 nt
Brightness	50 nt	Color	70 nt
Sharpness	50 nt	Audio output	20 W

in = inches; cm = centimeters; nt = nit; W = Watt.

The TV was placed at a height of 96 cm, and the center of the display height was at 129 cm from the lab floor. The viewing distances were set at 195 and 390 cm (3H and 6H). The background distance from the screen to the wall was 0.12 m. To avoid any disturbance to the participant's surrounding, black sheets were pasted on the wall around the TV, as shown in Figure 1. All experiments were performed in the human-factors laboratory with an average dry-bulb temperature and relative humidity of 23.8 °C and 30.6%, respectively. The experimental zone was ensured to have no vibrations or strong odors during the test. The lighting conditions were maintained constant in the sessions. The environmental illuminance at the screen center was approximately 250 lux. The participants were seated on an adjustable chair placed in front of the display and were asked to direct their heads toward the display. To control the head movement, each participant's back was laid on the chair, while the head was supported by the chair so that the center of the display would be at the same level as the participant's eye position at a viewing distance of 195 or 390 cm.

Before performing the experiment, participants were introduced and welcomed to the experimental room. The participants were instructed to have a rest of a full night and avoid cigarettes or caffeine for 6 h prior to the experiment. Each participant was asked to visit the lab four times on different days for approximately 2 h each time. This allowed the physical stress associated with the four films to be evaluated on separate days without interference from other movies. The time between visits was at least 48 h. The participants were then given the opportunity to ask any questions about the study. They were informed about their rights to refuse or stop participating before or during the experiment. Each participant was then given a consent form to read and sign. Each participant filled out a demographic questionnaire. Visual acuity and color blindness tests were then conducted. During the pre-test (first visit), each participant was briefed about the objective of the study and experimental procedure as well as the time required for each visit before the experiment was performed. During the test session, they were instructed to avoid moving their head as much as possible. In addition, to avoid low blood sugar levels that may influence the experiment outcomes, each participant performed the experiment of watching 2D/or 3D movies at least two hours after having a meal. The participants agreed to placing electrodes on their skin to record the EEG. To comparatively evaluate the visual fatigue, the 20 participants were randomly allocated to one of the four groups: "View the four videos in 3D from a distance of 195 cm (3D3H)"; "View the four videos in 3D from a distance of 390 cm (3D6H.)"; "View the four videos in 2D from a distance of 195 cm (2D3H.)"; and "View the four videos in 2D from a distance of 390 cm (2D6H)".

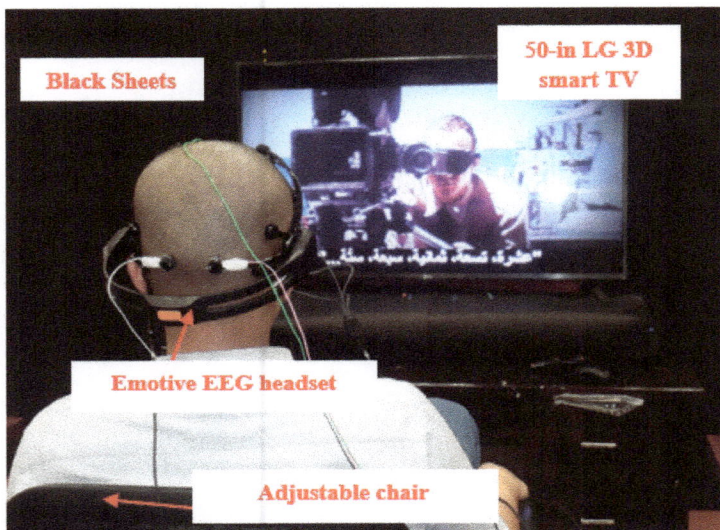

Figure 1. Experimental setup.

The four movies were randomly assigned to the participants. None of the participants were allowed to participate under more than one group to avoid bias and learning effect. In other words, if a participant watched the same movie in more than one group, the participant may become drowsy or bored, which could affect the EEG response. The configuration of the LG TV could be changed from 2D to 3D mode. The participants who watched the 3D movies wore passive 3D glasses with lenses having circular-polarizing filters. The participants were asked to answer a visual discomfort questionnaire (VDQ) involving 41 symptoms to rate their perceptions of visual fatigue twice [42,43]. The first VDQ was given before participating in the experiment, and the other was given after participation. Several measurements were performed on the participants before, during, and immediately after viewing the display. The EEG signal was recorded for 3 min (pre-test). The participants then watched the assigned movie from the designated distance in the assigned mode (2D or 3D) for 1 h while the EEG signal was continuously recorded for 60 min (this was later divided into six 10-min sessions). After the participants finished watching the movie, the EEG signal was recorded continuously for another 3 min (post-test). Finally, the participants were asked once again to rate their perceptions of visual fatigue.

2.5. Response Measures

2.5.1. Subjective Visual Discomfort Evaluation

Most early studies measured visual fatigue using questionnaires, which are not only affected by the feeling of visual fatigue but also by individual bias resulting from the emotional state of the participant. The subjective evaluation method used in this study included a total of 41 questions obtained from the studies conducted by Li [42] and Kennedy et al. [43]. The participants were asked to fill out a VDQ to rate their perceptions of visual fatigue. The total visual discomfort score was used to evaluate visual fatigue.

2.5.2. EEG Signal Response

The electrical activity in the brain is generated by billions of brain cells called neurons [44]. Hsu and Wang [10] suggested that the EEG data recorded from the occipital lobe, either left or right occipital lobes (either O1 or O2) are suitable to measure visual fatigue. Moreover, Iwasaki and Kurimoto [45] found that the inhibition function and the electrical activity of the cerebral cortex are relevant to visual fatigue. In this study, the EEG signals were recorded from the occipital lobes (O1 or O2). The positions of the occipital lobes (O1 and O2) were determined by calculating 10% for Oz from inion and at 5% from Oz on the left and right hemisphere of the head as O1 and O2, respectively, by adhering to the 10–20 international standard for EEG electrode placement [44]. Before placing the Ag/AgCl electrodes, the electrode sites (occipital lobes, forehead, and mastoid) were cleaned using 70% isopropyl alcohol swab, and subsequently, air dried for two minutes before performing the experiment. Thereafter, the Ag/AgCl disk electrodes were filled with super gel and placed on the prepared sites, which were held in place using the emotive headset. The reference and ground electrodes were placed over the mastoid region (behind right ear) and on the forehead of the participant, respectively [5]. Figure 2 shows the placement of the EEG electrodes. The ground, reference, O1, and O2 electrodes were connected to a four-channel EEG preamplifier for ME 6000; GND port, Ref port, ch-3, and ch-4, respectively. The four-channel EEG preamplifier was connected to ch1 and ch2 of ME 6000 device.

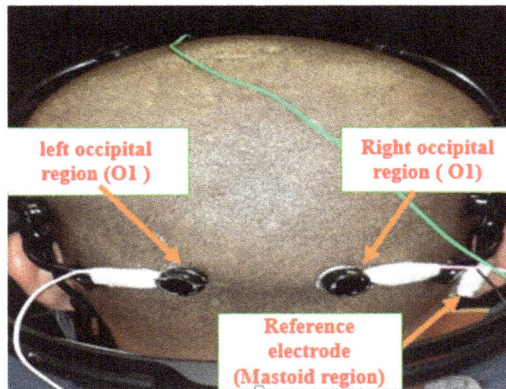

Figure 2. Electroencephalogram (EEG) electrodes positions.

The EEG data were amplified and recorded by Mega win 3.1b-13 system (Mega Electronics Ltd., Kuopio, Finland), with a high-quality four-channel digital EEG amplifier for ME 6000. The EEG data was recorded from occipital lobes (both O1 and O2) with a sampling rate of 1000 Hz. Before performing any analysis on the EEG waves, the noise was separated to eliminate undesired signals. The noise was filtered using a low-pass filter (four poles Elliptic filter was used) with a cut-off frequency of 32 Hz to remove the power-line noise and another high-frequency noise. Subsequently, the signal amplitude was carefully adjusted. The EEG waves were extracted by decomposition of the EEG data using multilevel discrete wavelet transform (DWT), where DWT contains sub-band of the signal. Four levels of DWT using Debauches 4 were implemented and the information about the frequencies of the EEG waves is divided into several wavelet levels. Moreover, a Fast Fourier transform (FFT) was employed to identify the frequency contained in each level. The relative powers of the β band and α/β power ratio were the responses selected as the indicators of visual fatigue, which are response variables related to the EEG based on previous studies [28,29]. The relative power of the β band was calculated using the integrated power of a particular band divided by the sum of the integrated powers of the θ, α, and β rhythms. For example, the basic index formula of β rhythm is expressed as a basic index $\beta = 100 \times$ (power of β)/(power of α + power of β + power of θ). The ratio indices are defined as the proportion of the basic indices, i.e., β/α, $(\alpha + \theta)/\beta$ [17,29].

2.6. Statistical Analysis

A four-way mixed repeated measures design (ANOVA) was implemented to examine the effects of the time, films, display, and distance on the response variables (EEG relative β power and α/β power ratio). The effects of watching 2D and 3D displays from particular distances on the total subjective visual discomfort score were evaluated using multivariate analysis of variance (MANOVA): a two-way multivariate analysis of variance helped compare the effect of independent variables on pre and post visual discomfort as dependent variables. Finally, pairwise comparisons were used to investigate the source of any significant effects. Statistical analysis was performed using SPSS Version 23. Specific assumptions should be met before performing the ANOVA, including homogeneity of variance in the data, normality, and continuous data (not dichotomous). Moreover, not only was the statistical significance calculated, but also, the effect size based on the partial eta-squared value (η^2). The partial eta-squared indicates the percentage of the variance in the dependent variables attributable to the particular independent variables [46].

3. Results

3.1. Subjective Visual Discomfort Rating

After the test, overall results showed that only the display type had a significant effect on the subjective visual discomfort as depicted in Table 2. The results revealed a significant difference in the display type pre- and post-test in term of the total subjective visual discomfort score: *Wilks' lambda = 0.864, F(2, 75) = 5.911, p = 0.004, partial eta squared = 0.136.* Given the significance of the overall tests, the univariate main effects were examined. The display type was significant for conducting pre- and post-tests. As presented in Table 2, after the participants watched videos on 2D and 3D displays, individuals under the 3D condition had higher scores [total subjective visual discomfort score: *F (1, 76) = 11.752, p = 0.001, partial eta squared = 0.134.* The scores before the participants watched any videos were not significantly different for those watching 2D and 3D displays.

Table 2. Score means (standard deviation) for total subjective visual discomfort score.

	Mean (SD)				Statistics			
Display	**2D**	**3D**	F [a]	p [a]	F [b]	p [b]	η^2	η^2
Pre	1.18 (0.958)	1.23 (1.03)	0.051	0.822				0.001
Post	8.30 (7.94)	16.20 (12.02)	11.752	0.001	5.911	0.004	0.136	0.134
Distance	**3H**	**6H**	F [a]	p [a]	F [b]	p [b]	η^2	η^2
Pre	1.35 (1.00)	1.05 (0.994)	1.826	0.181				0.023
Post	12.8 (11.3)	11.70 (10.5)	0.228	0.635	0.923	0.402	0.024	0.03

[a] Comparison between watching films at 2D or 3D TV for both pre and post (univariate test). [b] Comparison between watching films at pre and post (overall test).

3.2. EEG Signals Response Analysis

In this research work, only the signal from the occipital lobes (O1) was analyzed because both have the same values. The responses relative β-power band and α/β power ratio were selected as the indicators of visual fatigue.

3.2.1. Relative β Power

The relative β power band (β rhythm) is expressed as relative β power band = 100 × (power of β/(power of α + power of β + power of θ)). The results showed that the viewing time has a significant effect on the relative β power of the EEG, $F (7, 448) = 3.982$, $p < 0.0001$, partial eta squared = 0.059. Moreover, the display type has a significant effect on the relative β power of the EEG, $F (1, 64) = 9.332$, $p < 0.003$, partial eta squared = *0.127*. In addition, the viewing distance in terms of the display-type interaction had a significant effect on the relative β power of the EEG, $F (1, 64) = 8.526$, $p < 0.005$, partial eta squared = *0.118*.

Figure 3 shows the relationship between the EEG relative β power and viewing time. The results show that the relative β power of EEG decreases slightly with the increase in viewing time from previewing to thirty minutes (session 4), and subsequently, remains stable for up to fifty minutes (session 6). Thereafter, it significantly decreases from session six until the end of the movies, $p < 0.05$.

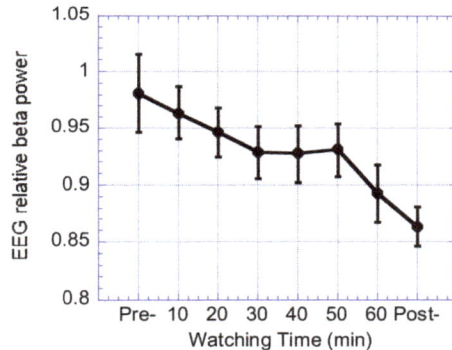

Figure 3. Effect of viewing time on relative β power (%).

Figure 4 shows the effect of the viewing distance (3H and 6H) with respect to the display type (2D and 3D) on the relative β power of the EEG. As shown in the Figure 4, the relative β power is significantly higher when watching videos in the 2D display from a short viewing distance when compared with watching the videos in 3D, $F (1, 38) = 23.662$, $p < 0.0001$, partial eta squared = *0.384*. Moreover, there is no significant difference between watching the videos in 2D or 3D from a long viewing distance (6H). Moreover, when the viewing distance is changed from 3H to 6H, the relative β power was increased in case of 3D-display, while it was decreased in case of 2D-display. The difference between watching the 3D display from near or far viewing distance is insignificant (3H or 6H), $F (1, 38) = 2.564$, $p = 0.118$, partial eta squared = *0.063*. In addition, there is a significant difference between watching the 2D display from different viewing distances (3H or 6H), $F (1, 38) = 7.097$, $p = 0.011$; partial eta squared = *0.157*. The results show that watching the 3D display from a short viewing distance resulted in a high visual fatigue while watching the 2D display from a short viewing distance resulted in a low visual fatigue.

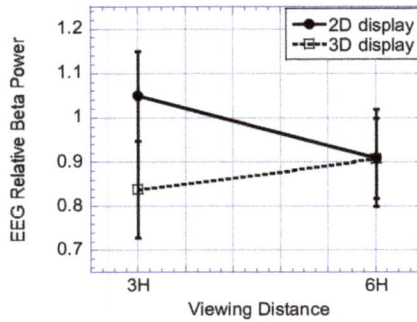

Figure 4. Effect of viewing distance by display type interaction on relative β power (%).

3.2.2. α/β Power Ratio

The results showed that the viewing time has a significant effect on α/β power ratio, $F_{(7, 448)} = 8.947$, $p < 0.001$, partial eta squared $= 0.123$. In addition, the viewing time with respect to the type of display has a significant effect on the α/β power ratio, $F_{(7, 448)} = 5.667$, $p < 0.001$.

Figure 5 shows the effect of the viewing time with respect to the display type on the α/β power ratio in decibel (dB). The results show that while watching the 3D display, the α/β ratio increases over viewing time while the α/β power ratio remains stable when the participants watched the 2D display over time. There is a significant difference between watching the 2D and 3D displays at time intervals of 4, 5, 6, 7, and 8 on the α/β power ratio at $p < 0.05$ based on the pairwise comparisons. When watching the 3D display, the differences between the time intervals 1 and 2; 2 and 3; 4 and 5; 5, 6, and 7; 6 and 7 are insignificant, $p > 0.05$. It can be concluded that when watching videos in 3D display, the participants were visually fatigued between ten and twenty minutes of watching the 3D display. Subsequently, the fatigue increases significantly for 10 min (at 30 min of watching). Thereafter, the fatigue increases gradually for the next 10 min (at 40 min of watching), and then, increases gradually for the next twenty minutes (at 60 min of watching).

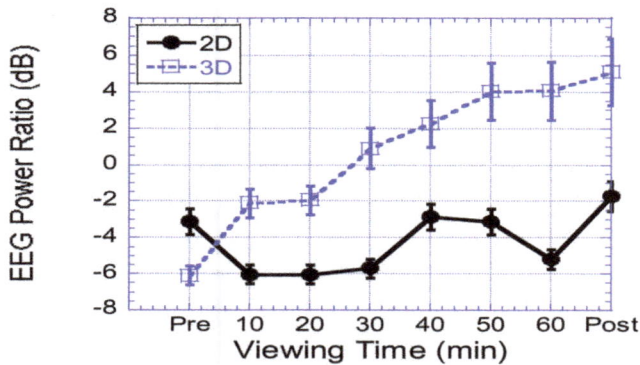

Figure 5. Effect of viewing time by display type on α/β power ratio.

4. Discussion

The subjective visual discomfort, relative β power of EEG, and α/β power ratio of EEG are used to evaluate the effect of watching videos on 2D and 3D displays associated with viewing distances

(3H vs. 6H) over a viewing time of 60 min. In terms of the visual discomfort, it is revealed that watching the 3D display clearly caused more visual discomfort than watching the 2D display, regardless of the viewing distance. These results are consistent with those of previous studies [9,33,38,42,47–49].

The mental responses of the participants were analyzed while watching different films on 2D and 3D displays over sixty minutes from long and short viewing distances. The results show that the relative β power of the EEG decreases slightly with the increase in the viewing time. Moreover, watching the 3D display contributes more to the decrease in the relative β power of the EEG than watching the 2D display, which causes more visual fatigue than that evident in the case of the 2D-display. These results are consistent with the study conducted by Chen et al. [6], who found that for most participants watching the 3D TV, the energy in the β frequency bands significantly decreases. Moreover, when the viewing distance was switched from 3H to 6H, the relative β power increased in the 3D-display case, while it decreased in the 2D-display case. In addition, the same levels of visual fatigue were observed when watching videos at 2D or 3D from a long viewing distance (6H). These results show that watching the 3D display from a short viewing distance resulted in a high visual fatigue. Viewing the 2D display from a short viewing distance resulted in a low visual fatigue as the viewing distance gets longer, resolution will be higher (pixels on the display will be imaged on a smaller region on viewer's retina, making the image harder to resolve), field of view will be smaller (hence the sense of immersion gets weaker), and binocular disparity will get smaller (even though the disparity is the same on the screen, the angular disparity with respect to the viewer will be smaller—therefore, the viewer will experience smaller amounts of vergence–accommodation conflict). Moreover, the type of display was found to have the most contribution to the visual fatigue compared to other variables (12.7%), followed by distance*display (11.8%), time (5.9%), and distance (1.4%).

The results of the α/β power ratio show that watching the 3D display contributes to increase in the α/β power ratio over viewing time while it remains stable when the participants watched the 2D display over time. Moreover, the type of display has the most contribution to the visual fatigue compared to other variables (13.1%), followed by time (12.3%), and time*display (8.1%). It can be concluded that when watching videos in 3D display, visual fatigue increased between ten and twenty minutes. Subsequently, it increased significantly for 10 min (at 30 min of watching). Thereafter, the increase was gradual for the next 10 min (at 30 min of watching), and then further increases gradually for the next 20 min (at sixty min of watching). Moreover, watching the 3D display contributes to increase the α/β power ratio more than watching videos in 2D display. This result was in agreement with that obtained by Hsu and Wang [10], where the increase in α/β was associated with visual fatigue. The results of this study showed that there were certain discrepancies between the subjective measure of visual fatigue and the objective metric. Because the subjective measure is one of the conventional methods which might be affected by the personal variation, practical experience, and familiarity, has a significant effect on conscious experience with both positive and negative consequence.

Shibata et al. [33] observed more discomfort and fatigue with a given vergence–accommodation conflict at longer distances. This finding is in agreement with the study conducted by Shibata et al. [33], wherein the viewing distance had a significant effect on visual fatigue. In stereoscopic 3D displays, the accommodation distance is fixed at a particular distance from the eyes to the display screen so the viewer needs to accommodate to the screen. However, the vergence distance varies depending on the distance being simulated on the display so the viewer needs to converge at a point located in front of or behind the display plane. Moreover, Wee et al. [50] and Wee and Moon [51] identified that accommodation and convergence may conflict in the participant viewing programs on 3D displays because of an increase in the near visual tasks, which in turn, may lead to eventual deterioration of the capability to accommodate and converge, thereby increasing the visual fatigue. Watching contents on 3D displays from short viewing distance contributes to eventual deterioration of the capability to accommodate and converge, which in turn, may lead to the development of visual fatigue. This result helps support the findings of the study conducted by Wee et al. [50] and Wee and Moon [51].

There are other methods of 3D displays such as integral photography, which is a promising method to display 3D optical images by reproducing exactly the same light rays as emitted from real objects [52]. This method duplicates the conditions of viewing real objects. Therefore, the convergence and accommodation responses have been predicted to be consistent with the depth position of the 3D target. The accommodation response to integral 3D displays has been theoretically investigated by many researchers [53–56]. The reports indicate that satisfying the super multi-view (SMV) condition is the most important requirement for obtaining a proper accommodation response. The SMV condition means that two or more light rays from the point lights of the reconstructed 3D objects reach the pupil of an observer. Hiura et al. [56] showed that six of the ten observers did not have an accommodation-convergence conflict in viewing integral photography in the range. Moreover, the required resolution was found to be 0.7 or more and less than 1.4 cycles per degree for inducing accommodation. Hiura et al. [56], concluded that integral photography can provide a natural 3D image that looks like a real object.

Alhaag and Ramadan [57] investigated the effects of display type (2D versus 3D), viewing distance (3H versus 6H) and viewing time on visual fatigue based on percentage of maximum electromyography (EMG) contraction (%MVC) of orbicularis oculi (OO) muscle activity. They reported that viewing time and distance had significant effects on the %MVC and OO muscle activity with the details depending on the display type. The OO muscle activity of the participants watching the 3D display from a long distance (6H) did not change over the viewing time. Alhaag and Ramadan [52] suggested that watching 3D display from a short viewing distance contributed to the eventual deterioration of the capability to accommodate and converge. This, in turn, likely caused development of greater visual fatigue watching the 3D display as compared to watching the 2D display from an equivalent short viewing distance. The reason offered for increasing visual fatigue while viewing 3D display from short viewing distance is that doing so increases the field of view.

Ukai and Howarththe [58], Suzuki et al. [59], and Fujikado [60] showed that the reason of conflict between the accommodation and vergence in 3D displays, which results in visual fatigue, is accommodation that should respond to the screen position but disparity in the two images for both the vergence stimulus of the eyes varies over time. These problems occur only for short viewing distances in the range of 31.1–53.6 cm in a volumetric display [49] because of the depth of field of the human eye [56]. The results of this study are in good agreement with those obtained by Alhaag and Ramadan [57], Ukai and Howarth [58], Suzuki et al. [59], and Fujikado [60]. In addition, it was explained that a wide field of view could cause more cyber sickness [61–64]. Visual fatigue increases while viewing 3D display from short viewing distance because the field of view increases, causing high strain. The result of this study is in good agreement with those obtained by Alhaag and Ramadan [56], Seay et al. [62] and Lin et al. [63]. Lee et al. [64] studied the effects of display type (2D versus 3D) and viewing distances (60 cm versus 90 cm) on the amount of eyestrain using a near-infrared pupil detection device. They found that watching the 3D display from a short distance (60 cm) caused high eyestrain. This result agrees with the one concluded in this study.

5. Conclusions

Watching 3D TV has different effects on the brain as compared against watching 2D TV. The brain is relaxed while watching 2D TV from short viewing distances, while watching 3D display from short viewing distances requires participants to utilize more cognitive loads for processing three-dimensional information compared to 2D displays. Hence, watching 3D displays from short viewing distances caused visual fatigue, which may decrease the cognitive capacity of the participants for processing visual information. In addition, the brain is relaxed while watching 3DTV from longer viewing distances. While watching 2D displays from long distances, the cognitive load increases because of low level of attentional focusing ability and regulates occipital lobe required by the brain to process the information. The results show that in case of 3D mode, the decrease in the relative β power of EEG and increase in the α/β power ratio were significantly higher from the start till the end of the test.

It was found that the "type of display" has most significant factor for visual fatigue compared to other variables (12.7%), followed by interaction of distance and display (11.8%), time (5.9%), and distance (1.40%). In addition, when the viewing distance was switched from 3H to 6H, the relative β power was increased in case of 3D-display, while it was decreased in case of 2D display. Moreover, the level of visual fatigue was the same when watching videos in 2D or 3D mode from a long viewing distance (6H).

Acknowledgments: The authors are grateful to the Raytheon Chair for Systems Engineering for funding. Also, the authors would like to acknowledge reviewers for their constructive comments and suggestions. Based on those comments, the manuscript is further enhanced and improved.

Author Contributions: Ramadan and Alhaag conceived the idea, and designed the experiments; Alhaag performed the experiments; Ramadan and Alhaag analyzed the data; all authors contributed materials/analysis tools; Ramadan and Alhaag organized the contents for the manuscript preparation and wrote the paper. All authors reviewed and approved the final version.

Conflicts of Interest: The authors declare no conflict of interest.

References

1. Kim, Y.J.; Lee, E.C. EEG based comparative measurement of visual fatigue caused by 2D and 3D. *Commun. Comput. Inf. Sci.* **2011**, *174*, 289–292.
2. Mun, S.C.; Park, M.C.; Park, S.I.; Whang, M.C. SSVEP and ERP measurement of cognitive fatigue caused by stereoscopic 3D. *Neurosci. Lett.* **2012**, *525*, 89–94. [CrossRef] [PubMed]
3. Jung, Y.J.; Kim, D.; Sohn, H.; Lee, S.I.; Park, H.W.; Ro, Y.M. Subjective and objective measurements of visual fatigue induced by excessive disparities in stereoscopic images. *Proc. SPIE* **2013**, *8648*, 86480M.
4. Cho, H.; Kang, M.-K.; Yoon, K.-J.; Jun, S.C. Feasibility Study for Visual Discomfort Assessment on Stereo Images Using EEG. In Proceedings of the International Conference on 3D Imaging, Liège, Belgium, 3–5 December 2012; pp. 1–6.
5. Park, S.J.; Subramaniyam, M.; Moon, M.K.; Kim, D.G. Physiological responses to watching 3D on television with active and passive glasses. In *International Conference on Human-Computer Interaction*; Springer: Berlin/Heidelberg, Germany, 2013; pp. 498–502.
6. Chen, C.; Li, K.; Wu, Q.; Wang, H.; Qian, Z.; Sudlow, G. EEG-based detection and evaluation of fatigue caused by watching 3DTV. *Displays* **2013**, *34*, 81–88. [CrossRef]
7. Chen, C.; Wang, J.; Li, K.; Wu, Q.; Wang, H.; Qian, Z.; Gu, N. Assessment visual fatigue of watching 3DTV using EEG power spectral parameters. *Displays* **2014**, *35*, 266–272. [CrossRef]
8. Chen, C.; Wang, J.; Li, K.; Liu, Y.; Chen, X. Visual fatigue caused by watching 3DTV: An fMRI study. *Biomed. Eng. Online* **2015**, *14*, S12. [CrossRef] [PubMed]
9. Li, H.C.; Seo, J.H.; Kham, K.T.; Lee, S.H. Measurement of 3D Visual Fatigue Using Event-related Potential: 3D Oddball Paradigm. In Proceedings of the 3DTV Conference: The True Vision—Capture, Transmission and Display of 3D Video, Istanbul, Turkey, 28–30 May 2008; pp. 213–216.
10. Hsu, B.W.; Wang, M.J.J. Evaluating the effectiveness of using electroencephalogram power indices to measure visual fatigue. *Percept. Mot. Skills* **2013**, *116*, 235–252. [CrossRef] [PubMed]
11. Wang, Y.; Liu, Y.; Zou, B.; Huang, Y. Study on issues of visual fatigue of display devices. In *Signal Recovery and Synthesis*; Optical Society of America: Washington, DC, USA, 2014.
12. Wang, Y.; Liu, T.; Li, S.; Wang, J. Using electroencephalogram spectral components to assess visual fatigue caused by sustained prism-induced diplopia. In Proceedings of the 2016 9th International Congress on Image and Signal Processing, BioMedical Engineering and Informatics (CISP-BMEI), Datong, China, 15–17 October 2016; pp. 1551–1556.
13. Bang, J.W.; Heo, H.; Choi, J.S.; Park, K.R. Assessment of eye fatigue caused by 3D displays based on multimodal measurements. *Sensors* **2014**, *14*, 16467–16485. [CrossRef] [PubMed]
14. Guo, M.; Liu, Y.; Zou, B.; Wang, Y. Study of electroencephalography-based objective stereoscopic visual fatigue evaluation. In Proceedings of the 2015 International Symposium on Bioelectronics and Bioinformatics (ISBB), Beijing, China, 14–17 October 2015; pp. 160–163.
15. Hou, C.; Yue, G.; Shen, L. Assessing the visual discomfort of compressed stereoscopic images using ERP. In *International Conference on Human Centered Computing*; Springer: Cham, Switzerland, 2016; pp. 127–137.

16. Frey, J.; Appriou, A.; Lotte, F.; Hachet, M. Classifying EEG signals during stereoscopic visualization to estimate visual comfort. *Comput. Intell. Neurosci.* **2016**, *2016*, 2758103. [CrossRef] [PubMed]

17. Cheng, S.; Lee, H.; Shu, C.; Hsu, H. Electroencephalographic study of mental fatigue in visual display terminal tasks. *J. Med. Biol. Eng.* **2007**, *27*, 124.

18. Xie, X.; Hu, J.; Liu, X.; Li, P.; Wang, S. The EEG changes during night-time driver fatigue. In Proceedings of the 2009 IEEE Intelligent Vehicles Symposium, Xi'an, China, 3–5 June 2009; pp. 935–939.

19. Lal, S.K.; Craig, A.; Boord, P.; Kirkup, L.; Nguyen, H. Development of an algorithm for an EEG-based driver fatigue countermeasure. *J. Saf. Res.* **2003**, *34*, 321–328. [CrossRef]

20. Liu, J.; Zhang, C.; Zheng, C. EEG-based estimation of mental fatigue by using KPCA–HMM and complexity parameters. *Biomed. Signal Process. Control* **2010**, *5*, 124–130. [CrossRef]

21. Kar, S.; Bhagat, M.; Routray, A. EEG signal analysis for the assessment and quantification of driver's fatigue. *Transp. Res. Part F* **2010**, *13*, 297–306. [CrossRef]

22. De Waard, D. *The Measurement of Drivers' Mental Workload*; Groningen University, Traffic Research Center: Groningen, The Netherlands, 1996.

23. Fisch, B.J.; Spehlmann, R. (Eds.) *Fisch and Spehlmann's EEG Primer: Basic Principles of Digital and Analog EEG*; Elsevier Health Sciences: Amsterdam, The Netherlands, 1999.

24. Cheng, S.-Y.; Hsu, H.-T. Mental Fatigue Measurement Using EEG. In *Risk Management Trends*; Intech: Rijeka, Croatia, 2011.

25. Tran, Y.; Thuraisingham, R.A.; Wijesuriya, N.; Nguyen, H.T.; Craig, A. Detecting neural changes during stress and fatigue effectively: A comparison of spectral analysis and sample entropy. In Proceedings of the 2007 3rd International IEEE/EMBS Conference on Neural Engineering, Kohala Coast, HI, USA, 2–5 May 2007; pp. 350–353.

26. Belyavin, A.; Wright, N.A. Changes in electrical activity of the brain with vigilance. *Electroencephalogr. Clin. Neurophysiol.* **1987**, *66*, 137–144. [CrossRef]

27. Subasi, A. Automatic recognition of alertness level from EEG by using neural network and wavelet coefficients. *Expert Syst. Appl.* **2005**, *28*, 701–711. [CrossRef]

28. Jap, B.T.; Lal, S.; Fischer, P.; Bekiaris, E. Using EEG spectral components to assess algorithms for detecting fatigue. *Expert Syst. Appl.* **2009**, *36*, 2352–2359. [CrossRef]

29. Eoh, H.J.; Chung, M.K.; Kim, S.H. Electroencephalographic study of drowsiness in simulated driving with sleep deprivation. *Int. J. Ind. Ergon.* **2005**, *35*, 307–320. [CrossRef]

30. Patterson, R. Human factors of 3D displays. *J. Soc. Inf. Disp.* **2007**, *15*, 861–871. [CrossRef]

31. Lambooij, M.; Fortuin, M.; Heynderickx, I.; IJsselsteijn, W. Visual discomfort and visual fatigue of stereoscopic displays: A review. *J. Imaging Sci. Technol.* **2009**, *53*, 1–14. [CrossRef]

32. Shibata, T.; Kim, J.; Hoffman, D.M.; Banks, M.S. The zone of comfort: Predicting visual discomfort with stereo displays. *J. Vis.* **2011**, *11*, 11. [CrossRef] [PubMed]

33. Shibata, T.; Kim, J.; Hoffman, D.M.; Banks, M.S. Visual discomfort with stereo displays: Effects of viewing distance and direction of vergence-accommodation conflict. *Proc. SPIE Int. Soc. Opt. Eng.* **2011**, *7863*, 78630P1–78630P9. [PubMed]

34. Park, M.C.; Mun, S. Overview of measurement methods for factors affecting the human visual system in 3D displays. *J. Disp. Technol.* **2015**, *11*, 877–888. [CrossRef]

35. Matthews, M.L.; Lovasik, J.V.; Mertins, K. Visual performance and subjective discomfort in prolonged viewing of chromatic displays. *Hum. Factors* **1989**, *31*, 259–271. [CrossRef] [PubMed]

36. American Optometric Association. *Guide to the Clinical Aspects of Computer Vision Syndrome*; American Optometric Association: St. Louis, MO, USA, 1995.

37. Kim, C.J.; Park, S.; Won, M.J.; Whang, M.; Lee, E.C. Autonomic Nervous System Responses Can Reveal Visual Fatigue Induced by 3D Displays. *Sensors* **2013**, *13*, 13054–13062. [CrossRef] [PubMed]

38. Park, S.; Won, M.J.; Mun, S.; Lee, E.C.; Whang, M. Does visual fatigue from 3D displays affect autonomic regulation and heart rhythm? *Int. J. Psychophysiol.* **2014**, *92*, 42–48. [CrossRef] [PubMed]

39. Assembly, I.R. *Methodology for the Subjective Assessment of the Quality of Television Pictures*; International Telecommunication Union: Geneva, Switzerland, 2003.

40. Park, S.J.; Oh, S.B.; Subramaniyam, M.; Lim, H.K. Human impact assessment of watching 3D television by electrocardiogram and subjective evaluation. In Proceedings of the XX IMEKO World Congress—Metrology for Green Growth, Busan, Korea, 9–14 September 2012.

41. Chen, C.Y.; Ke, M.D.; Wu, P.J.; Kuo, C.D.; Pong, B.J.; Lai, Y.Y. The influence of polarized 3D display on autonomic nervous activities. *Displays* **2014**, *35*, 196–201. [CrossRef]
42. Li, H.C.O. Human factor research on the measurement of subjective three dimensional fatigue. *J. Broadcast. Eng.* **2010**, *15*, 607–616. [CrossRef]
43. Kennedy, R.S.; Lane, N.E.; Berbaum, K.S.; Lilienthal, M.G. Simulator sickness questionnaire: An enhanced method for quantifying simulator sickness. *Int. J. Aviat. Psychol.* **1993**, *3*, 203–220. [CrossRef]
44. Andreassi, J.L. *Psychophysiology: Human Behavior & Physiological Response*; Taylor & Francis: Mahwah, NJ, USA, 2009.
45. Iwasaki, T.; Kurimoto, S. Objective evaluation of eye strain using measurements of accommodative oscillation. *Ergonomics* **1987**, *30*, 581–587. [CrossRef] [PubMed]
46. Brown, J.D.; Hilgers, T.; Marsella, J. Essay prompts and topics minimizing the effect of mean differences. *Writ. Commun.* **1991**, *8*, 533–556. [CrossRef]
47. Yano, S.; Ide, S.; Mitsuhashi, T.; Thwaites, H. A study of visual fatigue and visual comfort for 3D HDTV/HDTV images. *Displays* **2002**, *23*, 191–201. [CrossRef]
48. Park, S.; Won, M.J.; Lee, E.C.; Mun, S.; Park, M.C.; Whang, M. Evaluation of 3D cognitive fatigue using heart–brain synchronization. *Int. J. Psychophysiol.* **2015**, *97*, 120–130. [CrossRef] [PubMed]
49. Hoffman, D.M.; Girshick, A.R.; Akeley, K.; Banks, M.S. Vergence–accommodation conflicts hinder visual performance and cause visual fatigue. *J. Vis.* **2008**, *8*, 33. [CrossRef] [PubMed]
50. Wook Wee, S.; Moon, N.J.; Lee, W.K.; Jeon, S. Ophthalmological factors influencing visual asthenopia as a result of viewing 3D displays. *Br. J. Ophthalmol.* **2012**, *96*, 1391–1394.
51. Wook Wee, S.; Moon, N.J. Clinical evaluation of accommodation and ocular surface stability relevant to visual asthenopia with 3D displays. *BMC Ophthalmol.* **2014**, *14*, 1.
52. Lippmann, M.G. Epreuves reversibles donnant la sensation du relief. *J. Phys. Theor. Appl.* **1908**, *7*, 821–825. [CrossRef]
53. Jung, J.H.; Hong, K.; Lee, B. Effect of viewing region satisfying super multi-view condition in integral imaging. *SID Symp. Dig. Tech. Pap.* **2012**, *43*, 883–886. [CrossRef]
54. Maimone, A.; Wetzstein, G.; Hirsch, M.; Lanman, D.; Raskar, R.; Fuchs, H. Focus 3D: Compressive accommodation display. *ACM Trans. Graph.* **2013**, *32*, 1–13. [CrossRef]
55. Deng, H.; Wang, Q.-H.; Luo, C.-G.; Liu, C.-L.; Li, C. Accommodation and convergence in integral imaging 3D display. *J. Soc. Inf. Disp.* **2014**, *22*, 158–162. [CrossRef]
56. Hiura, H.; Komine, K.; Arai, J.; Mishina, T. Measurement of static convergence and accommodation responses to images of integral photography and binocular stereoscopy. *Opt. Express* **2017**, *25*, 3454–3468. [CrossRef] [PubMed]
57. Alhaag, M.H.; Ramadan, M.Z. Using electromyography responses to investigate the effects of the display type, viewing distance, and viewing time on visual fatigue. *Displays* **2017**, *49*, 51–58. [CrossRef]
58. Ukai, K.; Howarth, P.A. Visual fatigue caused by viewing stereoscopic motion images: Background, theories, and observations. *Displays* **2008**, *29*, 106–116. [CrossRef]
59. Suzuki, Y.; Onda, Y.; Katada, S.; Ino, S.; Ifukube, T. Effects of an eyeglass-free 3D display on the human visual system. *Jpn. J. Ophthalmol.* **2004**, *48*, 1–6. [CrossRef] [PubMed]
60. Fujikado, T. Asthenopia from the viewpoint of visual information processing-effect of watching 3D images. *J. Eye* **1997**, *14*, 1295–1300.
61. Patterson, R. Review Paper: Human factors of stereo displays: An update. *J. Soc. Inf. Disp.* **2009**, *17*, 987–996. [CrossRef]
62. Seay, A.F.; Krum, D.M.; Hodges, L.; Ribarsky, W. Simulator sickness and presence in a high field-of-view virtual environment. In *CHI'02 Extended Abstracts on Human Factors in Computing Systems, Proceedings of the CHI'02 Human Factors in Computing Systems, Minneapolis, MN, USA, 20–25 April 2002*; ACM: New York, NY, USA, 2002; pp. 784–785.

63. Lin, J.W.; Duh, H.B.L.; Parker, D.E.; Abi-Rached, H.; Furness, T.A. Effects of field of view on presence, enjoyment, memory, and simulator sickness in a virtual environment. In Proceedings of the IEEE Virtual Reality, Orlando, FL, USA, 24–28 March 2002; pp. 164–171.
64. Lee, E.C.; Heo, H.; Park, K.R. The comparative measurements of eyestrain caused by 2D and 3D displays. *IEEE Trans. Consum. Electron.* **2010**, *56*, 1677–1683. [CrossRef]

applied sciences

MDPI

Article

Estimation of Mental Distress from Photoplethysmography

Roberto Zangróniz [1], Arturo Martínez-Rodrigo [1], María T. López [2], José Manuel Pastor [1] and Antonio Fernández-Caballero [2,3,*]

[1] Instituto de Tecnologías Audiovisuales, Universidad de Castilla-La Mancha, 16071 Cuenca, Spain; roberto.zangroniz@uclm.es (R.Z.); arturo.martinez@uclm.es (A.M.-R.); josemanuel.pastor@uclm.es (J.M.P.)

[2] Instituto de Investigación en Informática, Universidad de Castilla-La Mancha, 02071 Albacete, Spain; Maria.LBonal@uclm.es

[3] Centro de Investigación Biomédica en Red de Salud Mental (CIBERSAM), 28029 Madrid, Spain

* Correspondence: antonio.fdez@uclm.es; Tel.: +34-967-599-200

Received: 11 November 2017; Accepted: 3 January 2018; Published: 5 January 2018

Abstract: This paper introduces the design of a new wearable photoplethysmography (PPG) sensor and its assessment for mental distress estimation. In our design, a PPG sensor obtains blood volume information by means of an optical plethysmogram technique. A number of temporal, morphological and frequency markers are computed using time intervals between adjacent normal cardiac cycles to characterize pulse rate variability (PRV). In order to test the efficiency of the developed wearable for classifying distress versus calmness, the well-known International Affective Picture System has been used to induce different levels of arousal in forty-five healthy participants. The obtained results have shown that temporal features present a single discriminant power between emotional states of calm and stress, ranging from 67 to 72%. Moreover, a discriminant tree-based model is used to assess the possible underlying relationship among parameters. In this case, the combination of temporal parameters reaches 82.35% accuracy. Considering the low difficulty of metrics and methods used in this work, the algorithms are prepared to be embedded into a micro-controller device to work in real-time and in a long-term fashion.

Keywords: distress estimation; wearable; heart rate variability; photoplethysmography

1. Introduction

Mental distress (or psychological distress) is a general term used to describe unpleasant feelings or emotions that impact your level of functioning. In other words, it is psychological discomfort that interferes with your activities of daily living. Mental distress can result in negative views of the environment, others, and the self. This is why it is important to investigate on devices and environments capable of recognizing and/or regulating negative emotions [1–5]. Sadness, anxiety, distraction, and symptoms of mental illness are manifestations of psychological distress.

Mental stress is accompanied by dynamic changes in activity of the autonomic nervous system (ANS). Although mental stress cannot be measured directly, the physiological response can be interpreted to assess the level of mental stress. Several physiological parameters (like electroencephalograph, heart rate variability, blood pressure, event-related potentials, and electromyography, among others) have been found sensitive toward any changes occurring in mental stress level [6]. Moreover, heart rate variability (HRV), the quantification of beat-to-beat variability in cardiac cycle over time, is one of the most determinant measures of ANS status [7,8].

The heart rate (HR) represents successive heart polarization and depolarization caused by the electrical impulses generated on the sinoatrial node and transmitted to the ventricles [9]. The sympathetic nervous system increases HR in response to stress, exercise or heart disease by acting on the accelerans nerve [10]. On the contrary, the parasympathetic nervous system decreases HR through acting on the vagus nerve [11]. During ventricular polarization, blood is pumped into the cells throughout the circulatory system. This process is reflected in an electrocardiogram as the QRS complex, where R-peaks are the most significant points within this wave. Accordingly, R-peaks are used as reference for computing HR. Indeed, HR is defined as a time series sequence of non-uniform RR intervals [12].

One of the most extended methodologies to measure HR consists in quantifying blood volume changes caused by the circulatory system functioning in veins or capillaries [13–15]. This technique is based in photoplethysmography (PPG), which consists of a low-cost optical technique capable of measuring small variations in reflected/transmitted light intensity, associated with changes in blood pumping function [16]. Technology has enabled to measure HR using inexpensive PPG sensors. By using robust software running a good algorithm it is possible to measure HR and HR related parameters like HRV, and in turn stress [17–19]. In this respect, blood movement in vessels goes from heart to fingertips in a wave-like motion, generating a lag by the time required for transmission of the pulse wave between heart pumping, depicted in EEG as QRS complex, and maximum blood volume in the vessels, depicted in PPG signals as PPG-peak (systolic peak). Although there is no time synchronization between R-peak and PPG-peak due to time lag, some authors have used peak-to-peak PPG signal interval instead of R-R interval in ECG signal. The use of pulse cycle interval is often called pulse rate variability (PRV) for this purpose. While there are many studies that analyze the use of PRV as an estimation of HRV, it is difficult to obtain quantitative conclusions due to the differences among experiments and methodologies. However, a high correlation between both metrics has been reported for subjects at rest [20].

Our research team has decided to design, build, and assess its own wearable photoplethysmography sensor aimed at exploiting HRV for the sake of estimating mental distress. In this case, estimating mental distress it is not conceived at this point of the design to be used in clinical applications. Although some commercial/research wearables support the possibility to calculate distress (e.g., Apple Watch (https://www.apple.com/watch/), Fitbit Ionic (https://www.fitbit.com/ionic); Garmin Forerunner (https://www.garmin.com/en-US), there are at least three reasons that support the decision to design our own wearable. The first one is that some commercial wearables (e.g., watches) stop being marketed after a short time, which prevents their use in future applications. The second justification is that virtually no commercial wearables allow access to their sensors' raw data [21], which is necessary for any further statistical analysis. The last rationale is that an own implementation provides enough freedom to extend the hardware with additional sensors in future implementations. Besides, this article gives constructive details of the wearable for reproducing the experiments. On the contrary, most wearables are commercial and closed.

The remainder of the paper is as follows. Section 2 introduces the design of the HR monitoring device, the experimental methodology employed, how photoplethysmogram data is processed in this proposal, which features have been selected for estimating mental distress and, finally, a description of statistical analysis. Then, Section 3 offers the most important results obtained in our work and the performance of the system. Lastly, Section 4 discusses the more relevant aspects of the design presented, and Section 5 the conclusions of the proposal.

2. Materials and Methods

2.1. Monitoring Pulse Rate Variability

In the proposed design, an optical plethysmogram technique is used to obtain blood volume information by means of a PPG sensor. In a PPG acquisition and signal conditioning circuit there are several key elements: light emitting diode (LED), LED driver circuitry, photo-detector (PD), and PD signal conditioning circuitry [22]. Although it is possible to address the development of these elements from scratch, nowadays there are commercial fully-integrated analog front-ends and optical sensors suitable for PPG signals. In this regard, an analog front-end AFE4400 (Texas Instruments Incorporated, Dallas, TX, U.S.A.) and an optical sensor NJL5310R (New Japan Radio Corporation Ltd., San Jose, CA, U.S.A.) have been selected. The AFE4400 integrates a LED driver and a PD signal conditioning circuitries, and a timing control module in a single package. The NJL5310R consists of two green LEDs and a high sensitivity PD. Hence, they are suitable to build a circuit capable of measuring blood volume through reflective photometry.

According to the manufacturer, the NJL5310R has been designed with an optimal LEDs and PD separation to get a high-quality PPG signal. In our design, the green LEDs have been connected front-to-front, so that they light up and off simultaneously. In reflective photometry, LEDs and PD are placed in the same plane as the body part, and the PD receives the reflected light from different depths under the skin (such as blood, tissue, and so forth).

Furthermore, the AFE4400 has two clearly differentiated stages. On the one hand, the transmit stage is divided into LED driver and LED current control. The LED driver sets a reference current of each LED and has been configured in push-pull mode to turn both LEDs on and off simultaneously (since LEDs in the optical sensor have been wired in common anode). LED current control regulates and ensures that LED current follows its reference. On the other hand, the receiver stage is split into transimpedance amplifier (TIA), conditioning section and analog-to-digital converter (ADC). The differential current-to-voltage TIA converts PD input current into a voltage. The differential voltage at the output of the TIA includes the component from ambient light. So, after TIA a programmable digital-to-analog converter (DAC) provides the current to cancel the ambient component. Then, the resulting signal is amplified, passed through a low-pass filter, and, finally, buffered before driving the 22-bit sigma-delta analog-to-digital converter (ADC).

In addition, power needs are different for transmit and receiver stages. For transmit stage, the supply voltage depends upon voltage drop in the photo sensor's LEDs and the transmit stage reference voltage (set as a compromise between low-power and better dynamic range). For these reasons, a supply voltage of 4.3 V has been selected for transmit stage. For receiver stage, the supply voltage is not so strongly warped by external elements, rather for the admissible analog front-end (AFE) supply voltage. Accordingly, a supply voltage of 3 V has been chosen for receiver stage.

As shown in Figure 1, a low-input DC-DC boost switching converter, TPS61093 (Texas Instruments Incorporated, Dallas, TX, U.S.A.), steps up incoming voltage to 6.5 V. The converter's input voltage range allows to power the design by two alkaline batteries, a single Li-ion battery or through a traditional 3.3 V and 5 V regulated power supply. A couple of 150 mA ultra-low-noise linear regulators, TPS7A4901 (Texas Instruments Incorporated, Dallas, TX, U.S.A.), generate 4.3 V and 3 V supply for transmit and receiver stages, respectively. The choice of a linear regulator as power supply final stage allows filtering out the output voltage ripple inherent to DC-DC switching conversion.

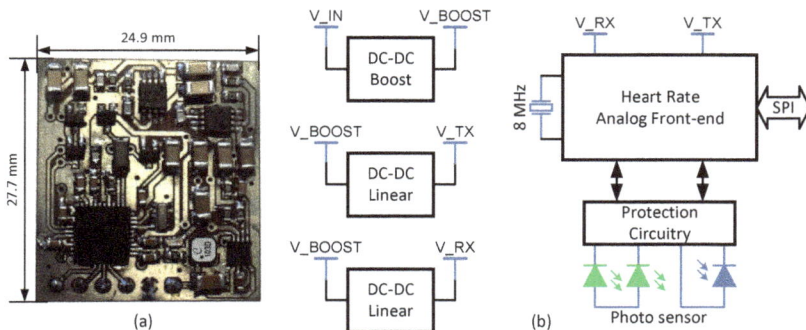

Figure 1. (**a**) Photoplethysmogram sensor, and (**b**) major building blocks: analog front-end, power rails, clock, and protection circuitry. DC: direct current; SPI: serial peripheral interface.

An 8 MHz external crystal feeds the AFE internal crystal oscillator that generates a 4 MHz internal master clock signal by means of a divide-by-2 block. AFE timer module uses this master clock to settle raising and falling edges of the different control signals. Lastly, the protection circuitry for the photo sensor consists of clamping diodes for each line in transmit and receiver stages (DC voltage restoration). Also, LED and PD tracks on the PCB (printed circuit board) are routed by means of differential pairs, because RF noise can attenuate photo sensor signals. Likewise, PD tracks have been guarded with the common-mode voltage signal from AFE, and common mode choke coils have been used.

The complete schematics of the design are attached as supplementary material.

2.2. Experimental Methodology

In order to classify distress versus calm condition we have used the well-known International Affective Picture System (IAPS) [23]. IAPS consists of a standard and categorized database of color photographs created to provide a wide range of affective stimuli. Moreover, the two primary dimensions recorded in the database are valence (ranging from pleasant to unpleasant) and arousal (from calm to excited). So, for each IAPS picture the mean and standard deviation of arousal and valence is provided in four different tables constructed from responses of men, women and children who responded to the emotion felt when exposed to pictures by means of the Self-Assessment Manikin (SAM), an affective rating system [24]. Thus, the idea is to use IAPS database to show a series of images to some volunteer participants. Each image used in the experiment should belong to one of two classes "high arousal-low valence" and "low arousal-high valence", corresponding to distress and calm, respectively.

Fifty healthy participants (twenty-eight men and twenty-two women; 20 to 28 years old) not suffering from evident mental pathologies were recruited to participate in the experiment. Moreover, the participants who agreed did not present cardiovascular or anatomic nervous system diseases that could alter their PRV. All participants were informed on the high emotional content of some pictures that they will be shown. All participants were students from Technical School at Cuenca, Spain. The students had to pass the PHQ-9 Depression Test Questionnaire to be accepted in the experiment. The exclusion criteria used to discard subjects from the experimental section were mainly based on the results of PHQ-9. In this regard, the exclusion threshold was fixed at scores greater than eight. Considering this prerequisite, four students were not welcomed, and one experiment was not valid due to technical problems. Thus, the number of valid experiments was forty-five (twenty-five males and twenty females). This study was approved by Universidad de Castilla-La Mancha institutional committee on human experimentation. All participants gave written informed consent in accordance with the Declaration of Helsinki.

The procedure for performing the experiment is described next (see Figure 2). The participant sits in front of the experimentation monitor to keep his/her movement minimal. The developed wearable sensor, described in Section 2.1, is put on his/her right wrist. In this regard, the experimentation monitor consists of a high resolution 28 inches screen. When the technician verifies the proper functioning of the wearable and its communication with the software, the experiment starts. Firstly, the participant has to carefully read the general instructions of the experiment. Next, ten pictures randomly chosen from a set of pictures that fulfill the condition to belong to negative stress (or distress) are shown consecutively to the participant during 6 s each. Silences consisting of blank images with a fixed duration of 1 s are inserted between two consecutive images. Afterwards, a distracting task is presented to the participant so that his/her emotional state comes to neutral. Next, the experiment continues by showing randomly another set of ten IAPS images that fulfill the condition to belong to calmness. Again, silences are used between each pair of images. Lastly, the distracting task is offered again.

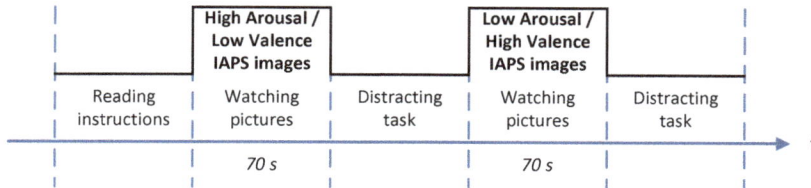

Figure 2. Experimentation time-line. IAPS: International Affective Picture System.

Pictures were classified into calm and stress subsets depending on their score of arousal and valence reported in IAPS database. Thus, the inclusion criterion for a picture to be added to the stress subset was that the picture rated an arousal level higher than 5 and a valence level lower than 3. Similarly, the inclusion criterion for a picture to be inserted into the calm subset was that it rated an arousal level lower than 4 and a valence level higher than 4 and lower than 6. Several studies have reported these ranges as those corresponding to negative stress and calmness, respectively [25].

Finally, and in agreement with Figure 2, the useful information extracted for each participant consists on two segments lasting 70 s for each stress and calm condition, respectively. Although a distracting task is deployed at the end of each image sequence, the PPG signal recorded in those segments is discarded for further analysis, since they are used exclusively to lead the participant to a neutral emotional state. Therefore, after experimentation, a total number of 90 PPG segments lasting 70 s were used in the subsequent analysis, where 45 belong to stress and 45 to calm condition.

2.3. Photoplethysmography Processing

The signal acquisition was held using the wearable PPG sensor at a sampling rate of 60 Hz and a 22-bit resolution. This sampling rate was chosen as minimum sampling frequency to prevent PPG signal distortion, since useful PPG information is located between 0 and 30 Hz [26]. However, in order to increase the reliability of the subsequent validation analysis and accuracy of PRV series computation, the acquired PPG was interpolated by using a cubic splines algorithm, increasing time resolution up to an equivalent sampling frequency of 1000 Hz. This type of interpolation has been used previously to increase the sampling rate in similar contexts [27].

Moreover, several factors, such as sensor location, electrical sources, ambient lights, skin properties or temperature, may affect quality of PPG signals [28]. These factors add different artifacts and noisy components to the waveform, which augments difficulty in signal characterization. Therefore, different filtering and processing techniques are applied to eliminate possible interferences and enhance the waveform before characterizing the signal. In this regard, power line interference is one of the most common noise sources. Indeed, ambient electromagnetic signals are present everywhere, modulating PPG signal over a sinusoidal component at its fundamental frequency. Moreover, variation in temperature or poor contact of photo sensor are only some of the causes of baseline in PPG signal [28]. Hence, baseline wander, high-frequency noise and power-line interference are removed by computing a forward/backward filtering approach. More concretely, baseline drift is removed by applying a 0.5 Hz cut-off high-pass, linear-phase FIR filter. Similarly, a 30 Hz cut-off low-pass, linear-phase FIR filter is applied to remove high-frequency noise and power-line interferences.

In this regard, approaches based on adaptive threshold have been extensively employed to detect peaks on PPG signals [29]. Nevertheless, PPG signals contain inherent noise (Gaussian noise), sudden amplitude changes or different morphologies caused by premature ventricular contractions or movements. Therefore, the use of a robust and reliable PPG peak detection algorithm is a key factor to face these difficulties. The algorithm has to deal with the detection of beats and artifacts while operating with a minimum computational burden in real-time. Considering all these premises, in this work an incremental-merge segmentation algorithm was used for PPG peak detection [30]. This algorithm extracts morphological features of PPG signal that are used as line segments. Next, these segments are classified as pulse or artifacts through using adaptive thresholds. More precisely, a PPG signal is converted into line segments by connecting the first and the last point of the line, depending of a tuning parameter *m* (length of line segments) which depends directly on the sampling rate. Then, line segments are classified such that pulse peaks are identified as endpoints of the validated up-slopes. On the other hand, horizontal lines are labeled as clipping or disconnection, so that up-slopes preceding and succeeding a horizontal line are labeled as artifacts. In this way, pulses from PPG corresponding to artifacts are identified and suppressed from beat-to-beat computation. Finally, PRV is estimated by measuring the time variation in consecutive PPG peaks.

2.4. Feature Extraction

Considering the short duration of each segment of analysis (70 s long), no additional windowing was set for the analysis, and the entire segment was used to extract the characteristics from the signal. It is important to highlight that all metrics were calculated retrospectively from PPG signals, rather than in real-time, as this work firstly intends to validate the proposed model. Similarly to other studies using short-term analysis of heart variability, classical time domain, frequency domain and morphological characteristics are computed from PRV and for both conditions (calm and stress) [31]. Table 1 shows the complete list of features used in the study.

Table 1. Temporal, frequency and morphological features computed for pulse rate variability (PRV) signals.

Signal	Analysis	Features
PRV	Temporal	*MNN, SDNN, SENN, DRNN, SDFD, RMSFD, pNN50*
	Morphological	*HRNN, TINN, SKNN, KUNN*
	Frequency	*LF, HF, NHF*

Concretely, adjacent normal cardiac cycles (*NN*) are computed and then, the mean of *NN* intervals (*MNN*), standard deviation of *NN* intervals (*SDNN*), standard error of successive differences of adjacent *NN* intervals (*SENN*) and the difference between the longest and shortest *NN* interval, that is the dynamic range, (*DRNN*) are calculated. Additionally, the ratio of pairs of successive *NN* intervals differing more than 50 ms (*pNN50*) are calculated by taking into consideration the total number of analyzed cardiac cycles. Finally, the first derivative (*FD*) of successive differences of adjacent *NN* intervals is computed and standard deviation (*SDFD*) and root mean square (*RMSFD*) are estimated.

Moreover, it has been reported that frequency parameters extracted from PRV may contribute with significant information that is not present in time-based methods [12]. Therefore, power spectral density (PSD) of the PRV is estimated in this work to obtain how power variance distributes as a function of frequency. A non-parametric method based on Fast Fourier Transform (FFT) is used. FFT is characterized by the simplicity of the algorithm and a high processing speed which are desirable specifications to be implemented in our design. According to the literature, three main spectral components are commonly used to assess PSD, namely very low frequency (*VLF*), low frequency (*LF*) and high frequency (*HF*), respectively [32]. The physiological meaning of *VLF* is quite diffuse, specially in short-term recordings and, consequently, it is out of this study [33]. On the contrary, *LF* and *HF* represent the control and balance of parasympathetic and sympathetic components [33]. More concretely, *LF* component ranges from 0.04 Hz up to 0.15 Hz and its increase is generally associated with a sympathetic activation [32]. On the other hand, *LF* component ranges between 0.15 Hz and 0.4 Hz and it is associated with parasympathetic modulation [32]. Finally, the relationship between the power found in *LF* and *HF* components is usually estimated, because it assess the sympatho-vagal balanve controlling the heart rate [34]. Thus, absolute values of potency (*LF* and *HF*), as well as relationship of *LF* and *HF* (*NHF*) regarding the total power (excluding *VLF* component) are calculated in this study.

With regard to morphological markers, HRV Triangular Index (*HRNN*) and Triangular Interpolation (*TINN*) are estimated. Morphological features are based on the fact that *NN* interval durations can be converted into a geometric pattern, such that *NN* density distribution is assessed by a simple equation to measure variability. In this work, the density distribution function (Δ) is constructed through assigning the number of equally long *NN* intervals to each value of their lengths. Then, the most frequent value of the distribution is calculated, i.e., $max(\Delta)$. *HRNN* is obtained by performing the ratio between the area integral of Δ by its maximum value. Similarly, *TINN* is computed by calculating the width of Δ, this way establishing the distribution boundaries, A and B, respectively. For the sake of performing this operation, Δ is transformed into a multi-linear function q, such that $q(t) = 0$ for $t \leq A$ and $t \geq B$. So basically, the Δ function with the best fit to the *NN* density distribution defines or identifies A and B boundaries. Then, *TINN* is calculated as temporal difference between A and B, that define the vertexes of the base of the triangle. Finally, the third and fourth moment of successive *NN* intervals, as well known as skewness (*SKNN*) and kurtosis (*KUNN*) from a distribution, were computed. Both parameters evaluate the asymmetry from PRV distribution caused by outliers or atypical pulse values within the series around the sample mean, thus assessing the shape of data distribution. Equation (1) shows the computation of *SKNN*, and Formula (2) represents the *KUNN* computation, where μ is the mean of successive *NN* intervals, σ is the standard deviation of successive *NN* intervals, and $E(t)$ corresponds to the expected value of quantity t.

$$SKNN = \frac{E(NN - \mu)^3}{\sigma^3} \tag{1}$$

$$KUNN = \frac{E(NN - \mu)^4}{\sigma^4} \tag{2}$$

2.5. Statistical Analysis

Shaphiro-Wilks and Levene tests have proved that distributions are normal and homoscedastic for all features studied. Consequently, the results are expressed in terms of mean $\pm$ standard deviation for all samples belonging to a same group. The statistical differences between both groups, calm and distress, are assessed by a *t*-Student test. A value of statistical significance $\rho < 0.05$ has been considered as significant.

Moreover, a ten-fold stratified cross-validation is used to assess the discriminant ability of each feature. This kind of cross-validation allows to obtain a highly reliable performance generalization of the metric under study [35]. Indeed, this approach makes use of all available data both for training and testing. This avoids the possibility of classification results to be highly dependent on the choice of a given training-test segmentation. Thus, the database is firstly partitioned into 10 equally sized folds, rearranging data to ensure that each fold is a good representative of the whole. Then, 10 training and validation iterations are performed, such that a fold of the data is held out for test, whereas the other ones are used for learning within each iteration. A receiver operating characteristic (ROC) curve is used to obtain the optimal discriminant threshold between calmness and distress for each learning set. The ROC curve is created by plotting true positive (TP) rate against false positive (FP) rate at various threshold settings. Here, TP rate (or sensitivity) is considered as the percentage of distress condition correctly classified. On the other hand, FP rate (or 1-specificity) corresponds to the rate of calm condition identified improperly. The optimal threshold is selected as the value which provides the highest accuracy, i.e., highest number of conditions correctly classified. Finally, global accuracy is obtained by averaging this procedure 5 times.

A decision tree (DT) classifier is used in order to assess the possible relationships among the different temporal, morphological and frequency features. This methodology is chosen due to its easy implementation and low computation burden when addressing a binary classification problem. Indeed, DT is based in consecutive if-else decisions. Regarding the DT configuration, each split is performed after considering the best optimization criterion, based on the Gini diversity index. Moreover, some rules are programmed to prevent an uncontrolled three overgrowth. Thus, the growth of every tree is always stopped when any node only contains samples from the same group (pure node) or less than 20% of all samples.

3. Results

Table 2 shows mean and standard deviation of the features under study. All characteristics calculated are shown, regardless of their statistical significance Only 4 out of 14 parameters show statistical differences when PRV is analyzed; and, all of them correspond to the time domain. More precisely, standard deviation (*SDNN*), dynamic range (*DRNN*) and standard error (*SENN*) of PRV series, as well as standard deviation of PRV derivative (*SDSD*), show statistical significance. In this regard, *DRNN* achieves the highest discriminatory power. It is worth noting that all significant parameters report an increasing temporal value when participants are elicited with stressing stimuli.

In order to estimate a reliable and robust power classification for each single parameter, a stratified 10-fold cross-validation is run five times. Figure 3 shows the ROC curves at a random iteration, together with the sensitivity (Se), specificity (Sp) and area under curve (AUC) for the four parameters that reported statistical significance. Additionally, average values of Se, Sp and accuracy (Ac), for both training and test subsets iterations, are shown in Table 3.

The single classification results are in agreement with the discriminatory power obtained previously. *DRNN* achieves the highest accuracy, correctly classifying 72.06% conditions. The rest of markers reach poorer performances, ranging from 66.18 to 67.65%. It is worth noting that features report a higher capability in discriminating true negatives, i.e., calmness than discriminating stress. In this regard, *SDNN* and *DRNN* achieve specificity values 76.47 and 82.35%, respectively.

Additionally, a series of tree-based classification models are calculated in order to study the potential relations among the different parameters. It is worth noting that all the parameters calculated are included in this analysis, regardless if they show statistical relevance, as there might exist underlying complementaries not revealed yet. Figure 4 shows the structure and the parameters composing the tree-based discriminant model more frequently obtained among the different iterations. As can be observed, the model is formed exclusively of temporal PRV parameters . Thus, the *SDNN* parameter is chosen as the most relevant in the model and a threshold of 0.5250 serves to divide the sample into two subgroups, calmness and stress.

Table 2. Results obtained. Mean and std values for emotional states of calm and distress and statistical significance (ρ), for all parameters are presented.

Physiological Analysis	Feature Acronym	Calmness Mean ± Std	Distress Mean ± Std	ρ
Temporal	*MNN*	0.7941± 0.1021	0.7765± 0.1064	0.4890
Temporal	*SDNN*	0.0596 ± 0.0243	0.1064 ± 0.0527	0.0046
Temporal	*DRNN*	0.2652 ± 0.1025	0.4522 ± 0.2453	1.15×10^{-4}
Temporal	*SENN*	0.0634 ± 0.0294	0.0934 ± 0.0500	0.0036
Temporal	*SDFD*	0.0589 ± 0.0361	0.1039 ± 0.0723	0.0018
Temporal	*RMSFD*	0.0853 ± 0.0570	0.1140 ± 0.0712	0.075
Temporal	*PNN50*	0.9997 ± 0.0018	0.9986 ± 0.0039	0.1495
Morphological	*HRNN*	5.2802 ± 1.4449	5.6542 ± 1.8026	0.3487
Morphological	*TINN*	0.3606 ± 0.1960	0.4894 ± 0.2373	0.053
Morphological	*SKNN*	−0.1009 ± 1.2990	−0.1335 ± 1.0867	0.9107
Morphological	*KUNN*	5.9136 ±6.8703	4.9149 ± 3.9008	0.4637
Frequency	*LF*	$0.0032 \pm 7.5112 \times 10^{-4}$	$0.0031 \pm 8.5693 \times 10^{-4}$	0.5710
Frequency	*HF*	0.0073 ± 0.0017	0.0071 ± 0.0019	0.5711
Frequency	*NHF*	$0.4401 \pm 1.7377 \times 10^{-4}$	$0.4401 \pm 1.7022 \times 10^{-4}$	0.9107

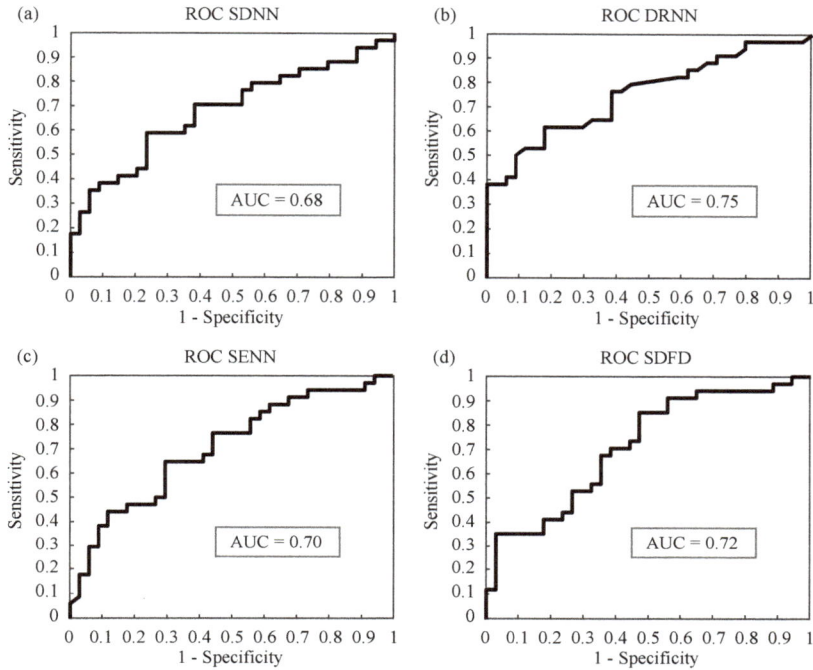

Figure 3. Receiver operating characteristic (ROC) curves for the four statistical significant parameters: (**a**) SDNN (standard deviation of NN intervals), (**b**) DRNN (dynamic range of NN intervals), (**c**) SENN (standard error of successive differences of adjacent NN intervals), (**d**) SDFD (the first derivative of standard deviation). AUC: area under curve.

Table 3. Sensitivity (Se), specificity (Sp) and accuracy (Ac) of significant parameters using ROC (receiver operating characteristic) analysis for training and test subsets.

Physiological Analysis	Feature Acronym	Learning			Test		
		Se (%)	Sp (%)	Ac (%)	Se (%)	Sp (%)	Ac (%)
Temporal	*SDNN*	61.30	84.02	72.34	58.82	76.47	**67.41**
Temporal	*DRNN*	69.57	85.10	77.32	61.76	82.35	**72.06**
Temporal	*SENN*	68.23	75.12	71.42	64.61	70.59	**67.65**
Temporal	*SDSD*	66.73	75.20	70.54	67.65	64.71	**66.18**

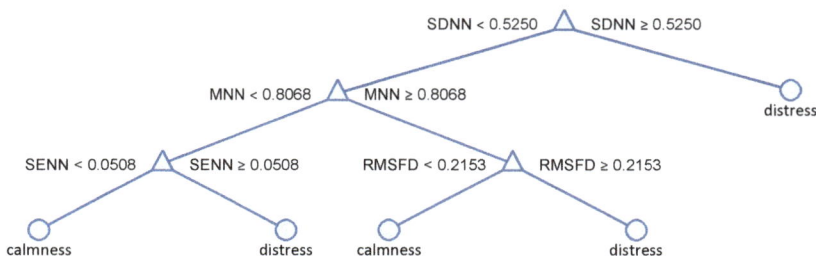

Figure 4. Tree-based discriminant models obtained by considering SDNN, MNN, SENN and RMSFD (the first derivative of root mean square) values from all PRV (pulse rate variability) signals.

In the next step, the samples labeled as calmness are newly partitioned into two subgroups using *MNN* and a threshold of 0.8068. Finally, in each remaining subgroup, *SENN* and *SDFD* are chosen to distribute the rest of samples by using thresholds 0.0508 and 0.2153, respectively. No further ramifications are formed with the criteria imposed. This tree-based model achieves a sensitivity, specificity and accuracy of 79.49, 85.29 and 82.35%, respectively. In this regard, the model improves global correctness more than 10% regarding the best single accuracy reported by *DRNN*. Finally, it is relevant to note that, while the single parameter with higher discriminatory power showed a limited sensitivity (i.e., ability to detect stress), when the parameters were combined by means of a tree-based classifier, sensibility increased more than 17%, achieving a more balanced discriminatory model.

4. Discussion

Negative stress (or distress) is one of the most important mental states due to its significant effects in health [1–3]. Distress is considered cause and consequence of failure and difficulties in a wide variety of daily situations. Thus, continuous monitoring of distress levels may prevent related health problems as well as unnecessary risks caused by suffering from stress. However, stress is a very complex subject and measuring it is not an easy task, as clearly stated recently [36]. Considering the increasing popularity of wearables such as continuous monitoring devices, photoplethysmography sensors have emerged as a reliable alternative to measure PRV. Although a number of works assessing stress condition are found in the literature, the most outstanding aspect of our contribution is the development of the necessary hardware and signal processing, as well as individual and global performance of the considered features. This enables deploying a wearable device with a high ability to discriminate between the two considered distress and calmness states. The lightness of our signal processing approach permits to work in real-time and in a long-term fashion.

In this work, stress and calmness conditions have been assessed by using temporal, frequency and morphological markers extracted from PRV. Interestingly, most of the temporal features reported statistical differences discriminating both conditions, while the rest of markers achieved low or no ability to differentiate between calmness and distress. In this respect, all classical metrics considered in this work have previously reported the ability to quantify changes in the ANS, responsible for regulating cardiac activity [32]. Moreover, they have been used before as good stress indicators [37]. However, only a few studies have evaluated this phenomenon from a short-time series viewpoint, which may lead to the described discrepancies in their performance achieved. Thus, temporal features like *MENN* or *SENN* have resulted statistically significant in the same context of stress when using ultra-short term analysis of HRV [31].

However, results from frequency parameters may variate considerably depending on the methodology used to perform PSD and length of the signals. Thus, some studies have reported that frequency analysis tend to produce better results for parametric instead of non-parametric methods when data length of the available signal is relatively short [38]. Moreover, non-parametric calculation (like FFT used in this study) is based in mathematical assumptions that severely limit frequency resolution. In this regard, considering the narrow bandwidth where the frequency parameters are computed, the quality of the calculated power spectrum could be affected. Furthermore, morphological parameters are based on the geometrical shape of NN time intervals distribution. Similarly, short time series may affect the shape of distribution, substantially altering the results. Indeed, some parameters like *TINN* have already been evaluated in the same context of stress and short-time series, and no statistical significance was found [31].

In recent works, it has been demonstrated that different ways to induce stress uses to trigger distinct cognitive processes [12]. This is why, comparison among approaches using stress detection should be discussed with caution. Nevertheless, it is interesting to note that the present study has reached better than, or comparable classification outcomes to, other similar analysis research. Firstly, it is worth noting that there are hardly any recent studies using short-term PRV analysis for assessing stress in the literature [39]. However, considering the correlation between HRV and PRV stated in this study, some comparison among works using short-term HRV series and mental stress can be provided.

In this regard, a combination of time, frequency and non-linear parameters computed over HRV signals using a wide range of classification algorithms has been studied very recently [40]. In the study, a global accuracy of 81.16% using Naive Bayes classifier was reported. Similarly, temporal, frequency and non-linear parameters on short-time HRV series, achieving a global performance of 64% using a support vector machine classifier, have been presented [41]. The same author reported an improved performance of 79% using additional non-lineal features using a complex tree classifier [42]. A global accuracy of 84.6% using exclusively RMSSD of ultra-short time HRV series and using a combination of binary tree classifiers has been reported [43]. It is mandatory to underline that these discriminant rates have been reached by combining parameters computed from different domains through advanced classifiers. Finally, there is also a number of recent works that have evaluated stress detection by using HRV in the context of long-term analysis, reporting stress detection classification rates ranging from 70 to 78% [36,44–47].

In view of these outcomes, our algorithm achieved a notably global accuracy of 82.35% using exclusively classical temporal parameters and a binary-tree classifier. Although all features were included in a multi-parametric analysis, only *SDNN*, *MNN*, *SENN* and *RMSFD* were chosen, showing an underlying complementarity of temporal features to classify stress patients. This interconnection is in agreement with other works previously published. Thus, a recent approach, proposed a methodology for stress detection based exclusively in the combination of time-domain features (*MNN*, *RMSSD* and *pNN50*) achieving global classification of 74.6% [36]. Just as in our approach, the authors state that it can be efficiently implemented on mobile devices, since the proposed method only uses time-domain features. In the same line, the arousal level has been assessed on patients, reporting that temporal parameters RMSSD, PNN12 and PNN20 showed statistical significance on ultra-short temporal series of 15 s length [48]. No global precision was reported in this study.

It is also interesting to notice that some recent works are exploring new areas and methodologies to enhance mental stress estimation. In this regard, some authors are computing standard and advanced non-linear analysis of HRV to recognize stress, among other emotions [49]. Unfortunately, some non-linear methodologies require a heavy computation cost, which makes the implementation of algorithms in real-time systems not viable. Furthermore, other authors have recently explored multivariate analysis to enhance distress classification. In these studies, some physiological variables like electro-dermal activity (EDA), electromyography (EMG) or skin temperature (SKT) are also combined with heart functioning to detect mental stress. Thus, multiple parameters, extracted from HRV, SKT and EDA with a support vector machine algorithm and K-means clustering to classify the obtained training data and index the user's stress level, which resulted in an overall 91.26% accuracy, were combined [44]. In the same line, a myriad of sensors, such as foot EDA, respiration, hand EDA, HR and EMG to improve the stress classification have been used [45]. Although our design is much easier and gets reasonable performance results with one single sensor, the aforementioned potential areas of research may benefit from the occurrence of new metrics and methodologies for mental distress estimation.

Finally, some comments about the suitability of PRV metric as substitute of HRV deserve consideration. At this point, several works have been published during last years, discussing if PRV series computed from PPG signals can be used instead of HRV computed from ECG recordings. Throughout the last years, some studies have claimed that PRV is a surrogate for HRV [50–52]. Furthermore, some recent works deserve a special mention, as they not only claim that PRV is suitable for HRV analysis, but they also provide details about which parameters extracted from HRV are most prone to show errors comparing both ECG and PPG methodologies. For instance, nineteen healthy subjects were enrolled in a recent study [50], where ECG and PPG signals were simultaneously recorded for each individual. ECG and PPG signals were recorded with sampling rate of 250 Hz and 500 Hz, respectively. Then HRV was computed from R-R and P-P series, and most typical parameters were extracted from both series. The results reported that the error for all PRV parameters was less than 6%, except for *pNN50*, which achieved a global error around 30% [50]. This could be the reason why PNN50 showed no relevance in our study. Similarly, ten healthy subjects were enrolled in another experiment [51], where ECG and PPG signals were recorded simultaneously at 1000 Hz. The results demonstrated an excellent correspondence between HRV parameters derived by ECG- and PPG-based methods. HRV was computed using R-R series and PP intervals, calculated from systolic peak of PPG waveform [51].

5. Conclusions

This paper has introduced a new wearable photoplethysmography sensor to be used in the domain of mental distress condition estimation. The International Affective Picture System database for inducing controlled arousal and valence has been applied to forty-five volunteers for the sake of assessing the proposal. The paper has introduced a complete description of the device capable of acquiring blood volume of a subject. The signals have been processed, a series of features have been extracted and it has been possible to classify calm and stress with a notably accuracy.

Although classical time, frequency and morphological analysis parameters are evaluated, our final proposal uses only time-domain PRV features. More concretely the mean, standard deviation and standard error of consecutive normal cardiac beats, and the root means square of the first derivative of consecutive normal cardiac beats are used. Moreover, these temporal metrics showed significant differences and acceptable single classification in the pulse variability generated by both emotional states. Nonetheless, in accordance with previous studies, their combination by means of a simple tree-based classifier revealed underlying complementarity that enhances the global accuracy of the discriminatory model up to 82.35%. This result outperforms or is comparable with other published works that use many parameters combined with complex classification algorithms.

It is also important to consider that all mathematical algorithms and processes used in this work can be embedded into a micro-controller. Thus, PPG peaks have been detected by using a real-time peak-detection algorithm [30]. The PRV-related features are defined in time-domain, extracted from short-time PPG signals, and the operations are based exclusively in sums and divisions. This fact opens the door for algorithms to work in real-time and in a long-term fashion.

Supplementary Materials: The following are available online at http://www.mdpi.com/2076-3417/8/1/69/s1. The complete schematics of the PPG sensor design are attached as supplementary material. Sheet 1: Cover page, Sheet 2: Analog front-end, Sheet 3: Power supply.

Acknowledgments: This work was partially supported by Spanish Ministerio de Economía, Industria y Competitividad, Agencia Estatal de Investigación (AEI) / European Regional Development Fund (FEDER, UE) under DPI2016-80894-R grant, and by the Centro de Investigación Biomédica en Red de Salud Mental (CIBERSAM) of the Instituto de Salud Carlos III. Arturo Martínez-Rodrigo holds an EPC 2016-2017 research fund from Escuela Politécnica de Cuenca, Universidad de Castilla-La Mancha.

Author Contributions: Roberto Zangróniz and Arturo Martínez-Rodrigo conceived and designed the study, programmed the experiments and drafted the manuscript. María T. López and José Manuel Pastor helped to interpret the results and reviewed the manuscript. Finally, Antonio Fernández-Caballero supervised the experiments, reviewed the manuscript and contributed to the final version. All authors have read and approved the final version of the manuscript.

Conflicts of Interest: The authors declare no conflict of interest.

Abbreviations

The following abbreviations are used in this manuscript:

ADC	analog-to-digital converter
AFE	analog front-end
ANS	autonomic nervous system
DAC	digital-to-analog converter

HF   high frequency
HR   heart rate
IAPS   international affective picture system
LED   light emitting diode
LF   low frequency
PCB   printed circuit board
PD   photo-detector
PPG   photoplethysmography
PRV   pulse rate variability
PSD   power spectral density
ROC   receiver operating characteristic
SAM   self-assessment manikin
SPI   serial peripheral interface
TIA   transimpedance amplifier
VLF   very low frequency

References

1. Zangróniz, R.; Martínez-Rodrigo, A.; Pastor, J.M.; López, M.T.; Fernández-Caballero, A. Electrodermal Activity Sensor for Classification of Calm/Distress Condition. *Sensors* **2017**, *17*, 2324.
2. Fernández-Caballero, A.; Martínez-Rodrigo, A.; Pastor, J.M.; Castillo, J.C.; Lozano-Monasor, E.; López, M.T.; Zangróniz, R.; Latorre, J.M.; Fernández-Sotos, A. Smart environment architecture for emotion recognition and regulation. *J. Biomed. Inform.* **2016**, *64*, 55–73.
3. Castillo, J.C.; Castro-González, A.; Fernández-Caballero, A.; Latorre, J.M.; Pastor, J.M.; Fernández-Sotos, A.; Salichs, M.A. Software architecture for smart emotion recognition and regulation of the ageing adult. *Cogn. Comput.* **2016**, *8*, 357–367.
4. Sokolova, M.V.; Fernández-Caballero, A. A review on the role of color and light in affective computing. *Appl. Sci.* **2015**, *5*, 275–293.
5. Fernández-Caballero, A.; Latorre, J.M.; Pastor, J.M.; Fernández-Sotos, A. Improvement of the elderly quality of life and care through smart emotion regulation. In *Ambient Assisted Living and Daily Activities*; Pecchia, L., Chen, L., Nugent, C., Bravo, J., Eds.; Springer: New York, NY, USA, 2014; pp. 348–355.
6. Yoo, K.-S.; Lee, W.-H. Mental stress assessment based on pulse photoplethysmography. In Proceedings of the 2011 IEEE 15th International Symposium on Consumer Electronics, Singapore, 14–17 June 2011; pp. 323–326.
7. Khan, N.A.; Jönsson, P.; Sandsten, M. Performance comparison of time-frequency distributions for estimation of instantaneous frequency of heart rate variability signals. *Appl. Sci.* **2017**, *7*, 221.
8. Heathers, J.A.J. Smartphone-enabled pulse rate variability: An alternative methodology for the collection of heart rate variability in psychophysiological research. *Int. J. Psychophysiol.* **2013**, *89*, 297–304.
9. Malik, M.; Bigger, J.T.; Camm, A.J.; Kleiger, R.E.; Malliani, A.; Moss, A.J.; Schwartz, P.J. Heart rate variability standards of measurement, physiological interpretation, and clinical use. *Eur. Heart J.* **1996**, *17*, 354–381.
10. Acharya, U.R.; Joseph, K.P.; Kannathal, N.; Lim, C.M.; Suri, J.S. Heart rate variability: A review. *Med. Biol. Eng. Comput.* **2006**, *44*, 1031–1051.
11. Schmidt-Nielsen, K. *Animal Physiology: Adaptation and Environment*; Cambridge University Press: Cambridge, UK, **1997**.
12. Valenza, G.; Lanata, A.; Scilingo, E.P. The role of nonlinear dynamics in affective valence and arousal recognition. *IEEE Trans. Affect. Comput.* **2012**, *3*, 237–249.
13. Sun, Y.; Thakor, N. Photoplethysmography revisited: From contact to noncontact, from point to imaging. *IEEE Trans. Biomed. Eng.* **2016**, *63*, 463–477.
14. Kumar, M.; Pachori, R.B.; Acharya, U.R. An efficient automated technique for CAD diagnosis using flexible analytic wavelet transform and entropy features extracted from HRV signals. *Expert Syst. Appl.* **2016**, *63*, 165–172.
15. Merone, M.; Soda, P.; Sansone, M.; Sansone, C. ECG databases for biometric systems: A systematic review. *Expert Syst. Appl.* **2017**, *67*, 189–202.
16. Allen, J. Photoplethysmography and its application in clinical physiological measurement. *Physiol. Meas.* **2007**, *28*, R1–R39.

17. Martínez-Rodrigo, A.; Pastor, J.M.; Zangróniz, R.; Sánchez-Meléndez, C.; Fernández-Caballero, A. ARISTARKO: A software framework for physiological data acquisition. In *Ambient Intelligence-Software and Applications*; Springer: New York, NY, USA, 2016; pp. 215–223.
18. Mohan, P.M.; Nagarajan, V.; Das, S.R. Stress measurement from wearable photoplethysmographic sensor using heart rate variability data. In Proceedings of the International Conference on Communication and Signal Processing, Melmaruvathur, India, 6–8 April 2016; pp. 1141–1144.
19. Sokolova, M.V.; Fernández-Caballero, A.; López, M.T.; Martínez-Rodrigo, A.; Zangróniz, R.; Pastor, J.M. A distributed architecture for multimodal emotion identification. In *Trends in Practical Applications of Agents, Multi-Agent Systems and Sustainability*; Springer: New York, NY, USA, 2015; pp. 125–132.
20. Schäfer, A.; Vagedes, J. How accurate is pulse rate variability as an estimate of heart rate variability?: A review on studies comparing photoplethysmographic technology with an electrocardiogram. *Int. J. Cardiol.* **2013**, *166*, 15–29.
21. Bonnici, T.; Orphanidou, C.; Vallance, D.; Darrell, A.; Tarassenko, L. Testing of Wearable Monitors in a Real-World Hospital Environment: What Lessons Can Be Learnt? In Proceedings of the 9th International Conference on Wearable and Implantable Body Sensor Networks, London, UK, 9–12 May 2012; pp. 79–84.
22. Kim, J.; Kim, J.; Ko, H. Low-power photoplethysmogram acquisition integrated circuit with robust light interference compensation. *Sensors* **2015**, *16*, 46.
23. Lang, P.J.; Bradley, M.M.; Cuthbert, B.N. *International Affective Picture System (IAPS): Technical Manual and Affective Ratings*; NIMH Center for the Study of Emotion and Attention: Gainesville, FL, USA, 1997.
24. Morris, J.D. Observations: Sam: The self-assessment manikin; an efficient cross-cultural measurement of emotional response. *J. Advert. Res.* **1995**, *35*, 63–68.
25. Hosseini, S.A.; Khalilzadeh, M.A.; Changiz, S. Emotional stress recognition system for affective computing based on bio-signals. *J. Biol. Syst.* **2010**, *18*, 101–114.
26. Nitzan, M.; Babchenko, A.; Khanokh, B. Very low frequency variability in arterial blood pressure and blood volume pulse. *Med. Biol. Eng. Comput.* **1999**, *37*, 54–58.
27. Gil, E.; Orini, M.; Bailón, R.; Vergara, J.M.; Mainardi, L.; Laguna, P. Photoplethysmography pulse rate variability as a surrogate measurement of heart rate variability during non-stationary conditions. *Physiol. Meast.* **2010**, *31*, 1271–1290.
28. Elgendi, M. On the analysis of fingertip photoplethysmogram signals. *Curr. Cardiol. Rev.* **2012**, *8*, 14–25.
29. Shin, H.S.; Lee, C.; Lee, M. Adaptive threshold method for the peak detection of photoplethysmographic waveform. *Comput. Biol. Med.* **2009**, *39*, 1145–1152.
30. Karlen, W.; Ansermino, J.M.; Dumont, G. Adaptive Pulse Segmentation and Artifact Detection in Photoplethysmography for Mobile Applications. In Proceedings of the 34th Annual International Conference of the IEEE EMBS, San Diego, CA, USA, 28 August–1 September 2012; pp. 3131–3134.
31. Lizawati, S.; Jaegeol, C.; Myeong, G.J.; Desok, K. Ultra Short Term Analysis of Heart Rate Variability for Monitoring Mental Stress in Mobile Settings. In Proceedings of the 29th Annual International Conference of the IEEE EMBS, Lyon, France, 22–26 August 2007; pp. 4656–4659.
32. Acharya, U.; Joseph, K.; Kannathal, N.; Min, C.; Suri, S. Heart rate variability: A review. *Med. Biol. Eng. Comput.* **2006**, *12*, 1031–1051.
33. Malik, M.; Xia, R.; Odemuyiwa, O.; Staunton, A.; Poloniecki, J.; Camm, A.J. Influence of the recognition artefact in automatic analysis of long-term electrocardiograms on time-domain measurement of heart rate variability. *Med. Biol. Eng. Comput.* **1993**, *31*, 539–544.
34. Malliani, A.; Pagani, M.; Lombardi, F.; Cerutti, S. Cardiovascular neural regulation explored in the frequency domain. *Circulation* **1991**, *84*, 482–492.
35. Jung, Y.; Jianhua, H. A k-fold averaging cross-validation procedure. *J. Nonparametr. Stat.* **2015**, *27*, 167–179.
36. Salai, M.; Vassányi, I.; Kósa, I. Stress detection using low cost heart rate sensors. *J. Healthc. Eng.* **2016**, *2016*, 5136705.
37. Mikuckas, A.; Mikuckiene, I.; Venckauskas, A.; Kazanavicius, E.; Lukas, R.; Plauska, I. Emotion Recognition in Human Computer Interaction Systems. *Elektron. Elektrotech.* **2014**, *10*, 51–56.
38. Sandhya, D. Parametric method for power spectrum estimation of HRV. In Proceedings of the International conference on Signal and Image Processing (ICSIP), Chennai, India, 15–17 December 2010; pp. 334–338.

39. Ham, J.; Cho, D.; Oh, J.; Lee, B. Discrimination of multiple stress levels in virtual reality environments using heart rate variability. In Proceedings of the Annual International Conference of the IEEE Engineering in Medicine and Biology Society (EMBS), Seogwipo, Korea, 11–15 July 2017; pp. 3989–3992.

40. Hwang, B.; Ryu, J.W.; Park, C.; Zhang, B. A novel method to monitor human stress states using ultra-short-term ECG spectral feature. In Proceedings of the Annual International Conference of the IEEE Engineering in Medicine and Biology Society (EMBS), Seogwipo, Korea, 11–15 July 2017; pp. 2381–2384.

41. Castaldo, R.; Montesinos, L.; Melillo, P.; Massaro, S.; Pecchia, L. To what extent can we shorten HRV analysis in wearable sensing? A case study on mental stress detection. In *IFMBE Proceedings*; Springer: New York, NY, USA, 2017; Volume 65; pp. 643–646.

42. Castaldo, R.; Xu, W.; Melillo, P.; Pecchia, L.; Santamaria, L.; James, C. Detection of mental stress due to oral academic examination via ultra-short-term HRV analysis. In Proceedings of the Annual International Conference of the IEEE Engineering in Medicine and Biology Society (EMBS), Orlando, FL, USA, 16–20 August 2016.

43. Mayya, S.; Jilla, V.; Tiwari, V.N.; Nayak, M.M.; Narayanan, R. Continuous monitoring of stress on smartphone using heart rate variability. In Proceedings of the IEEE 15th International Conference on Bioinformatics and Bioengineering (BIBE), Belgrade, Serbia, 2–4 November 2015.

44. Salafi, T.; Kah, J.C.Y. Design of unobtrusive wearable mental stress monitoring device using physiological sensor. In Proceedings of the 7th WACBE World Congress on Bioengineering, Singapore, 6–8 July 2015; pp. 11–14.

45. Haouij, E.I.N.; Poggi, J.-M.; Ghozi, R.; Sevestre-Ghalila, S.; Jaïdane, M. Random forest-based approach for physiological functional variable selection for driver's stress level classification. In Proceedings of the Conference of the Italian Statistical Society, Florence, Italy, 28–30 June 2017; pp. 393–398.

46. Zenonos, A.; Khan, A.; Kalogridis, G.; Vatsikas, S.; Lewis, T.; Sooriyabandara, M. HealthyOffice: Mood recognition at work using smartphones and wearable sensors. In Proceedings of the The Second IEEE International Workshop on Sensing Systems and Applications Using Wrist Worn Smart Devices, Sydney, Australia,14–18 March 2016; pp. 1–6.

47. Sandulescu, V.; Andrews, S.; Ellis, D.; Bellotto, N.; Mozos, O.M. Stress detection using wearable physiological sensors. In *Artificial Computation in Biology and Medicine*; Ferrández, J.M., Álvarez-Sánchez, J., de la Paz López, F., Toledo-Moreo, F., Adeli, H., Eds.; Springer: New York, NY, USA, 2015; pp. 526–532.

48. Schaaff, K.; Adam, M.T.P. Measuring emotional arousal for online applications: Evaluation of ultra-short term heart rate variability measures. In Proceedings of the Humaine Association Conference on Affective Computing and Intelligent Interaction (ACII), Geneva, Switzerland, 2–5 September 2013; pp. 362–368.

49. Nardelli, M.; Valenza, G.; Greco, A.; Lanata, A.; Pasquale, E. Recognizing Emotions Induced by Affective Sounds through Heart Rate Variability. *IEEE Trans. Affect. Comput.* **2015**, *6*, 385–394.

50. Jeyhani, V.; Mahdiani, S.; Peltokangas, M.; Vehkaoja, A. Comparison of HRV parameters derived from photoplethysmography and electrocardiography signals. In Proceedings of the 37th Annual International Conference of the IEEE Engineering in Medicine and Biology Society, Milan, Italy, 25–29 August 2015; pp. 5952–5955.

51. Selvaraj, N.; Jaryal, A.; Santhosh, J.; Deepak, K.K.; Anand, S. Assessment of heart rate variability derived from finger-tip photoplethysmography as compared to electrocardiography. *J. Med. Eng. Technol.* **2008**, *32*, 479–484.

52. Lu, S.; Zhao, H.; Ju, K.; Shin, K.; Lee, M.; Shelley, K.; Chon, K.H. Can photoplethysmography variability serve as an alternative approach to obtain heart rate variability information? *J. Clin. Monit. Comput.* **2008**, *22*, 23–29.

applied
sciences

MDPI

Article

Applicability of Emotion Recognition and Induction Methods to Study the Behavior of Programmers

Michal R. Wrobel

Department of Software Engineering, Faculty Of Electronics, Telecommunications and Informatics, Gdansk University of Technology, 80-233 Gdańsk, Poland; wrobel@eti.pg.gda.pl

Received: 22 December 2017; Accepted: 24 February 2018; Published: 26 February 2018

Abstract: Recent studies in the field of software engineering have shown that positive emotions can increase and negative emotions decrease the productivity of programmers. In the field of affective computing, many methods and tools to recognize the emotions of computer users were proposed. However, it has not been verified yet which of them can be used to monitor the emotional states of software developers. The paper describes a study carried out on a group of 35 participants to determine which of these methods can be used during programming. During the study, data from multiple sensors that are commonly used in methods of emotional recognition were collected. The participants were extensively questioned about the sensors' invasiveness during programming. This allowed us to determine which of them are applicable in the work of programmers. In addition, it was verified which methods are suitable for use in the work environment and which are only suitable in the laboratory. Moreover, three methods for inducing negative emotions have been proposed, and their effectiveness has been verified.

Keywords: affective computing; human–computer interaction; social computing; human aspects of software engineering; affective software engineering

1. Introduction

In the age of global information, it is of paramount importance to provide reliable and high-quality software at a reasonable cost and duration. Because of the continuous increase in the number of IT projects, the demand for information and communication technology (ICT) specialists is steadily growing [1]. Software development companies already have difficulties in recruiting specialists with the required knowledge and experience [2]. One of the solutions to the problem of an insufficient workforce in the ICT sector may be to increase employee productivity.

Introducing affect awareness in software development management may be one of the solutions to this problem. Recent research has already shown that, in IT projects, positive emotions increase productivity, while negative emotions can significantly reduce performance [3–7]. Affect-aware IT project management can help software developers stay productive as well as detect when emotions, such as frustration, reduce their performance [5]. However, to bring these ideas to life, tools to recognize the emotions of software developers while working are essential.

The affective computing domain has provided many methods and tools for recognizing the emotions of computer users. However, it has not been verified yet whether they can be used to monitor the emotions of programmers during their daily work and to what extent. Moreover, some of them, as a result of their invasiveness or cost, are not suitable for use in a work environment and can be used only in a laboratory.

Spontaneous emotions cannot be expected during laboratory experiments. Therefore, to study the emotional states of programmers in such an environment, they should be induced somehow. Only the induction of negative emotions was considered. Previous studies have shown that it is more efficient than in the case of positive emotions [8].

The aim of the study is to review the available emotion recognition methods for their use in a software development environment. In addition, selected methods of stimulating emotions during programming in a laboratory environment have been evaluated. Three research questions have been formulated:

RQ1 What methods and tools known from affective computing research can be used to recognize the emotions of programmers in a laboratory environment?
RQ2 Which of the identified methods are suitable for programming in a real working environment?
RQ3 How can the negative emotions of programmers be induced in a laboratory environment?

The rest of the paper is organized as follows: In Section 2, methods useful in recognizing emotions of software developers are described; Section 3 describes the experiment design, and Section 4 describes its execution and results; finally, Section 5 discusses the results, and Section 6 concludes.

2. Related Work

So far, several studies, which involved emotion recognition of members of IT teams, have been conducted in the field of software engineering. Numerous attempts were made to identify emotions using various available channels.

The most comprehensive research on utilizing physiological sensors during software developers' work was conducted by Müller and Fritz [5,9,10]. During their study [5] on 17 software developers, they collected the following data: electroencephalography (EEG) data using a Neurosky MindBand sensor, temperature, electrodermal activity (EDA) and blood volume pulse (BVP) using an Empatica E3 wrist band, and eye-tracking data using Eye Tribe. The results of the experiment showed that the EDA tonic signal, the temperature, the brainwave frequency bands, and the pupil size were the most useful predictive factors to classify the progress of software developers, and brainwave frequency bands, the pupil size, and heart rate were the most useful to classify their emotions. Nevertheless, they noted strong individual differences with respect to the correlation and classification of physiological data [5]. Similar differences have also been found in our other studies on the use of sensors to monitor the physiology of computer game players [11].

Müller and Fritz, along with Begel, Yigit-Elliott and Züger, also conducted an experiment to classify the difficulty of source code comprehension tasks using the same set of input channels. They stated that it is possible to use off-the-shelf physiological sensors in order to predict software developer task difficulty [12].

Facial electromyography (fEMG) is commonly regarded as a reliable method for measuring emotional reactions [13]. Ten et al. have proved, in an experiment with 20 participants, that fEMG activities are effective and reliable indicators of negative and positive emotions [14]. Bhandari et al. successfully used fEMG, along with EDA, to determine emotional responses during an evaluation of mobile applications [15].

Eye-tracking methods have previously been successfully used in other research in the software engineering domain [16–20]. For example, Bednarik and Tukiainen proved the usefulness of eye-movement tracking in a study of the comprehension processes of programmers [21]. An eye-tracking environment (iTrace) has been developed to facilitate eye-tracking studies in software systems [22].

One of the most popular methods of recognizing emotions is the analysis of facial expressions [23–25]. It has gained popularity mainly as a universal and non-invasive approach. Algorithms analyze video frames to identify face muscle movements and, on the basis of the Facial Action Coding System (FACS) [26], assess the user's emotional state. Successful attempts are also made to identify emotions on the basis of voice [27]. There are even frameworks that allow such an analysis to be performed [28]. A relatively new approach, which can be well suited for recognizing the emotions of programmers, uses keystroke dynamics and mouse movement analysis. It is completely non-intrusive and does not require any additional hardware [29]. There have already been attempts to use this method to monitor software developers [30].

The only channel used in previous research (e.g., [31,32]) that was excluded from the presented study was EEG. The Biometric Stand [33], on which the experiment was conducted, contains only a 3-channel EEG sensor that does not provide reliable data.

A number of studies have also been conducted on the use of sentiment analysis techniques to identify emotions on the basis of IT project artifacts (e.g., [34,35]). However, the purpose of this study was to check the ability to recognize emotions of developers while working, and therefore these methods have not been included.

3. Study Design

The aim of the study was to determine which of the methods can be used to detect the emotions of programmers. Emotion recognition methods are based on data received from one or more channels. For example, methods based on the analysis of facial expressions use video camera images, and methods based on the analysis of the physiological response of the human body use data from biosensors.

For the purpose of this study, the following input channels were selected on the basis of the analysis of methods used in the presented research in the field of software engineering:

- Video
- Audio
- Biosensors:
    - Skin conductance
    - Heart rate
    - Respiration
    - fEMG
- Computer keyboard and mouse movements

During the study, the participants were asked to solve simple algorithmic tasks in the Java language using a popular integrated development environment (IDE). While the participants were solving tasks, data were collected from multiple channels. At the same time, activities were performed to elicit the emotions of the developers. Before the study, the participants were informed about the purpose of the study but were not aware of attempts to influence their emotional states.

The study was designed to be conducted at a biometric stand in the Laboratory of Innovative IT Applications at Gdansk University of Technology (GUT). The room was divided into one part for the participant and one part for the observer, separated by an opaque partition (Figure 1). On the participant's desk there was a monitor, a keyboard and a mouse connected to the computer on which the tasks were performed. The computer itself was physically located in the second part of the room; it is labeled as Computer 2 on Figure 1. In addition, a video camera was located in front of the participant at the top of the monitor, followed by a lighting set, supplied with Noldus FaceReader software, which was used to recognize emotions on the basis of facial expressions [36]. Underneath the monitor, a myGaze Eye Tracker device was situated. A number of sensors were attached to the participant and were linked through the Coder FlexComp Infiniti by the Thought Technology analytical device with Computer 1, which was located in the observer's area. The BioGraph Infiniti application, developed by Thought Technology, was running on this computer and allowed visualization, pre-processing and exporting of the data from the physiological sensors. The observer also had a monitor, mouse, and keyboard connected to the participant's Computer 2. This allowed the observer to interfere in the activities of the participant. On Computer 2, Morae Recorder software was installed, which recorded the participant's desktop image, mouse movement, and fixation from the eye tracker. There was also a data acquisition program for keystroke analysis available on the same computer [30]. Computer 3 was used to collect all other data useful for the recognition of emotions, including recordings from the video camera and the microphone located in front of the participant.

Observer area

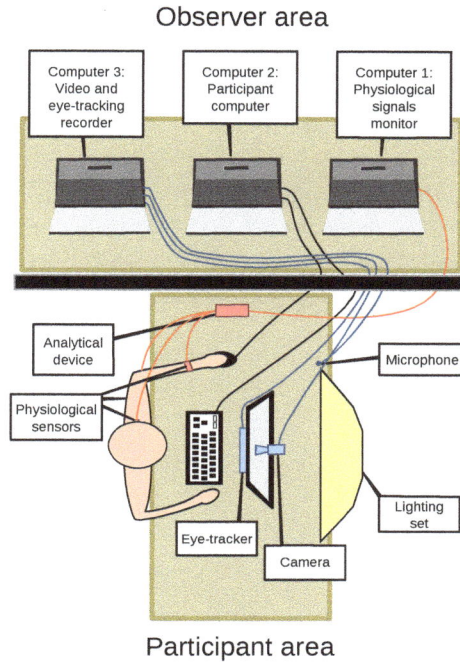

Figure 1. Study stand.

3.1. Plan of the Study

The study was organized in the form of consecutive sessions. During a session, the participant individually solved four programming tasks. Each participant took part in only one session; therefore, the number of sessions was equal to the number of participants. The purpose of each task was to solve one algorithmic problem. For each, the Java program was prepared, and then the key fragments of the source code were removed. The participant's goal was to complete the program code in the NetBeans environment and validate the solution by running a unit test, prepared for the purpose of this study.

During the session, the participant had to solve the following problems:

1. Sort the array using the bubble sort algorithm (Appendix A.1).
2. Return the indicated position within the Fibonacci sequence (Appendix A.2).
3. Check if the word is a palindrome (Appendix A.3).
4. Transpose a matrix (Appendix A.4).

To solve the first three tasks, the participants had a maximum of 5 min, and they had 3 min for the last task. Including the time necessary for the introduction and switching of the sensors, as well as the completion of the final questionnaire, the duration of the session was estimated at 40 min.

During the session, data was logged from channels that may be useful in the process of recognizing emotions. Before the participant started solving the tasks, the eye tracker was calibrated and the video camera was adjusted. During the session videos from the camera, eye-tracking data, microphone sound and mouse and keyboard patterns were constantly recorded. To verify the physiological sensors' obtrusiveness, they were switched on during the subsequent tasks. Only the respiration sensor, as the least onerous, was attached during all the tasks. During the first task, an EDA sensor was connected; during the second, fEMG was used; and during the third, BVP was used. The fourth task was carried out without any additional sensor.

After completing each task, the participants were asked to self-assess their emotional state using the Self-Assessment Manikin (SAM) [37]. Figure 2 presents the assessment form integrated with NetBeans, which was prepared for the purpose of the study. The top panel shows the happy–unhappy scale, which ranges from a smile to a frown. The middle panel corresponds to an excited-to-calm scale. Finally, the bottom scale reflects whether the participant feels controlled or in control. The SAM form is a recognized method of assessing the emotional state in the three-dimensional valence, arousal and dominance (VAD) scale.

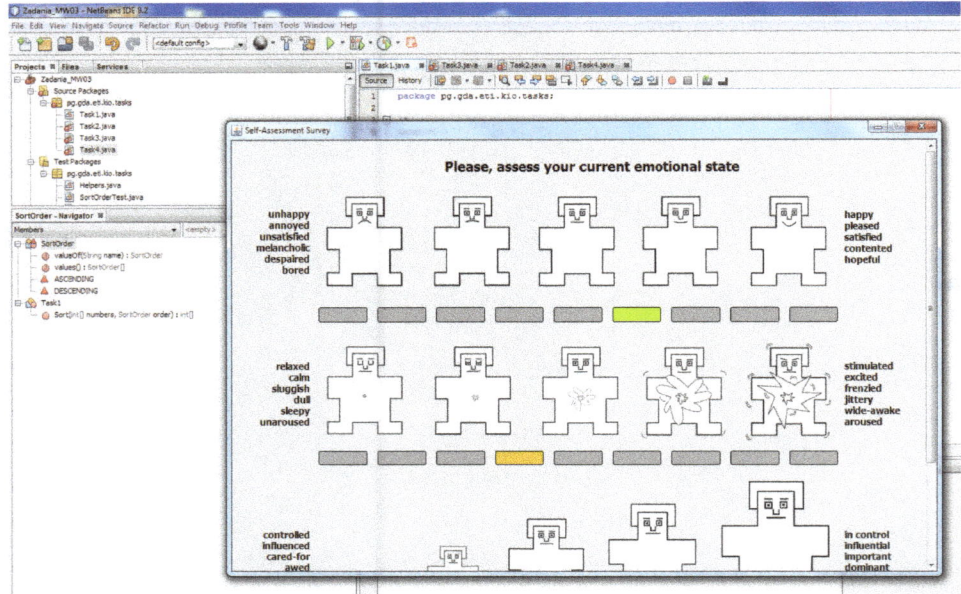

Figure 2. Self-Assessment Manikin used during the study.

3.2. Negative Emotion Induction

On the basis of the classification of emotion induction techniques proposed by Quigley et al. [38], the "real-world stimuli" technique was chosen. Three methods were applied that reflected the situations occurring in the working environment of software developers that are associated with negative emotions.

The participants performed their tasks using the NetBeans IDE. The functionality of this environment was enhanced with a plug-in called MaliciusIDE, which was developed for the purpose of this study. This allowed malice to be generated that would interfere with the participant during coding. During the study, malfunctions such as suspending a program for a specified number of seconds, duplicating the characters entered, or moving the mouse cursor were triggered manually via a Web interface running on Computer 3. This Wizard-of-Oz (WOZ) technique was implemented to ensure an appropriate number of events. Too few occurrences might not have induced emotions, but too many could have led to the disclosure of the malicious activity of the observer. Preliminary tests were conducted with automatically triggered malices. Their results revealed an insufficient number of malicious events that were noticed by the participants. For example, users did not notice that the content of the clipboard had been cleaned, because it was not used in a particular task.

For the second task, the goal of which was to return an indicated element of the Fibonacci sequence, an incorrect test case was prepared. Even in spite of the correct solution, the participants were always informed that the program had returned incorrect output. The purpose of such an action was to create confusion and consequently irritability and discouragement.

During the last task, an attempt was made to put time pressure on the participant. After 2 min, a beep signal was generated imitating the observer receiving the message. The participant was informed that the test had to be shortened and that he or she should try to finish the task within 1 min.

3.3. Questionnaire

After completing all tasks and disconnecting the sensors, the participant was asked to complete a survey implemented using the Google Forms service. The purpose of the questionnaire was to gather information on the participants' feelings about the methods of recognizing and inducing emotions.

The survey consisted of seven questions (Appendix B). In the first question, using the seven-level Likert scale, the participants assessed the nuisance of particular emotion recognition methods. A value of 1 corresponded to the claim that the sensor was unnoticeable, and a value of 7 corresponded to it having made the work completely impossible.

In the second question, the participants were asked to indicate which of the applied methods could be used in the daily work of programmers. In the next question, the participants reported which emotions were triggered by the emotion-inducing methods.

In the remaining questions, the participants answered how often they express emotions aloud, whether a wristwatch is intrusive during prolonged periods of typing, how often in real work an emotional self-assessment form could be used, and whether they would agree to investigate their emotional state during their daily work.

4. Execution and Results

The study was conducted in April and May 2017 at the Gdansk University of Technology, Poland. Altogether, 35 undergraduate computer science students, 6 women and 29 men, participated in the study. A single session lasted between 30 and 45 min, depending on the pace at which individual tasks were solved and the number of additional questions. Sample pictures of the participants during the study are shown in Figure 3.

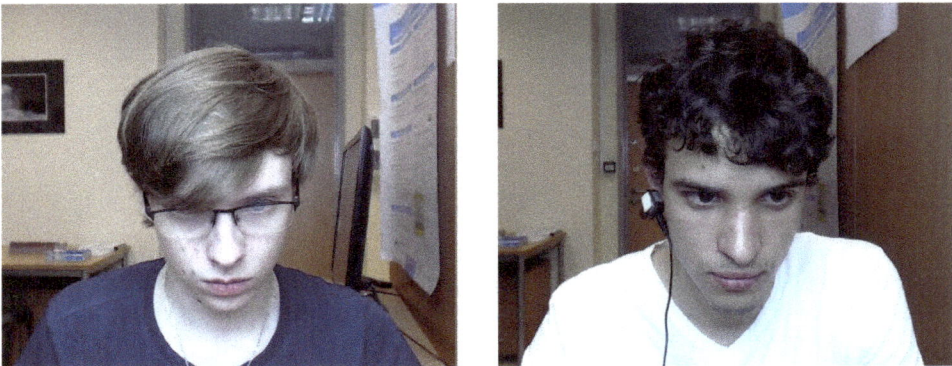

Figure 3. Images of the participants from the video camera: the first with glasses and a long fringe on the left, and another with the blood volume pulse (BVP) sensor attached to the earlobe on the right.

4.1. Availability

In order to check the possibility of using eye-tracking and video recording to recognize emotions of programmers, the availability metrics AV_EYE and AV_VIDEO, respectively, were introduced. For eye tracking, the AV_EYE metric was defined as the percentage of time for which the pupil's readings per minute were above the assumed sample quality threshold. Depending on the required accuracy of the measurements, four thresholds were presented, as shown in Table 1. Over most of the time (64.50%), the device recorded more than 29 readings per minute, with the sampling rate of the device at 30 Hz. Only 3.68% of the 1 min periods were without even a single detected fixation point, and 11.28% were with less than 10. The device did not recognize the position of the pupils when the head was tilted too far over the keyboard and also when the head was turned in one direction or the other. However, the collected data was sufficient to generate video clips with fixations and saccades during the solving of the tasks, as shown in Figure 4.

Table 1. Eye-tracker sample quality threshold and corresponding availability.

Sample Quality Threshold % (Readings per Minute)	Number of Samples Above Threshold (N = 40,352)	AV_EYE
80% ($\geq$24)	29,309	72.63%
85% ($\geq$26)	28,080	69.59%
90% ($\geq$27)	27,512	68.18%
95% ($\geq$29)	26,026	64.50%

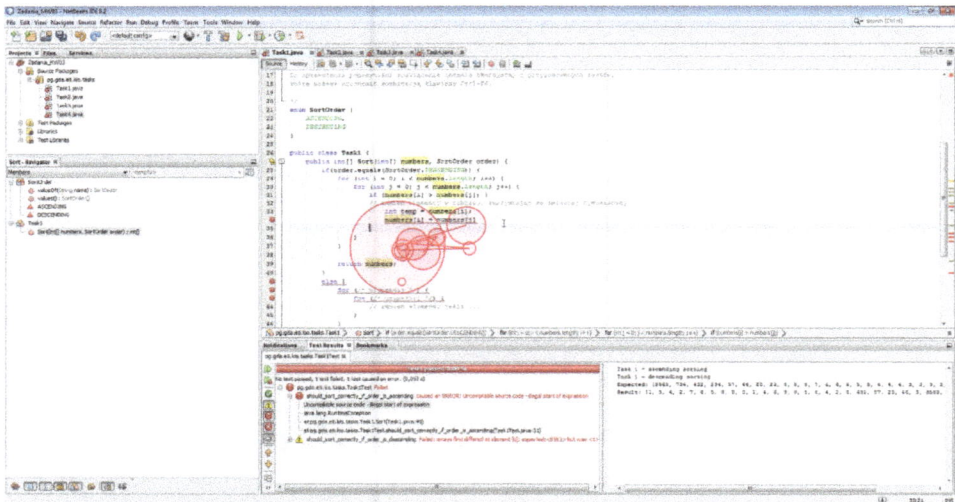

Figure 4. Fixations and saccades during Task 1.

Clips from the camera that recorded the faces of the participants were analyzed using Noldus FaceReader software. This recognizes emotional states on the basis of the FACS. For each video frame, the tool provides results as intensiveness vectors for the following emotions: joy, anger, fear, disgust, surprise, sadness and a neutral state. In the case of an error, instead of numerical values, the label FIND_FAILED is returned if the face cannot be detected on the frame, and FIT_FAILED is returned when the emotion cannot be recognized. In order to assess the accessibility of video-based emotion recognition of software developers during work, three metrics were proposed:

- AV_VIDEO—percentage of time for which emotion was recognized.
- FINDF—percentage of time for which a face was not detected.
- FITF—percentage of time for which the face was detected, but no emotion was recognized.

The results are shown in Table 2. The average availability across the samples exceeded 77%. However, a thorough analysis of the provided data has shown that the algorithm implemented in the Noldus FaceReader software had a major problem with recognizing the emotions of people with glasses. In this case, the availability decreased to just 55%, whereas for the remaining cases, it equaled 85% (Table 3). Other factors that reduced availability were fringes that partially covered the eyes, beards and moustaches (e.g., participant P21). Therefore, to obtain the best accuracy, the recognition of emotions on the basis of facial expressions can only be used for programmers without glasses or facial hair.

Table 2. Availability of video-based emotion recognition.

Participant	FINDF	FITF	AV_VIDEO	Glasses
P04	0.92%	3.73%	95.35%	No
P06	3.89%	16.52%	79.59%	No
P07	3.78%	2.76%	93.46%	No
P08	1.91%	5.47%	92.62%	No
P09	6.23%	25.67%	68.10%	No
P10	5.64%	22.47%	71.89%	Yes
P11	0.44%	2.84%	96.72%	No
P12	7.10%	33.83%	59.06%	Yes
P13	4.53%	28.16%	67.31%	Yes
P14	3.70%	11.72%	84.58%	No
P15	2.17%	0.86%	96.97%	No
P16	4.73%	13.58%	81.69%	No
P17	5.92%	5.79%	88.29%	No
P18	2.95%	5.10%	91.95%	No
P19	5.88%	36.41%	57.71%	Yes
P20	31.54%	19.96%	48.51%	No
P21	39.46%	7.87%	52.67%	No
P22	2.52%	4.77%	92.71%	No
P23	44.79%	50.23%	4.98%	Yes
P24	1.67%	13.18%	85.16%	Yes
P25	2.63%	4.22%	93.14%	No
P26	4.11%	50.73%	45.16%	Yes
P27	3.95%	3.46%	92.59%	No
P28	4.87%	3.23%	91.91%	No
P29	12.10%	10.44%	77.45%	No
P30	1.71%	2.99%	95.30%	No
P31	3.51%	16.19%	80.31%	No
P32	2.22%	8.02%	89.75%	No
P33	13.42%	30.58%	56.00%	No
P34	10.65%	54.71%	34.64%	Yes
P35	9.29%	20.27%	70.44%	Yes
P36	5.08%	7.49%	87.42%	No
P37	2.04%	4.41%	93.55%	No
P38	2.24%	3.72%	94.04%	No
P39	1.33%	0.96%	97.72%	No

Table 3. Average availability of video-based emotion recognition.

		FIND	FIT	AV
All	Mean	7.40%	15.21%	77.39%
	SD	10.28%	15.06%	21.35%
No Glasses	Mean	6.36%	8.55%	85.09%
	SD	9.15%	7.66%	13.97%
Glasses	Mean	10.41%	34.44%	55.15%
	SD	13.17%	14.86%	24.04%

4.2. Disturbance

On the basis of the questionnaire survey (Appendix B), the degree of disturbance of individual data collection methods was assessed. All methods were evaluated using the seven-level Likert scale, where a value of 1 corresponded to the claim that the method was unnoticeable, and a value of 7 indicated that it made the work impossible. Figure 5 shows the compilation of response distributions for all the examined channels.

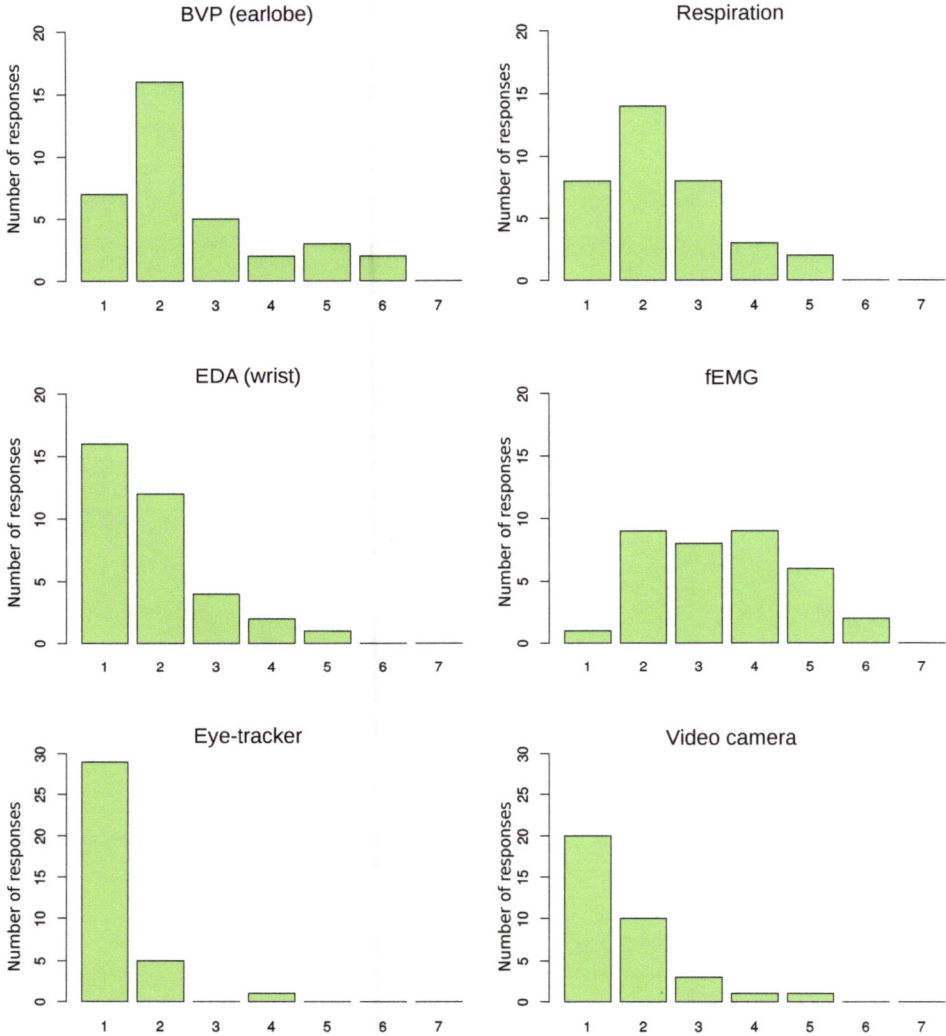

Figure 5. Disturbance rating distribution of channels.

Among the physiological sensors used in the study, the respondents pointed to the EDA sensor placed on the wrist as being the least cumbersome. As many as 16 respondents indicated that it was completely unnoticeable during coding, while only 3 were moderately disturbed.

The next two sensors, a respiration device placed on the chest and a BVP placed on the earlobe, were also rated as slightly intrusive. The last physiological sensor, the fEMG device, was considered the most cumbersome.

Other methods of collecting data for the purpose of emotion recognition were found by most respondents to be almost completely unnoticeable. Because of the bright light set, the camera that recorded the participant's face was evaluated slightly worse. However, the result of the assessment was still lower than the rating of the least intrusive physiological sensor.

Among all the tested methods of emotional recognition, the participants indicated the eye tracker as being the most acceptable in everyday work. Only one person did not indicate this method. Over half of the respondents reported that they would not be disturbed by collecting mouse movements and typing patterns (85.7%), by video camera recording (65.7%), by SAM (62.9%) or by EDA (62.9%). On the other hand, almost every respondent reported that the electromyographic sensor attached to the face would not usable in the work environment.

The respondents also revealed how often they thought the SAM questionnaire could be used in their daily work. The vast majority (71.4%) indicated that such data could be collected twice a day, for example, while starting and closing the IDE.

In view of the growing smart-watch market, the question as to whether a wrist watch interferes with the daily work of a software developer was raised. Over half of the respondents indicated that it is only slightly intrusive, and only four indicated otherwise. The detailed distribution of the response is shown in Figure 6.

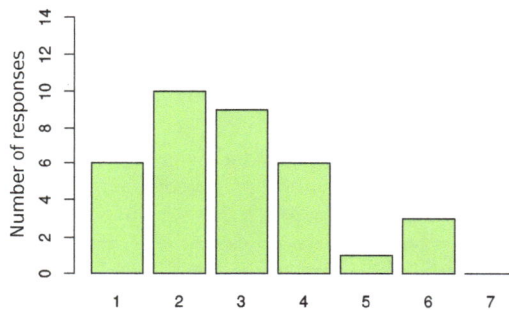

Figure 6. Evaluation of disturbance of the wrist watch in daily work.

Over 60% of the respondents stated that they often or very often express their emotions verbally while programming. Only seven claimed that they do so rarely or very rarely. However, during the study, no participants except one expressed their emotions this way. Therefore, voice recordings were not analyzed further.

4.3. Inducing Negative Emotions

According to the plan, attempts were made to induce emotions during each session. During all tasks, the observer disrupted the participant's work by causing malicious events in the NetBeans environment. The most commonly used events were adding additional characters while entering text, changing the position of the mouse pointer, freezing the environment for 7 s, clearing the contents of the clipboard, and temporarily hiding the IDE screen. These actions were carried out to disrupt work, but in a way that would seem to be natural behaviour of the application. The frequency of events was manually adjusted so that the users remained unaware of the intended actions of the observer. In addition, for task 2, an invalid test case was prepared, and the time for the last task was shortened.

In the questionnaire survey, the participants were asked to list which emotions were induced by specific actions. In the case of an unstable IDE, irritation most frequently appeared (42.86%), then anger (28.57%), followed by nervousness (25.71%) and frustration (11.43%). Four of the respondents (11.43%) indicated amusement (Figure 7). A post-study informal interview revealed that it was related to the fact that these participants had figured out that this unstable work was due to the deliberate actions of the observer. Other emotions were pointed out by only one or two participants and therefore were omitted from the analysis.

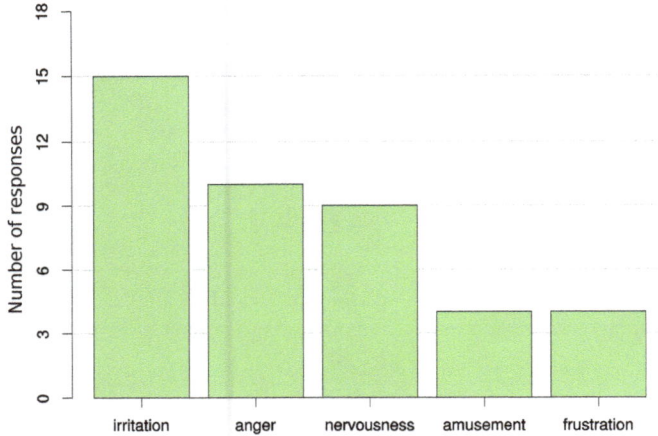

Figure 7. Negative emotions induced by the malicious behaviour of the integrated development environment (IDE).

An incorrect test case in one of the tasks had a lower impact on the emotional state of the programmer. Astonishment, the most commonly reported emotion, was indicated by only six people (17.14%). In addition, the respondents listed anger (14.29%), frustration (14.29%), uncertainty (11.43%) and irritation (11.43%). Other emotions were mentioned by fewer than three respondents.

Attempts to put time pressure on the participants almost completely failed. This had a negligible impact on the emotional state of the participants. Nearly half of the respondents indicated that this had no effect at all. On the other hand, this was the only action with a positive response—20% of the respondents indicated that the shortening of time was a mobilizing factor. Among the remaining responses, only five people listed negative emotions such as nervousness, irritation or fear.

The answers to the question about consent to monitor emotions in the work environment were not conclusive. The distribution of responses was similar to a normal distribution and is shown in Figure 8.

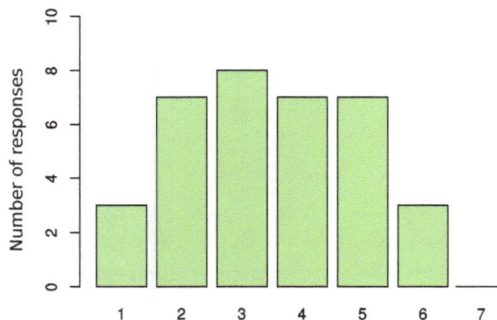

Figure 8. Distribution of answers on consent to monitor emotions at work.

5. Discussion

On the basis of the results of the study, the most appropriate methods for the recognition of the participants' emotions were those that were completely transparent to the subjects. Despite their low efficiency, keyboard- and mouse pattern-based methods were the most acceptable to the programmers.

Of course, the key factor in their implementation in the real work environment is to ensure privacy. The keylogger should not record which keys are pressed, but only patterns of typing, speed, the number of errors and, if possible, key pressures.

At first glance, the differences between the responses to the inconvenience when using the eye tracker and video camera were puzzling. As many as 11 participants pointed out that only the first device could be used in a working environment. Informal interviews conducted after the study revealed that this was related to lighting. During the study, a powerful light set (over 30,000 lm) was used, which was a prerequisite for obtaining high-accuracy results using the Noldus FaceReader software. Some respondents felt discomfort as a result of the very bright light.

Both the availability of eye-tracker data during programming and the user acceptance rating were high. However, studies conducted so far have shown that emotion recognition cannot be performed with high accuracy only on the basis of data from this channel. It can only be used in a multimodal approach. On the other hand, extended pupil movement pattern analysis, combined with keystroke dynamics or mouse movement [30], can reveal interesting results.

Although emotion recognition on the basis of facial expression is widely used, there are some major problems. The conducted study revealed that it can be used only in the case of people without glasses, a fringe or facial hair. For others, the availability is low; therefore, the recognition accuracy may be insufficient.

During the study, the results of the questionnaire on the expression of emotion vocally were not confirmed. Although the participants were informed that they could speak during the study, among all the participants, only one developer commented on his work, sometimes expressing emotions such as frustration or anger. This led to the surprising conclusion that the method of detecting emotion on the basis of audio analysis is not applicable in laboratory tests. However, the results of the questionnaire showed that it can likely be used in real work environments. To confirm this assumption, it is necessary to collect relevant data from the natural development environment.

Of all the physiological sensors used during the study, the EMG sensor located on the subject's face was recognized as the most intrusive. However, even this sensor was rated as moderately obstructive. This allowed us to conclude that from the point of view of work disruption, all studied sensors can be applied in a laboratory environment to monitor the physiology of software developers.

EDA is known as the physiological signal that allows emotions to be recognized with one of the highest accuracies [39]. However, the best locations for these sensors are the fingers. Clearly, because of the nature of the work of programmers, it is not possible to use this location. The research participant must be able to use the computer as in everyday work, and for programmers, the freedom to move the fingers is crucial. Therefore, an alternative location was chosen, and the sensor was attached to the participant's wrist. It has been shown in studies that this allows correct but less-accurate monitoring of EDA [40]. For similar reasons, the BVP sensor was placed on the ear lobe instead of the tip of the finger. However, it is necessary to be aware that such workarounds may lead to decreasing accuracy of collected data.

One of the possible solutions for the use of physiological sensors in everyday work is the smart watch. These are watches equipped with a set of sensors and software that allows it to collect, pre-analyze and transfer physiological data to a computer or smartphone. Commonly available devices are equipped with a BVP and some come with an EDA sensor. Among the participants, only four indicated that the watch bothered them significantly while typing. The widespread availability of smart watches, equipped with at least BVP and EDA sensors, would certainly allow for extensive monitoring of the emotional state of developers in their natural environment.

The most effective way to induce the participants' emotions was the manipulation of the IDE by the observer. Its unstable work evoked negative emotions in 32 participants (91.42%). The IDE is a basic tool for developers. Therefore, its unexpected behaviour over a prolonged time can lead to frustration and anger.

The second of the applied methods—an incorrect unit test—also elicited negative emotions, although in a lower number of participants. Over time, some developers began to suspect that the test was invalid. This method should be used for more complex tasks. In the study, it was used in a relatively simple task, which the participants solved in under half the time limit. The other half was spent on searching for an error—an error that was not there. One participant even opened the unit test code and modified it to complete the task.

Threats to Validity

Several threats to validity could have affected the results of this study. First of all, it was assumed that undergraduate students could participate in the study. They do not have as much experience working as a programmer as professional developers. Therefore, this threat may have had a particular impact on RQ2. On the other hand, the participants had used IDEs intensively when programming numerous student projects. Therefore, in the case of analyzing data related to availability and disturbance as well as methods of inducing emotions, the impact of this threat was rather low.

Another threat was the short time allocated for completing the tasks. Emotions related to the tasks performed may not have occurred in such a short time or could have been the result of previous activities. However, the observations of students during the sessions and the post-study interviews did not confirm this threat.

According to the study plan, each of the three physiological sensors was used in only one (the same) task for all participants. Such a study design may introduce a threat of confounding effects, in which the results are valid only for the particular sensor–task pair. However, because of the similar difficulty of each task, this was not believed to be the case for this study.

6. Conclusions

During the study, emotion recognition methods suitable for monitoring the emotional states of software developers in a laboratory environment were examined. Analysis of the collected data has allowed the research questions to be answered.

In response to RQ1, it can be stated that most of the tested channels can be used successfully during programming. Only audio channels are completely useless in a laboratory environment. Although in the survey the participants reported that they often express emotions verbally while programming, this was not confirmed during the study. In the case of tools for recognizing emotions on the basis of facial expressions, attention should be paid to the appearance of the subject. Studies have revealed that recognition results may be seriously compromised when a participant wears glasses, has a long fringe or has thick stubble.

Among the methods that can be used to monitor the emotion of programmers in the work environment (RQ2), non-invasive methods were indicated first and foremost. The suggested data channels that can be used in daily work include the eye tracker, typing patterns, mouse movements and video recordings. Most respondents also agreed to the use of the EDA sensor. Combined with the results on wearing a wrist watch while programming, it can be claimed that smart watches can be successfully used to monitor the emotions of developers.

Finally, emotion-inducing methods in a laboratory environment were evaluated (RQ3). The malicious plug-in to the IDE proved to be the best approach for triggering negative emotions. Of the remaining methods, shortening the task time did not meet the assumptions of creating time pressure. This incorrect test case, on the other hand, can be used in more complex tasks.

The study has revealed that, while most methods of recognizing the emotions of programmers can be used in laboratory tests, only those that are non-intrusive, such as an analysis of facial expressions or typing patterns, are accepted in the real working environment. In practice, this means that physiological sensors can be used to monitor the emotions of programmers only in the least invasive form—for example, biosensors built into smart watches. In addition, inducing the emotions in the laboratory environment proved to be challenging. Among the three evaluation methods, only the malicious behaviour of the IDE had an impact on the majority of the participants. To conduct research on the emotional states of programmers in a laboratory environment, it may be necessary to develop and validate more approaches to the problem of inducing emotions.

Conflicts of Interest: The author declares no conflict of interest.

Appendix A. Source Code of Provided Programming Tasks

Appendix A.1. Bubble Sort

```
/* Task 1
Complete the code so that it sorts the numbers with the bubble sort algorithm
according to the given order.

Examples:
1. For numbers 1,4,6,2,4 and order ASCENDING, the correct result is
   an array: 1,2,4,4,6
2. For numbers 1,5,7,2,3,6,4,8 and order DESCENDING, the correct result
   is an array: 8,7,6,5,4,3,2,1

Validate source code by pressing Ctrl + F6    */

enum SortOrder {
    ASCENDING,
    DESCENDING
}

public class Task1 {
    public int[] Sort(int[] numbers, SortOrder order) {
        if(order.equals(SortOrder.DESCENDING)) {
            for (int i = 0; i < numbers.length; i++) {
                for (int j = 0; j < numbers.length; j++) {
                    if (numbers[i] > numbers[j]) {
                        // swap elements in the array using the temporary variable
                    }
                }
            }
            return numbers;
        }
        else {
            for (/* complete */) {
                for (/* complete */) {
                    // swap elements if ...
                }
            }
            return numbers;
        }
    }
}
```

Appendix A.2. Fibonacci Sequence

```
/* Task 2
Complete the code to return the nth Fibonacci sequence number.

The formula for the Fibonacci sequence:
f(1) = 1
f(2) = 1
f(n) = f(n-1) + f(n-2)

Examples
1. n = 5
   Code returns 5
2. n = 7
```

```
    Code returns 13

Validate source code by pressing Ctrl + F6     */

public class Task2 {

    public int Fibonacci(int number) {
        return calculateFibonacci(number);
    }

    private int calculateFibonacci(int n) {
        if (n == 0) return 0;
        /* but if n is equal to 1, return 1 */
        /* in other cases, use recursion */
    }
}
```

Appendix A.3. Palindrome

```
/* Task 4
A palindrome is a word that reads the same backward as it does forward.
Your task is to check whether the given word is a palindrome. If so, return true,
otherwise return false. Ignore letter case.

Examples:
kayak   -> true
butterfly -> false

Validate source code by pressing Ctrl + F6     */

public class Task3 {
    public Boolean IsPalindrome(String phrase) {
        String reversedPhrase;
        reversedPhrase = reverse(phrase);
        if(/* Compare character strings, be case-insensitive */) {
            return true;
        }
        else {
            return false;
        }
    }

    /* Reverse the string*/
    private static String reverse(String input){
        char[] in = input.toCharArray();
        int begin = /* ... */
        int end = /* ... */
        char temp;
        while(end>begin){
            /* Replace the characters in the table using the temporary variable */
            end--;
            begin++;
        }
        return new String(in);
    }
}
```

Appendix A.4. Matrix Transponse

```
/* Task 4
Your task is to transpose the input matrix.

Examples:
A = |1,2,3|       A^T = |1,3|
    |3,5,6|             |2,5|
                        |3,6|

Validate source code by pressing Ctrl + F6    */

public class Task4 {
    public int[][] TransposeMatrix(int[][] A) {
        int row = /* .. */
        int col = A[0].length;

        int[][] T = new int[/*..*/][/*..*/];

        // rewriting values from matrix A to matrix T
        for(/* iterate the columns */) {
            for(/* iterate the lines */) {
                T[/*..*/][/*..*/]=A[/*..*/][/*..*/];
            }
        }
        return T;
    }
}
```

Appendix B. Questionnaire

1. How intrusive were the methods of data collection?

    (a)   Blood pressure sensor (ear):

        • (unnoticeable) 1 2 3 4 5 6 7 (made the work completely impossible)

    (b)   Breath sensor (belt on chest):

        • (unnoticeable) 1 2 3 4 5 6 7 (made the work completely impossible)

    (c)   Skin conductance (armband):

        • (unnoticeable) 1 2 3 4 5 6 7 (made the work completely impossible)

    (d)   Facial EMG (sensor on the face):

        • (unnoticeable) 1 2 3 4 5 6 7 (made the work completely impossible)

    (e)   Eye tracker (camera under the monitor):

        • (unnoticeable) 1 2 3 4 5 6 7 (made the work completely impossible)

    (f)   Video image (camera over the monitor + lighting set):

        • (unnoticeable) 1 2 3 4 5 6 7 (made the work completely impossible)

2. Choose methods that could be used during the daily work of the programmer according to comfort levels (you can select multiple options).

    • Blood pressure sensor (ear)
    • Breath sensor (belt on chest)
    • Skin conductance (armband)
    • Facial EMG (sensor on the face)

- Eye tracker (camera under the monitor)
- Video image (camera over the monitor + lighting set)
- Mouse movements and typing
- SAM self-assessment questionnaire

3. Does wearing a watch bother you during prolonged computer work?

- (it is unnoticeable) 1 2 3 4 5 6 7 (makes the work completely impossible)

4. Do you sometimes express your emotions (e.g., anger or frustration) aloud when programming?

- (never) 1 2 3 4 5 6 7 (always)

5. Which emotions were induced by:

- Unstable work of the IDE: _____
- Invalid test case in task 2: _____
- Shortening the time allocated for completing the last task: _____

6. How often could the SAM self-assessment form be displayed during the work so that it is not disruptive?

- Every 5 min
- Every 1 h
- Every commit
- When starting and closing the IDE
- Twice a day

7. Would you agree to monitor your emotional state while working?

- Yes
- No

References

1. ICT Specialists in Employment—Eurostat Report. Available online: http://ec.europa.eu/eurostat/statistics-explained/index.php/ICT_specialists_in_employment (accessed on 30 November 2017).
2. ICT Specialists—Statistics on Hard-To-Fill Vacancies in Enterprises—Eurostat Report. Available online: http://ec.europa.eu/eurostat/statistics-explained/index.php/ICT_specialists_-_statistics_on_hard-to-fill_vacancies_in_enterprises. (accessed on 30 November 2017).
3. Denning, P.J. Moods. *Commun. ACM* **2012**, *55*, 33–35.
4. Graziotin, D.; Wang, X.; Abrahamsson, P. Do feelings matter? On the correlation of affects and the self-assessed productivity in software engineering. *J. Softw. Evol. Process* **2015**, *27*, 467–487.
5. Müller, S.C.; Fritz, T. Stuck and frustrated or in flow and happy: Sensing developers' emotions and progress. In Proceedings of the 2015 IEEE/ACM 37th IEEE International Conference on Software Engineering (ICSE), Florence, Italy, 16–24 May 2015; Volume 1, pp. 688–699.
6. Fountaine, A.; Sharif, B. Emotional awareness in software development: Theory and measurement. In Proceedings of the IEEE/ACM 2nd International Workshop on Emotion Awareness in Software Engineering, Buenos Aires, Argentina, 21 May 2017; pp. 28–31.
7. Graziotin, D.; Fagerholm, F.; Wang, X.; Abrahamsson, P. Consequences of unhappiness while developing software. In Proceedings of the 2nd International Workshop on Emotion Awareness in Software Engineering, Buenos Aires, Argentina, 20–28 May 2017; IEEE Press: Piscataway, NJ, USA, 2017; pp. 42–47.
8. Uhrig, M.K.; Trautmann, N.; Baumgärtner, U.; Treede, R.D.; Henrich, F.; Hiller, W.; Marschall, S. Emotion elicitation: A comparison of pictures and films. *Front. Psychol.* **2016**, *7*, 180.

9. Fritz, T.; Müller, S.C. Leveraging biometric data to boost software developer productivity. In Proceedings of the 2016 IEEE 23rd International Conference on Software Analysis, Evolution, and Reengineering (SANER), Suita, Japan, 14–18 March 2016; Volume 5, pp. 66–77.
10. Müller, S.C.; Fritz, T. Using (bio) metrics to predict code quality online. In Proceedings of the 38th International Conference on Software Engineering, Austin, TX, USA, 14–22 May 2016; pp. 452–463.
11. Landowska, A.; Wróbel, M.R. Affective reactions to playing digital games. In Proceedings of the 2015 8th International Conference on Human System Interactions (HSI), Warsaw, Poland, 25–27 June 2015; pp. 264–270.
12. Fritz, T.; Begel, A.; Müller, S.C.; Yigit-Elliott, S.; Züger, M. Using psycho-physiological measures to assess task difficulty in software development. In Proceedings of the 36th International Conference on Software Engineering, Hyderabad, India, 31 May–7 June 2014; pp. 402–413.
13. Van Boxtel, A. Facial EMG as a tool for inferring affective states. In Proceedings of Measuring Behavior. Noldus Information Technology Wageningen, Eindhoven, The Netherland, 24–27 August 2010; pp. 104–108.
14. Tan, J.W.; Walter, S.; Scheck, A.; Hrabal, D.; Hoffmann, H.; Kessler, H.; Traue, H.C. Facial electromyography (fEMG) activities in response to affective visual stimulation. In Proceedings of the 2011 IEEE Workshop on Affective Computational Intelligence (WACI), Paris, France, 11–15 April 2011; pp. 1–5.
15. Bhandari, U.; Neben, T.; Chang, K.; Chua, W.Y. Effects of interface design factors on affective responses and quality evaluations in mobile applications. *Comput. Hum. Behav.* **2017**, *72*, 525–534.
16. Sharif, B.; Maletic, J.I. An eye tracking study on camelcase and under_score identifier styles. In Proceedings of the 2010 IEEE 18th International Conference on Program Comprehension (ICPC), Braga, Portugal, 30 June–2 July 2010; pp. 196–205.
17. Rodeghero, P.; McMillan, C.; McBurney, P.W.; Bosch, N.; D'Mello, S. Improving automated source code summarization via an eye-tracking study of programmers. In Proceedings of the 36th International Conference on Software Engineering, Hyderabad, India, 31 May–7 June 2014; pp. 390–401.
18. Sharif, B.; Falcone, M.; Maletic, J.I. An eye-tracking study on the role of scan time in finding source code defects. In Proceedings of the Symposium on Eye Tracking Research and Applications, Santa Barbara, CA, USA, 28–30 March 2012; pp. 381–384.
19. Sharafi, Z.; Shaffer, T.; Sharif, B.; Guéhéneuc, Y.G. Eye-tracking Metrics in Software Engineering. In Proceedings of the 2015 Asia-Pacific Software Engineering Conference (APSEC), New Delhi, India, 1–4 December 2015; pp. 96–103.
20. Kevic, K.; Walters, B.; Shaffer, T.; Sharif, B.; Shepherd, D.C.; Fritz, T. Eye gaze and interaction contexts for change tasks—Observations and potential. *J. Syst. Softw.* **2017**, *128*, 252–266.
21. Bednarik, R.; Tukiainen, M. An eye-tracking methodology for characterizing program comprehension processes. In Proceedings of the 2006 Symposium on Eye Tracking Research & Applications, San Diego, CA, USA, 27–29 March 2006; pp. 125–132.
22. Shaffer, T.R.; Wise, J.L.; Walters, B.M.; Müller, S.C.; Falcone, M.; Sharif, B. Itrace: Enabling eye tracking on software artifacts within the ide to support software engineering tasks. In Proceedings of the 2015 10th Joint Meeting on Foundations of Software Engineering, Bergamo, Italy, 30 August–4 September 2015; pp. 954–957.
23. Fragopanagos, N.; Taylor, J.G. Emotion recognition in human—Computer interaction. *Neural Netw.* **2005**, *18*, 389–405.
24. Gunes, H.; Piccardi, M. Affect recognition from face and body: Early fusion vs. late fusion. In Proceedings of the 2005 IEEE International Conference on Systems, Man and Cybernetics, Waikoloa, HI, USA, 12 October 2005; Volume 4, pp. 3437–3443.
25. Majumder, A.; Behera, L.; Subramanian, V.K. Emotion recognition from geometric facial features using self-organizing map. *Pattern Recognit.* **2014**, *47*, 1282–1293.
26. Sayette, M.A.; Cohn, J.F.; Wertz, J.M.; Perrott, M.A.; Parrott, D.J. A psychometric evaluation of the facial action coding system for assessing spontaneous expression. *J. Nonverbal Behav.* **2001**, *25*, 167–185.
27. Ooi, C.S.; Seng, K.P.; Ang, L.M.; Chew, L.W. A new approach of audio emotion recognition. *Expert Syst. Appl.* **2014**, *41*, 5858–5869.
28. Eyben, F.; Wöllmer, M.; Schuller, B. OpenEAR—Introducing the Munich open-source emotion and affect recognition toolkit. In Proceedings of the 3rd International Conference on Affective Computing and Intelligent Interaction and Workshops, Amsterdam, The Netherlands, 10–12 September 2009; pp. 1–6.

29. Kołakowska, A. A review of emotion recognition methods based on keystroke dynamics and mouse movements. In Proceedings of the 2013 The 6th International Conference on Human System Interaction (HSI), Sopot, Poland, 6–8 June 2013; pp. 548–555.
30. Kołakowska, A. Towards detecting programmers' stress on the basis of keystroke dynamics. In Proceedings of the 2016 Federated Conference on Computer Science and Information Systems (FedCSIS), Gdansk, Poland, 11–14 September 2016; pp. 1621–1626.
31. Khezri, M.; Firoozabadi, M.; Sharafat, A.R. Reliable emotion recognition system based on dynamic adaptive fusion of forehead biopotentials and physiological signals. *Comput. Methods Programs Biomed.* **2015**, *122*, 149–164.
32. Crk, I.; Kluthe, T.; Stefik, A. Understanding programming expertise: An empirical study of phasic brain wave changes. *ACM Trans. Comput.-Hum. Interact. (TOCHI)* **2016**, *23*, 2.
33. Landowska, A. Emotion monitor-concept, construction and lessons learned. In Proceedings of the 2015 Federated Conference on Computer Science and Information Systems (FedCSIS), Lodz, Poland, 13–16 September 2015; pp. 75–80.
34. Novielli, N.; Calefato, F.; Lanubile, F. Towards discovering the role of emotions in stack overflow. In Proceedings of the 6th International Workshop on Social Software Engineering, Hong Kong, China, 17 November 2014; pp. 33–36.
35. Jurado, F.; Rodriguez, P. Sentiment Analysis in monitoring software development processes: An exploratory case study on GitHub's project issues. *J. Syst. Softw.* **2015**, *104*, 82–89.
36. Brodny, G.; Kołakowska, A.; Landowska, A.; Szwoch, M.; Szwoch, W.; Wróbel, M.R. Comparison of selected off-the-shelf solutions for emotion recognition based on facial expressions. In Proceedings of the 2016 9th International Conference on Human System Interactions (HSI), Portsmouth, UK, 6–8 July 2016; pp. 397–404.
37. Bradley, M.M.; Lang, P.J. Measuring emotion: The self-assessment manikin and the semantic differential. *J. Behavior Ther. Exp. Psychiatry* **1994**, *25*, 49–59.
38. Quigley, K.; Lindquist, K.A.; Barrett, L.F. Inducing and measuring emotion and affect: Tips, tricks, and secrets. In *Handbook of Research Methods in Social and Personality Psychology*; Cambridge University Press: Cambridge, UK, 2014; pp. 220–252.
39. Picard, R.W. Affective computing: From laughter to IEEE. *IEEE Trans. Affect. Comput.* **2010**, *1*, 11–17.
40. Landowska, A. Emotion monitoring—Verification of physiological characteristics measurement procedures. *Metrol. Meas. Syst.* **2014**, *21*, 719–732.

applied
sciences

MDPI

Concept Paper

A User-Centred Well-Being Home for the Elderly

Nuno Rodrigues [1] and António Pereira [1,2,*]

[1]   School of Technology and Management, Computer Science and Communications Research Centre,
       Polytechnic Institute of Leiria, Leiria, 2411-901 Leiria, Portugal; nunorod@ipleiria.pt
[2]   INOV INESC INOVAÇÃO Institute of New Technologies of Leiria, Leiria, 2411-901 Leiria, Portugal
*   Correspondence: apereira@ipleiria.com; Tel.: +351-244-820-300

Received: 13 March 2018; Accepted: 18 May 2018; Published: 23 May 2018

Abstract: Every single instant a person generates a large amount of information that somehow is lost. This information can assume a large diversity of means, such as an oral word, a sneeze, an increase in heartbeat or even facial expressions. We present a model which promotes the well-being of the elderly in their homes. The general idea behind the model is that every single experience may mean something, and therefore may be recorded, measured and even have adequate responses. There is no device that provides a more natural interaction than a human body and every one of us, sends and receives useful information, which sometimes gets lost. Trends show that the future will be filled with pervasive IoT devices, present in most aspects of human life's. In this we focus on which aspects are more important for the well-being of a person and which devices, technologies and interactions may be used to collect data directly from users and measure their physiological and emotional responses. Even though not all the technologies presented in this article are yet mainstream, they have been evolving very rapidly and evidence makes us believe that the efficiency of this approach will be closely related to their advances.

Keywords: ambient intelligence; human-computer interaction; internet of things; smart homes

1. Introduction

This article describes a universal model for a home well suited for the well-being of the elderly and suggests how some technologies and types of interaction may be used to improve their lives. Some of the technologies and solutions proposed are already mainstream and present in some modern devices. Others, although are not yet largely disseminated, may become plausible solutions in the future. The model is universal and may be adapted for different types of people, but our particular case of study is mostly focuses on the elderly. These are undeniably a group of the society who can enrich their lives by integrating implementations of some of the proposed topics addressed.

The model assumes communications to several electronic devices in a home in order to monitor and keep track of user activities. This is truly one of the major challenges since it is important to gather some information which allows us to determine the condition of an elder living in her/his home. This can involve several aspects such as the health, the activities carried out and even the emotional state. Ways of measure these aspects are then compulsory to define a model which can interact with an elder in a positive and effective manner.

In the field of health, there is a lot of previous work in IoT intelligent devices intended for the elderly [1–4] and several solutions were applied with success. Actually, even with some complexity, some implementations of these are really straightforward. Physical activities and emotional states, however, are a little more complex than the previous (health).

1.1. Measuring Physical Activity

To measure physical activities the Metabolic Equivalent (MET) may be used. MET is a physiological measure expressing the energy cost of physical activities. This measurement is the one used in "The Adult Compendium of Physical Activities", conceptualized by Dr. Bill Haskell from Stanford University and developed for use in epidemiologic studies to standardize the assignment of MET intensities in physical activity questionnaires [5]. We opted for using the MET since it has been used extensively and effectively in a large variety of studies related to the measurement of physical activities e.g., [6–8].

If a person is adopting a sedentary behaviour such as simply sitting quietly or additionally watching television or listening to music, it will have an energy expenditure within the range 1.0 to 1.5 MET. In practical terms, this means that the MET ranges from 1.0 to 1.5 kcal/kg/hour. If, instead, the person is performing more active tasks such as moderately bicycling on a stationary bike it may have a MET of about 6.8 (a little bit different then going uphill on an outdoor bike which can take as much as 16.0 MET).

1.2. Detecting Emotions

Another aspect that may be measured concerns the emotional state of a person. One way of reaching such goal is to determine the user emotions at different times. For this purpose, the Facial Action Coding System (FACS) may be useful, which has been used successfully in a large variety of fields to categorize facial expressions e.g., [9–12]. The FACS is a system which allows for describing facial movement, based on an anatomical analysis of observable facial movements, and it is often considered a common standard to systematically categorize the physical expression of emotions. This system, which has proven to be useful to psychologists and to animators, encodes movements of individual facial muscles from slight different instant changes in facial appearance. It has been established as an automated system which detects faces (e.g., in videos and most recently using automatic real-time recognition using 3D technology [13]), extracting their geometrical features and then producing temporal profiles of each facial movement.

It was first published by Ekman and Friesen in 1978 [14]. Since then, it has been revised and updated several times, such as in 2002, by the one from Paul Ekman, Wallace V. Friesen and Joseph C. Hager [15] and more recently in 2017 (Facial Action Coding System 2.0—Manual of Scientific Codification of the Human Face from Freitas-Magalhães [13]). This recent instrument to measure and scientifically understand the emotion in the human face presents 2000 segments in 4K, using 3D technology and automatic and real-time recognition.

The FACS use Action Units (AU) which represent the fundamental actions of individual muscles or groups of muscles. Emotions can, therefore, be represented by AU which can also be scored by their intensity ranging from minimal to maximal.

We identify two fields were the FACS may be used in the context of the purposed model. One is to recognize human expressions to measure the user mood. The other is to apply human expressions in avatars that may be used to communicate with users. So much as a virtual caregiver or friend which may accompany the user and provide feedback including different expressions—those that all of us learn to recognize since our early years. Indeed it may reveal to be a very useful enhancement in systems such as the virtual butler [16].

Although the FACS may be used through image processing, using algorithms such as the Face Landmark Algorithm [17], extracting characteristic points of the face from an image, the use of 3D scanning broadens the possibilities of the hardware and algorithms which may be exploited. If not so long ago, 3D scanning technology was too expensive to be considered in domestic solutions, the fact is that this is rapidly evolving. Thinking about recent advances, it is not so eccentric to think that it is probable to be present for everyday uses in a near future. Indeed, it is as already starting to appear in small and mainstream devices such as smartphones. An example is the TrueDepth camera included in the "new" iPhone X [18], which allows analysing over 50 different facial muscles. This is achieved with the help of a built-in infrared emitter, which projects over 30,000 dots onto a user's face to build a unique facial map. These dots are projected in a regular pattern and are then captured by an infrared camera for analysis. This system is somewhat similar than the one found on the original version of the Kinect sensor [19] from Microsoft but built-in a much smaller scale. This sensor, especially the Windows version, has proven to be an effective solution for the low-cost acquisition of several real-world geometries in several projects [20–22].

One of the advantages of this system is that it can also be used for facial recognition, one of the features integrated into the iPhone X (designed by Face ID). This also has some advantages in the field of security, especially when considering more traditional facial recognition systems based on image analysis, which can often be fooled with photographs and masks.

Such methods are evolving fast and trends indicate that they will become mainstream and be used effectively as secure authentication systems, in several applications such as unlocking systems, retrieving personal data, make use of social media, making payments, amongst others. Indeed, most systems already use advanced machine learning to recognize changes in the appearance of the users.

Even the multimedia ISO standard MPEG-4 includes tools such as the Face and Body Animation (FBA) [23,24], which enables a specific representation of a humanoid avatar and allows for very low bitrate compression and transmission of animation parameters [25]. One component of this FBA is the Face Animation Parameters (FAP), to manipulate key feature control points in a mesh model to animate facial features. These were designed to be closely related to human facial muscles and are used in automatic speech recognition [26] and biometrics [27]. The FAP, representing several displacements and rotations of the feature points from the neutral face position, allows the representation of several facial expressions which may indeed be applied to Human-Computer Interaction (HCI) [28].

Undeniably, there are several studies related to human emotions in HCI which represent different strategies or computational models for affect or emotion regulation e.g., [29–35]. These also range a large multitude of domains, e.g., sports [36], virtual storytelling [37], games [38] and even therapeutic fields [39]. It is unarguable that emotions are important. In any single interaction between two (or more) humans both monitor and interpret each other's emotional expressions.

2. A Model of Wellbeing for the Elderly

One of the foremost questions considered when trying to define a model for which an elders' home IoT solution might be unified is which facets of a person's life it should touch. Consequently, the model is composed of three dimensions: health, activity and emotional, as represented in Figure 1.

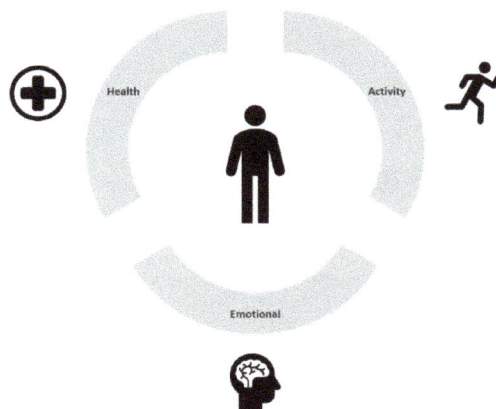

Figure 1. The three dimensions of well-being.

Even if sometimes these are observed isolated from each other, in literature as well as in some solutions, the fact is that on some occasions they are intrinsically related. Hence a change in one of the dimensions will affect other(s). Certainly, a chronic health problem, for many persons, will affect their emotional state. Similarly, it may prevent them from performing some type of activities. This dependency which is observed sometimes isn't by any means always negative. More positive relations may also occur. For instance, someone that is emotionally strong and is in a good physical shape (e.g., because of some daily activity) may also be less affected by some health complications and/or more easily surpass them.

Although other or more dimensions might have been selected, we selected these because of their noticeable contribution to the well-being of the elderly. Moreover, they were also chosen considering that some signals may be attained from sensors to determine some aspects related to these three dimensions. For example, a sensor measuring the heart rate can tell us something about the health. Sensors in appliances and sports machines can identify the activity that is being carried out and for how long and the type of activity (e.g., sports activity, entertainment—such as watching the favourite TV show). Finally, a camera capturing facial expressions can tell us something about the emotional state. For sure, all the factors that may influence a person's well-being, their relative importance and what can actually be measured and/or influenced with today's technology justifies further investigation.

Considering these three dimensions, it is possible to identify all the entities which may be present in a home that in some way can intervene with the resident. We classify these entities into three categories: agents, devices and geography elements. These may be interpreted as an overview for much of the different types of objects that compose a home in the vast literature e.g., [40–43] that can be found nowadays about smart homes, smart environments or ambient assisted living (just to name a few of the most common terms found). Analogous to Figure 1 these are represented as user-centred graphics as in Figure 2.

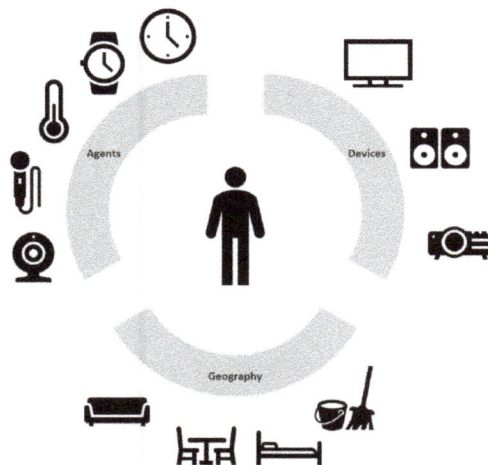

Figure 2. Entities of a home.

On this classification, the agents represent all the entities that may be present in a home intended to collect or transmit information to the resident. Therefore, the agents may be classified as input agents, output agents or both. Examples of input agents are cameras, panic buttons, heartbeat reader modules, etc. Output modules are projectors, stereos, notification clocks, amongst others.

Agents may also be simultaneous input and output agents. An example of an output agent can be a bracelet which reads the user pulse and gives him information about eventual irregularities sensed.

Devices are the appliances of the house, i.e., the pieces of equipment operated electrically for performing domestic chores (e.g., a TV, a fridge, a washing machine, etc.). These can also act as agents, or interact with them, by collecting or transmitting information. For example, a fridge can act as an input agent, collecting information through integrated technology, which allows detecting all the goods inside it. This same fridge can also act as an output agent by emitting alerts when some specific item is missing or almost over. It can also emit alerts when expirations dates are near.

Finally, geography elements, designation inspired by the animation field, are all the home entities, non-electric, which may even become props (using the same animation analogy) if the user somehow intervenes with them (e.g., home furniture). For example, the user may seat on a *couch*, sleep on a *bed* or sit at the *table* having dinner.

2.1. The Model of a Wellbeing Home

Based upon the classification of the several entities of a home, a simple model may then be defined intended to improve the well-being of an elder's life. Although this article is written taking into consideration the wellbeing of the elderly, the model may be used universally.

To notice also that it isn't our intention to replace existing architectures or frameworks already presented in other projects ex: [1,2,4,16,40–43]. Truly some can "fit" inside this model, have their own model or represent also other models for smart homes. Our goal is to present a simplified model of a home well suited for the wellbeing of the elderly.

The model, represented in Figure 3, besides the three types of entities (agents, devices and geography), also includes a Home Processing Unit (HPU) which is responsible for managing all the data received, managing and taking all the decisions. This element is thus something with computing capabilities (such as a personal computer). It is our belief that an intelligent home should centralize all the information in a central module of the home, which uses this information to perform informed actions.

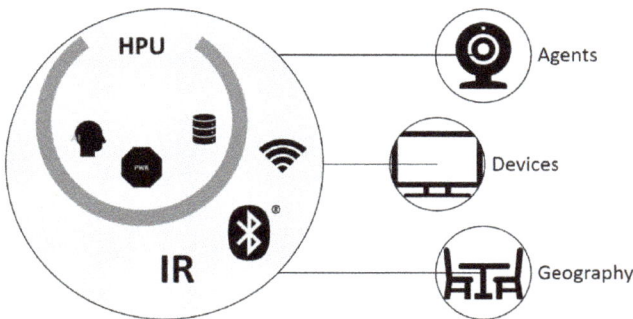

Figure 3. The model of a wellbeing home.

Homes will then include a type of HPU which is going to be responsible for interacting with all the intelligent devices of the home in an integrated form. If one of the innovations of modern houses, promoted by home sellers a few years ago, was a central vacuum system in a house (composed by a central cleaner in the house and several connections points, where one could simply connect and still have access to the function expected: vacuum cleaning) it is most likely that future houses will integrate and be promoted as having some kind of HPU.

In some major appliance retail stores categories such as "Smart Home" are now appearing, which include several sensors, controllers, connectable devices and several other types of devices promoted as "smart". These mainly work on a stand-alone basis, i.e., not within an integrated solution. Indeed, IoT devices seem to be a future trend, were around 20 billion of connected IoT related devices are forecast by 2023 [44]. Nonetheless, a universal platform for them to work in an integrated manner doesn't exist. This platform may take advantage of their full potential and even broaden their possibilities.

In order to have all the agents and/or devices communicating with a HPU, some sort of communication technology (or several) is necessary (e.g., Wi-Fi, Bluetooth, Infrared). This HPU will have the capability of communicating with all the home agents and/or devices. It may even use these to collect information of geography elements in the house (e.g., perform object detection and use segmentation algorithms on images or videos gathered from cameras). As such it can receive data from these entities (agents and devices) as well as sending data (including commands). To store all the data, one or more databases will then be necessary. Similarly, to allow to take decisions accordingly to the data received, it is necessary to assure some processing capabilities which may even include Artificial Intelligence (AI) algorithms.

The HPU presented in this article is centred on the well-being of the elder and will be driven by a Personal Well-being Rating (PWR) constantly being updated from the information gathered from the several entities present in the home. The PWR will be determined by the Personal Wellbeing Equation (PWE):

$$PWE(p) = K_h \, h + K_a \, a + K_e \, e, \tag{1}$$

where,

h = health
a = activity
e = emotional
K_h = health coefficient
K_a = activity coefficient
K_e = emotional coefficient

plus,

$$K_h + K_a + K_e = 1$$

The PWE is a function of a precondition (p). In practice, what this means is that it depends on a previous condition for a certain person (hence the "Personal" in the name of the Equation). It is not expected that every elder will have the same health, would perform the same activities or would have the same emotional responses. In other words, the Equation assumes that the wellbeing of a person depends on the three dimensions presented (health, activity and emotional) that can be monitored by agents and, ideally, somehow influenced.

Considering that we have the well-being rating that, based on the data collected by the agents, permits to discover when some dimension is affecting the general wellbeing of a person, the next step is to determine what to do to improve it. This may be achieved on a dimension approach, i.e., determining the dimension which is below some threshold and simply trigger an event. The challenge resides in to determine how to measure each dimension and what actions may be triggered.

Independently of the dimension there are two distinct ways of triggering an action: seamless (where the user doesn't have to confirm any possible action before it happens—which doesn't prevent her/him to cancel it, if desired) or with user intervention (where the user is explicitly solicited to manually confirm that he agrees with some action). Either way may be the most adequate depending on the specific scenario.

The next sections provide some insights about these concerns and give some ideas for possible future implementations in each dimension.

2.2. Activity

The first thing to consider in this dimension is how to determine the type of activities being carried out by the user on every single instant of time. This can be accomplished using IoT solutions but not only. It is easy to include agents to determine which devices are in use. It is trivial to determine if a TV or Hi-Fi system is turned on. Likewise, it is also easy to determine if a stationary bike is in use. One thing that simplifies this problem is the fact that all of these systems are electronic devices (even the more modest domestic stationary bikes include electronic components to monitor the activity—or electronic components can be easily coupled).

The aforementioned, however, is not although exactly true for any type of activity, which may not include electronic devices. For example, it may not be the most natural thing to ask someone to use some kind of wearables while dancing. The fact that music may be playing isn't a synonym that someone is dancing—it may simply be seated listening to music, which takes a little bit less than the 4.5 MET that takes salsa dancing (estimated value). For such cases, further work may be required by using data collected from cameras or 3D sensors, and most likely using advanced imaging techniques, 3D techniques, or both and even Artificial Intelligence algorithms. One important component in all of these is a clock which measures the amount of time spent in each activity.

The strategy can be defined as an assembly of IoT devices, which keep track all the user activities and the time spent in each activity. All this information is sent to the HPU. This is then responsible for quantifying the total amount of MET spent in each period of time. Considered a hypothetic elder's day, it could result in something such as what is represented (for simplification purposes several activities were omitted—e.g., taking medication, sitting on a toilet, etc.—and the table summarized) in Table 1.

Table 1. Activities and corresponding MET of a hypothetic elder's day.

Activity	MET	Minutes Spent
Cleaning the house (i)	2.5	30
Self-feeding (b)	1.5	80
Dressing (b)	2.5	20
Preparing meals (i)	3.5	110
Functional mobility: walking (b)	3.5	60
Personal hygiene: showering (b)	2.0	20
Personal hygiene: brushing teeth (b)	2.0	10
Personal hygiene: grooming (b)	2.0	10

The aforementioned activities correspond to recognized Activities of Daily Living (ADL) [45,46], where the (b) activities correspond to basic ADL, i.e., those that must be accomplished every day by everyone who wishes to thrive by their one. The (i) activities correspond to instrumental ADL, i.e., those not necessary for fundamental functioning, but still very important to give someone the ability to live independently.

There are no devices specifically conceived to determine the activities performed by the user. However, even the 3D sensors found in many modern entertainment devices, such as the Kinect sensor, can cope with several users at the same type and be programmed to know at each time which user is doing what. We do believe that a well-considered MET measurement may integrate solutions to promote more healthy, active and fulfilling lifestyles.

Actions

Determining the activities performed by the user and the time spent in each one, is just one part of the problem. The second part of the problem is determining the actions that can be carried out when some dimension is below some threshold value. In the activity dimension, it wouldn't be reasonable to force someone to perform an exercise, but there are some more subtle ways of improving the value of this activity. Knowing that the MET may be used to identify the energy expenditure of each activity, it is possible, for instance, to set an agent to play a music known for frequently make the home habitant dance and therefore spend 3.0 MET (estimated value of dancing a waltz) which is significantly more than the 1.3 MET from sitting quietly.

Another way is to simply inform a family member, a caregiver or a friend (represented in Figure 4) that the value of this dimension is low, and this person would beneficiate from some physical activity.

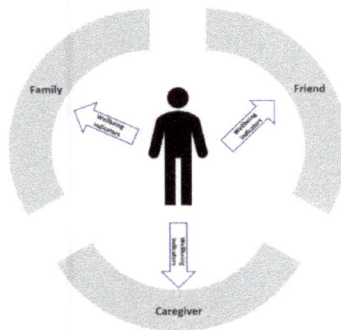

Figure 4. User circle.

Thus, what could be an eventual boring sedentary day, could become an outdoor day simply because a friend appeared at the door inviting for a walk accompanied by a good talk. This could also have some impact on the health and emotional dimensions. In addition, who knows if the friend who appeared wasn't also with a low value in the activity value and even got alerted by its HPU (which suggested to find a friend for a walk). This makes us imagine a world of connected HPU collaborating to improve the life of everyone.

2.3. Emotional

One thing to consider regarding the emotions is exactly which emotions do we want to recognize. Actually, different authors define different emotions and we could go as far in this field at least until the ancient Greek treatise on the art of persuasion "Aristotle's Rhetoric", dating from 4th century BC [47], passing through "The Expression of the Emotions in Man and Animals" of Charles Darwin in 1872 [48], or even the eight basic emotions Robert Plutchik's theory in 1980 [49], which were represented in a "wheel of emotions" which includes different intensities.

The FACS also considers several emotions such as those in the list represented in Table 2.

Table 2. FACS Emotions.

Emotions
Happiness
Sadness
Surprise
Fear
Anger
Disgust
Contempt

For now, from the list of Table 2, the emotion we are most interested in is happiness. One could also consider sadness—which may be considered the opposite of happiness, but this could bring us a few problems. For example, it is normal for someone to feel sad for long periods of time, such as when mourning a loved one. In such case, however, it is a typical reaction and it doesn't always prevent someone from showing happy feelings eventually during the day when facing some situations—e.g., when a grandchild gets a new achievement or makes a joke. The problem is that according to some studies, such as in [50], sadness can last, on average, for 120 h. This is significantly more than other feelings. For these reasons, we are concentrated on happiness by now.

One way of knowing if a person is happy is for example if she/he is smiling. However, a person can't be smiling all the time. As such, instead of measuring the amount of time a person is smiling, the approach consists of detecting the number of smiles per day. Nowadays there are several devices which can detect smiles. Modern cameras and even the latest smartphones can take pictures automatically when they detect smiles. Indeed, there are a lot of approaches that can be used to detect smiles e.g., [51–55].

The approach may then be based on a count of smiles per day. It is evident that this depends on many factors, including the country of origin and the age of a person. It is well known, for example, that, generally speaking, children smile a whole lot more than adults. For example, a study says that an UK adult smiles 11 times a day [56].

Actions

Knowing a way of recognizing that someone is happy is from their smile, the issue here is to trigger an emotional response which can be measured with a smile. It's easy to find lists of things which cause people to smile e.g., [57,58]. An excerpt of ten things that can make people smile is in Table 3, but there are many others.

Table 3. Things that can make people smile.

10 Things That Can Make People Smile
Sunny weather
Looking back at old photographs
Getting a nice message from a loved one
When your favourite song comes on the radio
Seeing any type of baby animal
Watching your favourite film
Tax rebates
Your football team winning
Getting a nice comment on social media
Being greeted by your dog

The challenge resides in to determine which things may be triggered and controlled by the HPU, and which are more effective. For example, some are infeasible such as controlling the weather, tax rebates, making the favourite football team winning or making a pet greet his owner. Other is feasible and strategies may be defined to achieve these. It is possible to ask the user if she/he wants to watch photographs of their good moments, hear her/his favourite song or watch her/his favourite movie or TV show.

The next step, assuming there was a positive response from the user, is to choose the corresponding agent or device which is responsible for performing the corresponding task. This may be as simple as, for example, turn on a display or projecting photographs are known to make the user happy. While the activity is taking place, and even after, the HPU should keep collecting input data from the agents, in order to discover if the triggered reaction caused the desired response from the user (monitoring the PWR and taking proper actions in case of need). Another example is, when detecting a low emotional state, the HPU contact a close relative or loved one, which may then send a message to the elder or post some nice comment on social media.

2.4. Health

If for one hand this dimension is probably the one that may more severely impact the quality of life of the majority of the elders, on the other hand, it is the simplest to monitor (depending of course of the type of health indicators that we want to observe). Indeed, a lot of previous work exists in the field of intelligent devices for the elderly (e.g., in [4] Chan et al. several smart home projects are presented), where health monitoring systems have been developed, such as: pulse, blood pressure, heart rate, fall detection devices and panic buttons. These essentially are made of wearable devices to monitor in real-time the health status of the elderly. These devices must have versatile functions and be user-friendly to allow elders to perform tasks with minor intrusion and disturbance, pain, inconvenience or movement restrictions [1].

In [2] an Android-based smartphone with a 3-axial accelerometer was used to detect falls of its holder in the context of a smart elderly home monitoring system. In this device, a panic button was also implemented. In [3] Gaddam et al. use intelligent sensors with cognitive ability to implement a home monitoring system that can detect abnormal patterns in the daily home activities of the elderly. This also includes a panic button.

Certainly, previous work allows us, in a simple manner, to monitor several health indicators and send them to the HPU. The action is carried out, however, is noticeable a more sensible matter.

Actions

The action will always be dependent on the health status of every person. It may not be reasonable to have an HPU acting as a doctor and recommending medication when it detects an unusual variation in this dimension. Amongst all the possible negative consequences that this could represent, it would be meaningful knowing why there was a diminishing in this dimension. For example, a rise in the heartbeat may result from watching a scary movie (something that "smart" agents eventually could determine). A more reachable and feasible solution would simply send a message to a familiar, caregiver or family's doctor. This person, knowing the medical history of the home resident, would be in the best condition to evaluate the situation and determine the best action to take.

More simple actions are simple alerts when the elder simply forgets to take some medication. In this situation, audio and/or visual information could be provided to alert the user to take their medicaments at certain times of the day.

A little less obvious is a different scenario where agents act in a preventive manner, in order to prevent some circumstances, which could lead to a situation where the user health could be compromised. One example is preventing accidental falls. Geography elements can be mapped inside a house and tracked when they are moved. This is essentially important for the elderly. For example, as time goes by some elders may develop diseases, such as cataracts, which lead to a decrease in vision. Agents can keep a map of all the house entities, keeping track of their location and eventually warn the user for some obstacles that may found when traversing the home and may lead her/him to fall. This can be achieved with cameras or 3D sensors, strategically collocated to avoid occlusion problems and with the aid of complex algorithms that follow the user in real time and detect potential risks of collision. In such scenarios, voice interfaces may also be included, where several aspects (e.g., type of language, on/offline system, context-aware, etc.) must be observed in order for them to become useful (in [59] several of these aspects are discussed).

2.5. Privacy

Another question is if people would accept being monitored in the context of a smart home. On a UK study [60] the perception of the purpose and benefits of smart home technologies, from the majority of participants, is: "the control of several devices in the house"; "improving security and safety"; "supporting assisted living or health"; "improve quality of life"; "provide care". Though a significant amount (around 40 to 60% in each of the following items) also agree that "there is a risk that smart home technologies . . . ": "monitor private activities"; "are an invasion of privacy"; "are intrusive". Nevertheless, the clear majority agree that "for there to be consumer confidence in smart home technologies, it is important that they . . . ": "guarantee privacy and confidentiality"; "securely hold all data collected"; "are reliable and easy to use".

The benefits expected of technology are also presented in [61] where one of "the most frequently mentioned benefit is an expected increase in safety". Similarly, the importance the participants in studies presented also mention "that they expect that the use of technology for aging in place will increase their independence or reduce the burden on family caregivers" and that "The desire to age in place sometimes leads to acceptance of technology for aging in place".

The issues concerning privacy are also of great concern for us. While we believe that most people would agree on a system that would improve their quality of life, there are some studies where some people show some discomfort. For example, a study conducted in [62], showed that the majority of aged persons have no problem with audio, but refused the video modality.

While we believe that most elders won't object to transmit some information to someone of their close circle of friends or family, to achieve a greater good, we don't deny that in some cases this could represent a problem. It is impossible to make predictions about if this will become an issue, but the truth is that the model doesn't prevent the users from using most of the solutions only in the privacy of their homes as personal monitor systems. Foremost, as we described in Section 2.1, independently of the dimension of wellbeing there are two distinct ways of triggering an action: seamless (where the user doesn't have to confirm any possible action before it happens—which doesn't prevent her/him to cancel it, if desired) or with user intervention (where the user is explicitly solicited to manually confirm that he agrees with some action). This allows for a desired level of control and to the development of scenarios which may parameterize with only the options allowed by the users.

The next chapter describes further a study that we have conducted to gain more insights about most of these topics.

3. Results

We have conducted a formative evaluation to identify possible improvements of the model, to gather some insights about the main areas of concern and focus and gain more awareness about the needs of the elderly. This evaluation was accomplished through the means of seven interviews: a CEO of a company in the field of healthcare; an occupational therapist; a technical director and phycologist of an elders' home; a doctor; a nurse; a direct-action and coordinator of physical activity program for the elderly; a professor in a higher school of health sciences specialized in people in their final stages of life. The common aspect with all of these is that all of them have or had close contact with elder within the scope of their professional activity.

The interviews included several questions whereas we present the most relevant for the evaluation of this study and summarize the main findings.

When asked if the dimensions considered for the study could represent the general well-being of the elders, none of the interviewed opposed against them. Nevertheless, other terms also come along, such as "social", "psychological" (terms also referred in other literature [63]) or even several other terms which represent several sub-areas of health dimensions. Indeed, although even other organizations were also considered, it was consensual that most of the new areas of well-being presented actually "fit" with the ones presented in the model. For example, the socialization contributes to improving the emotional state and the psychological, emotional and even spiritual aspects may be considered as contributors to the health dimension. It was also recognized that most dimensions are correlated, which confirms our initial presumptions (example: the physical activity may contribute to improving the health condition).

For such reasons it is our conviction that the proposed dimensions seem appropriate, a fact acknowledged by the several experts consulted, which doesn't exclude possible future additional dimensions or organizations from being included in future work.

When asked if elderly preferred to live in their own homes, if they had the possibility of having a system that would be monitoring their condition and act in case of need, everyone confirmed that this is an undeniable truth. Similarly, they all agreed that it is difficult for an elder to accept a stranger in their home, i.e., accept social assistance from someone they don't know. It was also referred that when willing to accept assistance they would prefer a familiar. After the initial resilience and well explained why the care is needed and the advantages for the elder, they would end up accepting the assistance. All the interviewed also agreed that once an elder finally accepts a stranger in their houses it is hard for them to accept new persons. Generally, all recognized that the elder would feel safer if they knew that they were being monitored at distance. The occupational therapist also referred, that the elders' days are made of routines, so if, some routines are included at specific periods (e.g., video calls 2–3 days a week), after little time it would be the elders asking for these activities. It was also referred that one possible way of knowing something may eventually be wrong with an elder, is exactly by analysing these routines. For example, an elder which typically wakes up at 9 am and performs the same chores, and suddenly fail these routines may have a strong reason for such. This leads us to consider in the model the possibility of monitoring the daily chores.

Considering that technology may help the elderly to live more independently, the possibility of being monitored without having a stranger in their homes and, when possible, the person being someone they know, is a clear benefit. Similarly, the use of technology to include the routines in the several dimensions is easy to achieve. For example, video calls for the health (e.g., talking with their doctor, nurse, etc.) and emotional dimension (talking to a close one, something acknowledged by most of the interviewed as an important aspect, sometimes the elder only feel the need of someone to talk with) and alerts when it is time to perform specific activities (activity dimension).

When asked about the main concerns of the elder that live alone, several aspects were mentioned: the fear of dying alone; the fear of needing help and no one is present to help them; the fear of being unable to help their partners when needed; the concern about bothering other persons; safety issues (being robbed, accidents); loneliness; lack of security;

These are all aspects where technology based on the described dimensions can help. The agents can gather the necessary information to monitor each dimension, reacting properly in each situation.

A sensible matter is also if elders are willing to provide share their personal information (e.g., physiological data) with others. Although the general opinion amongst the experts was that they will be willing if the benefits were well explained, several concerns were also pointed out, such as: the need to have several profiles of access to the information; privacy concerns; data protection concerns; the persons to which the information will be shared with; the order in which the information flows (e.g., who would be contacted first).

Conversely, when asked about the possible use of video cameras to monitor the elders' activities the majority of the interviewed (4 persons) mentioned that would be difficult to achieve, as the elderly would interpret it as an invasion of privacy. One of the interviewed (nurse) declared that it would be only possible with some of the elders. Only the direct-action specialist said that after some initial resilience, and after explaining all the benefits and purpose of the videos cameras, the majority of the elderly would agree with these devices. During one of the initial interviews two ideas came up: the possible use of video cameras only on a specific place in the house, known by the elders, and the possibility of only using the cameras by explicit evocation by the elders or eventually after some occurrence which implies urgent help. When faced with this possibility all agreed that it would be accepted by the elders.

For such reasons, and considering the importance of private data, possible implementations would have to be created supported in privacy and security concerns.

As for the reason why the elders don't perform more physical activities, the responses were: no reason why—normally their physical activity was related within the context of some type of job/work, since they don't work anymore, they simply don't see the need; lack of motivation; absence of specific programs; lack of time; lack of information about the benefits of periodic physical activity; because they are tired; the cost of activities. Some also mentioned that they would do it if it was within the context of social activities or if recommended by the doctor. These recommendations can be integrated within the context of social games or virtual assistants which may improve both the activity and emotional dimension and consequently the health dimension.

Generally, all the interviewed considered that the model would be of much use to monitor and improve the well-being of the elderly. The nurse even denoted that some elder is already prone to technology since there are several which receive home support which has alarm buttons and GPS bracelets (which allows them to be located outside of their homes when they press a button).

4. Discussion

The model presented for a home conceived for the well-being of the elderly introduced in this article is the foundation of a series of projects which will be presented in the future literature. However, we can also backtrack the application of the model, with the vast list of solutions we have been working in the last few years. Although the model was presented considering the elders best interests it is observable that this may be considered a universal model.

Regarding the use of the MET of "The Adult Compendium of Physical Activities" there are limitations since it doesn't account for differences in body mass, adiposity, age, sex, efficiency of movement, geographic and environmental conditions in which the activities are performed. However, this may serve as a reference and there is also always the possibility to use another type of measurements or even calculate measurements for each person. None of these alternatives poses obstacles to the correct application of the model.

There are also some constraints related to the FACS. For example, it deals with what is clearly visible in the face. There are also some features excluded from the FACS, such as facial sweating, tears, rashes, amongst others. Yet, most expressions are possible to determine with modern technologies in most of the cases. It is true that there will always be some situations where expressions can't be determined based on the FACS (e.g., in a case of facial paralysis), but these are the exception. For such cases, or where someone is in more severe conditions, probably a caregiver or similar will always be present. The user wouldn't then be taking the full advantage of a HPU equipped home (pretty much as a person who has a home equipped with a central vacuum cleaner, but which is no longer in conditions to vacuum clean). The model therefore assumes that there is a minimum extent of things that a person can do, which eventually allows her/him to live alone.

In some areas, those directly related with the dimensions of well-being considered (health, activity and emotional), some more future studies amongst specialists (e.g., doctors, nurses, psychologists, etc.) will benefit further developments of the model. Concerning the modern technology and our previous projects, there are strong evidence that the insights and solutions presented are plausible. The dimensions presented have noticeable contributions to the well-being of everyone, and we strongly believe that future projects founded on this model will allow an improvement of the quality of life of the elderly. Several aspects were discussed with a specialist which brought us several aspects to account for in future implementations. Moreover, the dimensions of well-being and human-computer interaction methods proposed are intended for the implementation of "more natural" interfaces which can effectively enrich the well-being of the elderly.

As mentioned in this article, the monitoring of the well-being status of the elder will be evaluated by the Personal Wellbeing Equation, where we are presently working.

Author Contributions: N.R. and A.P. contributed equally to this work with significative work in the review of the literature, in the specification of the dimensions of wellbeing, in the definition of the model of wellbeing, in the formative evaluation and in the final discussion. The article also benefits from several insights from previous projects from these authors. Both authors also read and approved the final version of this article.

Acknowledgments: This work was supported by national funds through the Portuguese Foundation for Science and Technology (FCT) under the project UID/CEC/04524/2016.

Conflicts of Interest: The authors declare no conflict of interest.

References

1. Ferreira, G.; Penicheiro, P.; Bernardo, R.; Mendes, L.; Barroso, J.; Pereira, A. Low cost smart homes for elders. In *Universal Access in Human–Computer Interaction. Human and Technological Environments*; Lecture Notes in Computer Science (LNCS); Springer: Cham, Switzerland, 2017; Volume 10279, pp. 507–517.
2. Lee, J.V.; Chuah, Y.D.; Chieng, K.T.H. Smart elderly home monitoring system with an android phone. *Int. J. Smart Home* **2013**, *7*, 17–32.
3. Gaddam, A.; Mukhopadhyay, S.C.; Gupta, G.S. Elder care based on cognitive sensor network. *IEEE Sens. J.* **2011**, *11*, 574–581. [CrossRef]
4. Chan, M.; Estève, D.; Escriba, C.; Campo, E. A review of smart homes-Present state and future challenges. *Comput. Methods Programs Biomed.* **2008**, *91*, 55–81. [CrossRef] [PubMed]
5. Ainsworth, B.E.; Haskell, W.L.; Herrmann, S.D.; Meckes, N.; Bassett, D.R., Jr.; Tudor-Locke, C.; Greer, J.L.; Vezina, J.; Whitt-Glover, M.C.; Leon, A.S. 2011 Compendium of Physical Activities: A second update of codes and MET values. *Med. Sci. Sports Exerc.* **2011**, *43*, 1575–1581. [CrossRef] [PubMed]
6. Kohler, S.; Behrens, G.; Olden, M.; Baumeister, S.E.; Horsch, A.; Fischer, B.; Leitzmann, M.F. Design and Evaluation of a Computer-Based 24-Hour Physical Activity Recall (cpar24) Instrument. *J. Med. Internet Res.* **2017**, *19*, e186. [CrossRef] [PubMed]
7. Lyden, K.; Kozey, S.L.; Staudenmeyer, J.W.; Freedson, P.S. A comprehensive evaluation of commonly used accelerometer energy expenditure and MET prediction equations. *Eur. J. Appl. Physiol.* **2011**, *111*, 187–201. [CrossRef] [PubMed]
8. Welk, G. *Physical Activity Assessments for Health-Related Research*; Human Kinetics: Champaign, IL, USA, 2002.
9. Kohler, C.G.; Martin, E.A.; Stolar, N.; Barrett, F.S.; Verma, R.; Brensinger, C.; Bilker, W.; Gur, R.E.; Gur, R.C. Static posed and evoked facial expressions of emotions in schizophrenia. *Schizophr. Res.* **2008**, *105*, 49–60. [CrossRef] [PubMed]
10. Hamm, J.; Kohler, C.G.; Gur, R.C.; Verma, R. Automated Facial Action Coding System for dynamic analysis of facial expressions in neuropsychiatric disorders. *J. Neurosci. Methods* **2011**, *200*, 237–256. [CrossRef] [PubMed]
11. Kohler, C.G.; Turner, T.; Stolar, N.M.; Bilker, W.B.; Brensinger, C.M.; Gur, R.E.; Gur, R.C. Differences in facial expressions of four universal emotions. *Psychiatry Res.* **2004**, *128*, 235–244. [CrossRef] [PubMed]
12. Anolli, L.; Mantovani, F.; Confalonieri, L.; Ascolese, A.; Peveri, L. Emotions in serious games: From experience to assessment. *Int. J. Emerg. Technol. Learn.* **2010**, *5*, 7–16. [CrossRef]
13. Freitas-Magalhães, A. *Facial Action Coding System 2.0: Manual of Scientific Codification of the Human Face*; FEELab Science Books: Porto, Portugal, 2017.
14. Ekman, P.; Friesen, W.V. *Facial Action Coding System: A Technique for the Measurement of Facial Movement*; Consulting Psychologists Press: Palo Alto, CA, USA, 1978.
15. Ekman, P.; Friesen, W.V.; Hager, J.C. *Facial Action Coding System: The Manual*; Research Nexus: Salt Lake City, UT, USA, 2002.
16. Costa, N.; Domingues, P.; Fdez-Riverola, F.; Pereira, A. A mobile virtual butler to bridge the gap between users and ambient assisted living: A smart home case study. *Sensors* **2014**, *14*, 14302–14329. [CrossRef] [PubMed]
17. Huang, Y.; Yao, H.; Zhao, S.; Zhang, Y. Towards more efficient and flexible face image deblurring using robust salient face landmark detection. *Multimedia Tools Appl.* **2017**, *76*, 123–142. [CrossRef]
18. Apple. iPhone X. Available online: https://www.apple.com/iphone-x/ (accessed on 25 December 2017).
19. Kinect Desenvolvimento de Aplicações do Windows. Available online: https://developer.microsoft.com/pt-pt/windows/kinect (accessed on 25 December 2017).
20. Correia, V.H.; Rodrigues, N.; Gonçalves, A. Reconstructing Roman Archaeological Sites: Theory and Practice—The Case of Conimbriga. *Open J. Soc. Sci.* **2016**, *4*, 122–132. [CrossRef]
21. Ferreira, C.; Rodrigues, N.; Gonçalves, A.; Hipólito-Correia, V. Reconstruindo Conimbriga—Medianum Absidado Digital. In Proceedings of the Interação 2013 5ª Conferência Nacional sobre Interação, Vila Real, Portugal, 7–8 November 2013.
22. Hipólito-Correia, V.; Gonçalves, A.; Rodrigues, N.; Ferreira, C. Reconstructing Conimbriga: Collaborative work between archaeology, architecture and computer graphics. In Proceedings of the 19th EAA—European Association of Archaeologists, Pilsen, Czech Republic, 4–8 September 2013.

23. Pandzic, I.S.; Forchheimer, R. *MPEG-4 Facial Animation: The Standard, Implementation and Applications*; John Wiley & Sons, Ltd.: Hoboken, NJ, USA, 2002; Volume 13.

24. Video. Available online: https://mpeg.chiariglione.org/standards/mpeg-4/video (accessed on 27 December 2017).

25. Tao, H.; Chen, H.H.; Wu, W.; Huang, T.S. Compression of MPEG-4 facial animation parameters for transmission of talking heads. *IEEE Trans. Circuits Syst. Video Technol.* **1999**, *9*, 264–276. [CrossRef]

26. Liew, A.W.C. *Visual Speech Recognition: Lip Segmentation and Mapping: Lip Segmentation and Mapping*; IGI Global: Hershey, PA, USA, 2009.

27. Aleksic, P.S.; Katsaggelos, A.K. Audio-visual biometrics. *Proc. IEEE* **2006**, *94*, 2025–2044. [CrossRef]

28. Petajan, E. MPEG-4 face and body animation coding applied to HCI. In *Real-Time Vision for Human-Computer Interaction*; Springer Science & Business Media: Berlin, Germany, 2005; pp. 249–268.

29. Bosse, T.; Gerritsen, C.; de Man, J.; Treur, J. Learning emotion regulation strategies: A cognitive agent model. In Proceedings of the 2013 IEEE/WIC/ACM International Conference on Intelligent Agent Technology (IAT), Atlanta, GA, USA, 17–20 November 2013; Volume 2, pp. 245–252.

30. Hudlicka, E. From Habits to Standards: Towards Systematic Design of Emotion Models and Affective Architectures. In *Emotion Modeling*; Lecture Notes in Computer Science; Springer: Cham, Switzerland, 2014; Volume 8750, pp. 3–23.

31. Manzoor, A.; Treur, J. An agent-based model for integrated emotion regulation and contagion in socially affected decision making. *Biol. Inspired Cogn. Archit.* **2015**, *12*, 105–120. [CrossRef]

32. Marsella, S.; Gratch, J.; Petta, P. Computational models of emotion. In *A Blueprint for an Affectively Competent Agent: Cross-Fertilization between Emotion Psychology, Affective Neuroscience, and Affective Computing*; Oxford University Press: Oxford, UK, 2010; pp. 21–41.

33. Mehrabian, A. Framework for a comprehensive description and measurement of emotional states. *Genet. Soc. Gen. Psychol. Monogr.* **1995**, *121*, 339–361. [PubMed]

34. Treur, J. Displaying and Regulating Different Social Response Patterns: A Computational Agent Model. *Cognit. Comput.* **2014**, *6*, 182–199. [CrossRef]

35. Jeon, M. Emotions and Affect in Human Factors and Human–Computer Interaction: Taxonomy, Theories, Approaches, and Methods. In *Emotions and Affect in Human Factors and Human-Computer Interaction*; Academic Press: Cambridge, MA, USA, 2017; pp. 3–26.

36. Hanin, Y.L. Emotions in Sport: Current Issues and Perspectives. In *Handbook of Sport Psychology*, 3rd ed.; John Wiley & Sons, Inc.: Hoboken, NJ, USA, 2012; pp. 31–58.

37. Bosse, T.; Pontier, M.; Siddiqui, G.F.; Treur, J. Incorporating Emotion Regulation into Virtual Stories. In Proceedings of the 7th International Workshop on Intelligent Virtual Agents (IVA 2007), France, Paris, 17–19 September 2007; pp. 339–347.

38. Bosse, T.; Pontier, M.; Treur, J. A computational model based on Gross' emotion regulation theory. *Cogn. Syst. Res.* **2010**, *11*, 211–230. [CrossRef]

39. Campbell-Sills, L.; Barlow, D.H. Incorporating emotion regulation into conceptualizations and treatments of anxiety and mood disorders. In *Handbook of Emotion Regulation*; Guilford Press: New York, NY, USA, 2007; pp. 542–559.

40. Pereira, A.; Felisberto, F.; Maduro, L.; Felgueiras, M. Fall Detection on Ambient Assisted Living using a Wireless Sensor Network. *ADCAIJ Adv. Distrib. Comput. Artif. Intell. J.* **2012**, *1*, 62–77.

41. Magaña-Espinoza, P.; Aquino-Santos, R.; Cárdenas-Benítez, N.; Aguilar-Velasco, J.; Buenrostro-Segura, C.; Edwards-Block, A.; Medina-Cass, A. WiSPH: A wireless sensor network-based home care monitoring system. *Sensors* **2014**, *14*, 7096–7119. [CrossRef] [PubMed]

42. Palumbo, F.; Ullberg, J.; Stimec, A.; Furfari, F.; Karlsson, L.; Coradeschi, S. Sensor network infrastructure for a home care monitoring system. *Sensors* **2014**, *14*, 3833–3860. [CrossRef] [PubMed]

43. Hanke, S.; Mayer, C.; Hoeftberger, O.; Boos, H.; Wichert, R.; Tazari, M.; Wolf, P.; Furfari, F. universAAL—An Open and Consolidated AAL Platform. In *Ambient Assited Living*; Springer: Berlin/Heidelberg, Germany, 2011; pp. 127–140.

44. Jonsson, P.; Carson, S.; Sethi, J.S.; Arvedson, M.; Svenningsson, R.; Lindberg, P.; Öhman, K.; Hedlund, P. *Ericsson Mobility Report*; Niklas Heuveldop: Stockholm, Sweden, 2017.

45. Lawton, M.P.; Brody, E.M. Assessment of Older People: Self-Maintaining and Instrumental Activities of Daily Living. *Gerontologist* **1969**, *9*, 179–186. [CrossRef] [PubMed]

46. Katz, S. Assessing self-maintenance: Activities of daily living, mobility, and instrumental activities of daily living. *J. Am. Geriatr. Soc.* **1983**, *31*, 721–727. [CrossRef] [PubMed]

47. Cope, E.M.; Sandys, J.E. *Aristotle: Rhetoric*; Online Resource (356 TS-WorldCat T4-Volume); Cambridge Library: Cambridge, UK, 1877; Volume 2, p. 1.

48. Darwin, C.R. *The Expression of the Emotions in Man and Animals*, 1st ed.; John Murray: London, UK, 1872. Available online: http://darwin-online.org.uk/content/frameset?itemID=F1142&viewtype=text&pageseq=1 (accessed on 6 December 2017).

49. Plutchik, R. A General Psychoevolutionary Theory of Emotion. In *Theories of Emotion*; Academic Press: Cambridge, MA, USA, 1980; pp. 3–33.

50. Verduyn, P.; Lavrijsen, S. Which emotions last longest and why: The role of event importance and rumination. *Motiv. Emot.* **2015**, *39*, 119–127. [CrossRef]

51. Deniz, O.; Castrillon, M.; Lorenzo, J.; Anton, L.; Bueno, G. Smile detection for user interfaces. In *Advances in Visual Computing*; Lecture Notes in Computer Science; Springer: Berlin/Heidelberg, Germany, 2008; Volume 5359, pp. 602–611.

52. Murthy, V.; Sankar, T.V.; Padmavarenya, C.; Pavankumar, B.; Sindhu, K. Smile Detection for User Interfaces. *Int. J. Res. Electron. Commun. Technol.* **2014**, *2*, 21–26.

53. Whitehill, J.; Littlewort, G.; Fasel, I.; Bartlett, M.; Movellan, J. Toward practical smile detection. *IEEE Trans. Pattern Anal. Mach. Intell.* **2009**, *31*, 2106–2111. [CrossRef] [PubMed]

54. Shan, C. Smile detection by boosting pixel differences. *IEEE Trans. Image Process.* **2012**, *21*, 431–436. [CrossRef] [PubMed]

55. Li, P.; Phung, S.L.; Bouzerdom, A.; Tivive, F.H.C. Automatic recognition of smiling and neutral facial expressions. In Proceedings of the 2010 Digital Image Computing: Techniques and Applications (DICTA 2010), Sydney, NSW, Australia, 1–3 December 2010; pp. 581–586.

56. Brits Smile 11 Times Every Day and 232,000 Times in Their Lifetime. Available online: https://www.thesun.co.uk/news/4670575/brits-smile-11-times-every-day-and-232000-times-in-their-lifetime/ (accessed on 28 December 2017).

57. Express.co.uk. Life. Life & Style. REVEALED: The Top 50 Things Most Likely to Make YOU Smile. Available online: https://www.express.co.uk/life-style/life/718333/Top-50-things-make-you-smile (accessed on 28 December 2017).

58. 40 Things That Will Make You Smile. Available online: https://www.theodysseyonline.com/40-things-that-will-make-you-smile?utm_expid=.oW2L-b3SQF-m5a-dPEU77g.0&utm_referrer= (accessed on 7 December 2017).

59. Vacher, M.; Caffiau, S.; Portet, F.; Meillon, B.; Roux, C.; Elias, E.; Lecouteux, B.; Chahua, P. Evaluation of a context-aware voice interface for Ambient Assisted Living: Qualitative user study vs. quantitative system evaluation. *ACM Trans. Access. Comput.* **2015**, *7*. [CrossRef]

60. Wilson, C.; Hargreaves, T.; Hauxwell-Baldwin, R. Benefits and risks of smart home technologies. *Energy Policy* **2017**, *103*, 72–83. [CrossRef]

61. Peek, S.T.M.; Wouters, E.J.M.; van Hoof, J.; Luijkx, K.G.; Boeije, H.R.; Vrijhoef, H.J.M. Factors influencing acceptance of technology for aging in place: A systematic review. *Int. J. Med. Inform.* **2014**, *83*, 235–248. [CrossRef] [PubMed]

62. Vacher, M.; Portet, F.; Fleury, A.; Noury, N. Development of Audio Sensing Technology for Ambient Assisted Living: Applications and Challenges. *Int. J. E-Health Med. Commun.* **2011**, *2*, 35–54. [CrossRef]

63. Dewsbury, G.; Dewsbury, G.; Qb, A. The social and psychological aspects of smart home technology within the care sector. *New Technol. Hum. Serv.* **2001**, *14*, 9–17.

applied sciences

MDPI

Article

Large Scale Community Detection Using a Small World Model

Ranjan Kumar Behera [1,*,†], Santanu Kumar Rath [1], Sanjay Misra [2,3], Robertas Damaševičius [4,*] and Rytis Maskeliūnas [4]

[1] Department of Computer Science and Engineering, National Institute of Technology, Rourkela 769008, India; skrath235@gmail.com
[2] Department of Computer Engineering, Atilim University, Incek, Ankara 06836, Turkey; sanjay.misra@covenantuniversity.edu.ng
[3] Department of Electrical and Information Engineering, Covenant University, Ota 1023, Nigeria
[4] Department of Multimedia Engineering, Kaunas University of Technology, Kaunas 51368, Lithuania; rytis.maskeliunas@ktu.lt
*   Correspondence: jranjanb.19@gmail.com (R.K.B.); robertas.damasevicius@ktu.lt (R.D.); Tel.: +91-943-959-2352 (R.K.B.)
†   Current address: Department of Computer Science and Engineering, NIT Rourkela, Rourkela 769008, Odisha, India.

Received: 27 September 2017; Accepted: 2 November 2017; Published: 15 November 2017

Abstract: In a social network, small or large communities within the network play a major role in deciding the functionalities of the network. Despite of diverse definitions, communities in the network may be defined as the group of nodes that are more densely connected as compared to nodes outside the group. Revealing such hidden communities is one of the challenging research problems. A real world social network follows small world phenomena, which indicates that any two social entities can be reachable in a small number of steps. In this paper, nodes are mapped into communities based on the random walk in the network. However, uncovering communities in large-scale networks is a challenging task due to its unprecedented growth in the size of social networks. A good number of community detection algorithms based on random walk exist in literature. In addition, when large-scale social networks are being considered, these algorithms are observed to take considerably longer time. In this work, with an objective to improve the efficiency of algorithms, parallel programming framework like Map-Reduce has been considered for uncovering the hidden communities in social network. The proposed approach has been compared with some standard existing community detection algorithms for both synthetic and real-world datasets in order to examine its performance, and it is observed that the proposed algorithm is more efficient than the existing ones.

Keywords: small world network; six degrees of separation; map reduce; community detection; modularity; normalize mutual information

1. Introduction

In a real world, various categories of networks play different roles in the society for different purposes viz. social networks, which represents social interactions among human beings in society, citation networks that represent the articles of various authors published in the particular field and their associated citations in other papers, technological networks that represent the distribution of resources, biological networks that represent protein–protein interaction in the network, etc. Social networks are considered as having interesting research domains due to their characteristics of involving human social activities. Network evolution [1], network modeling [2], centrality analysis [3], information diffusion [4], link prediction [5], and community detection [6] are some of the interesting research directions in social networks. Power-law degree distributions [7], small world networks [8], and community structures are some of the important properties observed in the social network.

Communities are found to be one of the most important features of large-scale social networks. Uncovering such hidden features enables the analysts to explore the functionalities in the social network. There exists quite a good number of definitions of community depending on the contexts pertaining to different applications. However, as per few number of commonly accepted definitions, they are considered to be a group of nodes that have a dense connection among themselves as compared to sparsity outside the group. Communities in the social network represent a group of people who share common ideas and knowledge in the network. Hidden communities can be explored through learning from social dynamics in the network [9]. Their identification helps in getting insight into social and functional behavior in the social network. However, due to unprecedented growth in the size of social networks, it is quite a hard task to discover subgroups in the network within a specified time limit. Real-world social networks are observed to follow the power-law in both degree-distribution and community size distribution [7].

A distributed framework like Hadoop may be considered as a better alternative for processing a large volume of data in complex and heterogeneous social networks. Hadoop internally uses a Map-Reduce algorithm for processing computation in multiple nodes in a cluster. It uses a dedicated file system known as Hadoop Distributed File System (HDFS) for storing data across multiple nodes in the cluster. A network is said to have small-world properties if the geodesic distance between any two nodes is small. In a small world network, for a fixed average degree, the average path length between pairs of the node in the network increases logarithmically with the increase in number of nodes or, in other words, small world network exhibits pure exponential growth with respect to walk-length in the network [10]. These inherent properties of real-world networks make it difficult for graph mining. There exist a plethora of community detection algorithms in the literature, where most of them emphasize maximizing the quality parameter in order to detect communities in a large-scale network. Sometimes, they are insignificant in exploring communities in a reasonable amount of time, due to the resolution limit of modularity [11]. The community detection algorithm is said to be faster and efficient, only if it follows a small world network phenomenon. The small world network concept is based on the six degrees of separation principle [12].

In this study, the Map-Reduce approach has been used to uncover the hidden communities in a large-scale network. Map-Reduce algorithms always follow two crucial phases: one is mapper and another one is the reducer. In this work, mapper phase has been used in mapping the nodes to their corresponding communities. An effort has been made to discover the communities using a small world model. In reducer phase, nodes are being clustered based on their walk length and similarity index (η) with the source node. Random walk based similarity index is introduced to measure the strength of social ties.

The subsequent sections of this paper is organized as follows: in Section 3, some preliminaries about community structure, small world network, power-law degree distribution and clustering coefficient has been discussed. Section 4 presented the random walk process in the network. In this section, a new similarity index has been devised based on a random walk in the network. The proposed methodology has been presented in Section 5. Community detection and clustering phase have been discussed in this section. Section 6 presented the implementation part of the work. Experimental results have been discussed in Section 7. Section 8 presents the possible threat to validation of the work. Conclusions and future work have been discussed in Section 9.

2. Related Work

Community detection is similar to a graph partitioning problem. Most of the graph partition methods are based on optimizing a quality function. Girvan and Newman have proposed the first community detection algorithm, which is based on a hierarchical partitioning problem in a graph.

In their work, modularity has been chosen as the objective function for accessing the quality of obtained partition [13]. In this algorithm, edges are removed iteratively in the order of their edge-betweenness value until it reaches the maximum modularity. Edge-betweenness value of an edge can be expressed as the number of shortest paths between a pair of nodes that passes through the edge. Modularity taken in this paper does not consider the information about unrelated pairs inside the network. A new modularity known as MIN-MAX modularity has been devised by R. Behera and M. Jena to optimize the community partitions in the paper [14]. It not only considers dense connections within the group, but it also gives the penalty to unrelated pairs within the group.

Random graph is a kind of graph where edges are distributed randomly among the nodes, but, unfortunately, it does not resemble a real-world network. A random graph generation model was proposed by Watts and Stogatz, which helps in generating the random graph with small world properties like average short path length and high clustering coefficient [15]. It has been observed that a small world network lies in between regular and random graphs. Communities are frequently observed in small world networks.

Ego network is the group of nodes consisting of a central actor and other nodes that are directly connected to it. They resemble with the properties of the small-world network. The central node in the ego network may have highest influential ability as compared to other nodes in the network. Exploring focal nodes may helps in modeling influence propagation in the network [16]. The social network allows users to make a group based on the common interest or common event happening in their social life. However, automatically, group construction is a difficult task when friends are added or removed dynamically in their social life. Authors McAuley and Leskovec have proposed an efficient model that enables detecting ego circles in large-scale networks that capture both structural and user profile information [17]. In this paper, the author has developed a model where a user can belong to different ego circles. This model allows the user to detect an overlapping community as well as hierarchically nested circles in a large-scale social network. Social circles in ego networks have a great impact on the evolution of the network.

Three fundamental network models come into the picture while discussing structural parameters of the network. The first one is the random network where nodes and edge distribution is random over the network. Degree distribution in the network follows binomial or Poisson distribution in a random network [18]. It is similar to the homogeneous network where most of the vertices are having the same degree. The small world model is another kind of network model that lies in between random and lattice network models. It exhibits a high clustering coefficient like a lattice network and smaller average path length like the random network. Degree distribution of a small world network follows the binomial distribution. A real-world network follows the power-law degree distribution that resembles the scale-free network model. A small world network model may have the scale-free distribution like a real-world social network. Chopade and Zhan have discussed the structural and functional characteristics for community detection process in the complex social network in their paper [19]. Community detection based on the structural parameter of the network topology has attracted an interest of research as compared to community detection based on the functional parameter of the network.

Several methods for community detection techniques have been developed and each has its own strength and weakness [6,13,20,21]. An efficient community detection method that considers both local and global information about topological structure has been explained by De Meo et al. [22]. Global information about the network topology helps to yield good results about community; however, it is not suitable for the large-scale complex network. Large scale network needs to be preprocessed through dimension scaling in order to map the global information to local one [23]. Local information about network topology may lead to faster community detection but are less accurate in nature. In this paper, the community detection process is based on optimizing the modularity value based on global information about the structure and yet is able to compare scalability of the network with local methods. In this work, communities have been detected in two phases. In the first phase, walk length for each node from a source node has been detected using an information propagation model, which is based on the random walk in the network. In the second phase, Euclidean distance between the nodes has been used for clustering process to partition the network.

Steve Gregory proposed label propagation algorithm for community detection in linear time complexity [24]. The main idea behind the algorithm is that a node is more likely be a part of that community, to which its maximum neighboring nodes belong. Labeling of a node is propagated through its neighboring nodes in multiple iterations until a label is confined to a group of nodes. It is the fastest available community detection method, which has been claimed to have linear time complexity. The community detection algorithm spends most of the time measuring the similarity values between a pair of nodes, especially in the case of unweighted graphs.

Community detection using random walk has been discussed by Pons et al. [25]. The algorithm discussed in this paper is well known as the Walktrap algorithm. The intuition behind the Walktrap algorithm is that a walker more likely gets trapped inside the dense region if it moves randomly inside the network. In this paper, the author has made an effort in discovering clusters by observing the movement of the walker inside the network. The time complexity of Walktrap algorithm is found to be $O(mn^2)$ in the worst case, where m is the number of edges and n is the number of nodes in the network. In this work, similarity between nodes has been calculated based on the random walk in the network.

Spin-Glass is a unique community detection algorithm that is based on the statistical mechanics of spin around the network [26]. The expected number of communities has been overestimated in the community detection and It has the worse approximation for the community when complexity and size of the network increases. However, it works fine for a small world model but too much expensive. Similarity between objects is determined by the spin associated with objects in graph configuration. Similarity between nodes is higher if their spins are of the same order. Communities are detected based on Pott's spin-glass model.

In the literature, most of the community detection algorithms deal with the undirected network. However, the real-world complex network often resembles the directed graph. Agreste et al. have made an extensive comparison of community detection algorithms for the directed network [27]. Infomap and Label propagation algorithms are the first ones to implement in the directed network. However, for the sake of simplicity, we have implemented these two algorithms on the undirected network. Peng and Lill have proposed a framework for mapping the community detection algorithm from undirected to the directed network. They have applied modularity optimization technique for obtaining optimal partitioning of the network [28].

Xiaolong et al. have proposed an optimized community detection algorithm, which is based on the vector influence clustering coefficient and directed information transfer gain of vertices in the network. They have implemented their algorithm on the directed network. In their work, they have also proposed an efficient optimization parameter (target optimal function) to evaluate the community partition in the network [29].

Rosvall and Bergstrom have developed an elegant community detection algorithm for discovering modules in a large-scale network, which is known as Infomap [30]. It is based on an optimizing a map function. This algorithm is similar to the Louvain algorithm, where, initially, each node in the network is assigned to a module, and, in each iteration, nodes in the modules are migrated themselves into the nearest module in order to minimize the map function. The proposed algorithm is quite similar to this algorithm where nodes are migrated into the modules based on the detected walk length in which nodes are discovered. Nodes belong to the same module are forced to migrate into another module to optimize the map function at the time of rebuilding the network structure. As a result, nodes assigned to one module at one point may differ from the assigned module at a later point in time. The Infomap algorithm is well suited for a small network. Its accuracy is found to be best as compared to other standard community detection algorithms [31].

Since the real world network follows the power-law degree distribution and quite complex in nature, traditional algorithms are practically inefficient unless it is implemented in some parallel architecture. The proposed work is similar to random walk community detection algorithm proposed by Pons et al. [25]. However, we have considered the concept of a small world model to evaluate the similarity between vertices in the network. Unlike the Walktrap algorithm, which processes the whole network for quantifying the similarity between the vertices, the proposed similarity measure has been calculated in less number of steps in a recursive manner that improves the performance of community detection. The proposed algorithm is further improved by implementing it in a Hadoop distributed platform. The proposed algorithm may surpass previously discussed community detection algorithms in terms of accuracy, as most of them are based on either a regular graph model or a random graph model [15]. Both of these models have less resemblance with the real-world network. However, a small world model closely resembles the real-world network.

The proposed algorithm behaves in a more consistent manner as compared to Infomap and the Spin-Glass community detection algorithm. The Infomap algorithm is based on the information flow in the network, which is calculated through random walk probability in the network. Huffman coding is used to generate the two-level encoding schema for the network. Community partitioning is identified by optimizing the mapping function that tries to compress the encoded schema by simulated annealing. This approach seems to be unrealistic and inconsistent for a large-scale network. The optimizing criteria in the Spin-Glass algorithm are similar to the Infomap algorithm. The proposed algorithm outperforms these two algorithms due to its efficient optimizing criteria for community evaluation.

3. Preliminaries

3.1. Structural Definition of Community

Social network may be represented in the form of a graph $G = (V, E)$, where V is the set of nodes and E is the set of edges in graph G. A group of nodes $c \in V$ is said to form a community if it satisfies the following condition:

$$\frac{2e_{in}}{n_c(n_c - 1)} > \frac{2E}{V(V - 1)} > \frac{e_{out}}{n_c(V - n_c)}, \tag{1}$$

where e_{in}, e_{out} are the number of edges existing inside the community and number of edges existing from a node in the community to a node outside the community respectively. n_c is the number of nodes within the community c. First and third part of Equation (2) corresponds to the fraction of number of links within the communities and between the communities in the graph, respectively. The middle part corresponds to the density of the graph. For a graph with community structure, a fraction of intra-community links are expected to be larger than graph density and graph density is expected to be larger than a fraction of inter-community links in the graph.

3.2. Small World Phenomenon

Small world phenomenon is one of the inherent principles behind the analysis of today's large-scale social network, which indicates that any two people in the network can be linked by a small number of acquaintances [10]. It may be observed that there always exists a path of short length that can be discovered using local information. A small world network is often associated with a high clustering coefficient and its characteristic path length decreases more rapidly than the clustering coefficient as the randomness increases.

Definition 1. *A graph is said to be a small world network if average path length L_{avg} is less than or equal to the path length L_{rand} and average clustering coefficient C_{avg} is strictly less than clustering coefficient C_{rand} in a random degree distribution of the graph:*

$$L_{avg} \preceq L_{rand} \text{ and } C_{avg} \prec C_{rand}. \tag{2}$$

This phenomenon is based on the six degrees of separation principle.

Definition 2. *Six degrees of separation is the principle, which indicates that every two people in the world is connected with a chain of no more than six acquaintances.*

Small world network exhibits the following important characteristics:

- Short Average Path Length;
- High Clustering Coefficient;
- Exhaustive search using local information.

3.3. Power Law Degree Distribution

The degree of a user in a social network is the number of relationships that the user maintains. Distribution of relationships among users is known as degree distribution. A network is said to be scale-free if it follows power-law degree distribution [12].

Definition 3. *A network is said to have power-law degree distribution if a fraction of nodes having degree k in the network depends on the power of k with some constant. Social network often follows power-law degree distribution as indicated by the following equation:*

$$p_d = kd^{-\gamma}, \quad 2 \le \gamma \le 3, \tag{3}$$

where p_d is the fraction of nodes having degree d, k and γ are power-law intercept and power-law exponent, respectively. Usually, γ ranges from 2 to 3.

3.4. Clustering Coefficient

Clustering coefficient is a measure used to define the network as the small world network. Clustering coefficient defines the friendship transitivity in a network [32]. This measure can have two versions: one is local and another one is global. Local clustering coefficient is associated with each node in the network where global clustering coefficient represents the clustering density for the whole network.

3.4.1. Global Clustering Coefficient (GCC)

Small world network is often observed to have high global clustering coefficient. GCC of the network is defined as the ratio of a number of triangles and the possible number of connected triplets in the network. A triplet is the set of three connected nodes in the network. Each triangle in a network contributes to three triplets. GCC of a network may be framed as below:

$$GCC = \frac{3 \times Number\ of\ triangle\ in\ the\ network}{Number\ of\ triplets\ in\ the\ network}. \tag{4}$$

3.4.2. Local Clustering Coefficient (LCC)

LCC signifies the tendency of a node to form a cluster. The higher the clustering coefficient of a node, the more chances to be involved in a cluster. It is defined as the fraction of edges existing between the neighboring nodes to the total number of edges possible. LCC for a node i can mathematically be represented as follows:

$$LCC_i = \frac{2\,|(u,v):(u,v) \in E\ and\ u,v \in Neighbor(i)|}{n_i(n_i-1)}. \tag{5}$$

Here, (u, v) is the edge between neighboring nodes, E is the set of edges in the network, Neighbor (i) is the set of neighboring nodes of i and n_i is the number of elements in the set. The average of local clustering coefficients for all of the nodes may be considered as the global clustering coefficient.

4. Random Walk

Community detection in a large complex network can be carried out by capturing the topological structure using random walk in the network. The intuition behind random walk is that the network tends to be trapped inside a denser region (community) for a longer period of time. This idea can be used for inclusion of nodes in the community. In this paper, an efficient similarity metric based on the random walk has been proposed to include a node in the community. This metric may provide the following features:

- Structure of the network is well captured in the process of community detection;
- It can be used in an agglomerative hierarchal clustering;
- Computation to find community may be more efficient.

Social network can be represented in the form of a graph, where nodes represent social entities and edges represent relationships between the entities. The graph can be stored in the form of adjacency matrix A, where $A_{ij} = 1$, if there is an edge, existing between nodes i and j, and 0 otherwise.

Random walk in a graph is the process of visiting a neighboring node randomly from the source node and continuing the process of visiting throughout the graph. Random walk process is well explained on the basis of Markov chain in which each node corresponds to vertices in the visited path [33]. In this paper, transition matrix obtained by Random walk in the graph has been considered for detection of communities. Transition matrix describes the probability of visiting each node from every other node in k number of steps i.e., $T_{i,j}^k$ corresponds to the probability of visiting node j from i in k number of steps. T^1, T^2, T^3 and T^k are the transition matrices for random walk corresponding to 1, 2, 3 and k walk length, respectively. Transition probability from vertex i to vertex j in one length random walk is defined by the following equation:

$$T_{ij}^1 = \frac{A_{ij}}{d_i},$$

(6)

where A_{ij} is the adjacency matrix of the network and d_i is the degree of vertex i.

Property 1. *Probability for a random walker to visit a node j from node i in walk length that tends to infinity depends only on the degree of j, rather than degree of the source or the intermediate node. It may be represented as:*

$$\lim_{k \to \infty} T_{i,j}^k = \frac{d_j}{2E},$$

(7)

where d_i is the degree of i and E is the total number of edges in the network.

Property 2. *Ratio of probabilities for a random walker that visits a node from i to j and j to i through a fixed walk length depends only on the degree of i and j. It may be represented as follows:*

$$\frac{T_{i,j}^k}{T_{j,i}^k} = \frac{d_j}{d_i}.$$

(8)

Vertex Similarity Based on Random Walk

Vertices belonging to the same community seems to have similar behavior as compared to vertices outside the communities. Any two nodes inside a community look the same way as other nodes in the network. We may consider an example of a random walk of length k in graph $G(V, E)$, which represents a social network. The probability of visiting all nodes from all other nodes in the network through k length random walk is represented by transition matrix T^k. Each tuple in the transition matrix corresponds to probabilities of visiting all other nodes from node i in k walk length. These probabilities are based on structural information in the network. From the structure of the network, the following inferences may be drawn:

- If two nodes i and j, are in the same community, the probability of visiting node j from i would be higher as compared to visiting a node outside the community. In addition, converse may not be true i.e., if the probability is high, it does not mean that they belong to the same community.
- The probability $T_{i,j}^k$ depends on the degree of j because the walker tends to visit towards vertices, where the degree is high.
- Two vertices belonging to the same community tend to see all other vertices in the same way:

$$\text{i.e.,} \quad T_{i,m}^k \approx T_{j,m}^k \ , \quad \forall i,j \in \ \text{same community} \quad \text{and} \quad m \in [1,n]. \tag{9}$$

In this paper, similarity between two vertices are identified from the transition matrix $T_{i,j}^k$ based on the walk length k. Probability of reaching one node from another would be different for different walk lengths. Similarity between i and j for k walk length can be computed by the Euclidean distance between row vectors corresponding to nodes i and j, in matrix T^k:

$$Sim(i,j) = \sqrt{\sum_{l=i}^{n} \frac{(T_{i,l}^k - T_{j,l}^k)^2}{d_l}}, \tag{10}$$

where d_l is the degree of vertex 1.

5. Proposed Methodology

Detecting communities in a social network having a good number of nodes in a reasonable amount of time is a challenging task due to its size and structure. In this work, distributing computing environment has been considered for processing large-scale networks. Hadoop is a framework applied to solving the complex problem by distributing the computation in multiple nodes in the cluster. Hadoop implicitly schedules the pieces of task on the different computing node. It automatically takes care of load balancing and resource scheduling over the cluster of nodes. Users need not worry about the internal execution policy. However, users can provide the application specific mapper and reducer program to the Hadoop framework. Users have the control over the Map-Reduce program structure but does not have control over execution environment. In this work, the community detection process in Hadoop has been carried out in the following two phases:

- mapper phase,
- reducer phase.

Prior to the computation in Hadoop, the random walk process has been carried out in order to visit all the nodes in the network. In this work, similarity values between all pair of nodes have first been calculated from the transition probability matrix. Transition matrix has been generated separately for random walk lengths 1 to 3.

In the mapper phase, community memberships of nodes are being evaluated by using the local information available at each node. In a Small world network model, quick navigation is possible through two fundamental processes: one is to find the short chain of acquittance and the other is to use only local information regarding structure around the node. The community detection process starts from a randomly selected source node. Neighboring nodes who have similarity value less than the threshold value, η, are identified and then included in the same community, to which the source node belongs. Here, η is treated as the clustering parameter for a small world model that determines the probability of having a connection between two nodes in the network. Small world model parameters have been determined based on the random walk in the network. Once a node has been mapped with a community number, the node is temporarily deleted from the graph in order to reduce the computation. For the sake of simplicity, the undirected graph has been taken into consideration. It is observed that a real world social network follows a small world network, or, in other words, by following a small number of steps, one can visit all other nodes in the network. Mapper phase nodes are mapped into their community in the order of their reachability from the source node. Steps adopted in the mapper phase are shown in Algorithm 1. The execution flow for community detection is presented in Figure 1.

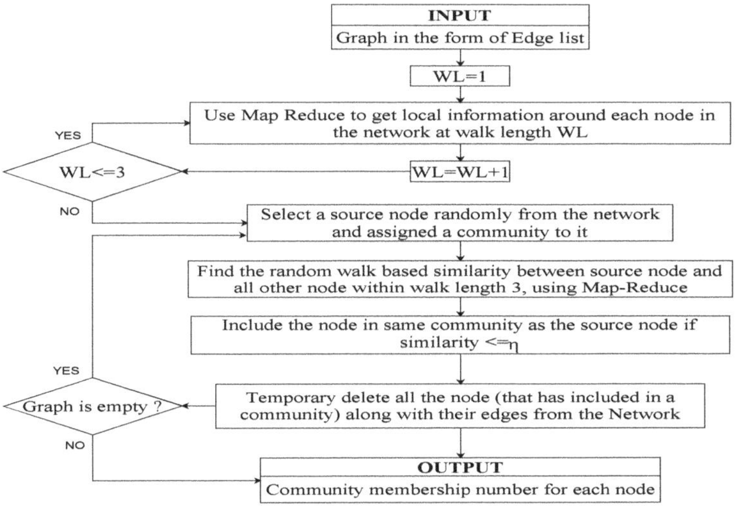

Figure 1. Execution flow for community detection using small world phenomenon.

Algorithm 1: Community detection using small world phenomenon (CDSW).

Mapper (Source Node, Community Membership, Graph G)

Input : The social network graph in the form of edge lists.
Output : Community Membership for each node in the form of (Key,Value) pair.

Here Community Membership is the Key and (Walk Length (WL), Node ID) is the Value.
Initialization :
Select a node randomly as the Source Node.
C,append (Source Node)
$C' = $ NULL
WL=0, Community Membership=1
Emit(Community Membership, WL, Source Node, NULL)
for WL=1 to 3
 for all v in C
  for each neighbor $\notin C'$ and Sim(Node v, Neighbor Node) $\leq \eta$
   Emit (Community Membership, WL, Neighbour Node)
  endfor
  Delete v from both Graph and C
 endfor
 $C = C'$
 Delete the nodes in C', because they are already visited.
end for
Community Membership++
Source Node = Random source node for next community
if graph is empty
  exit()
else
  Mapper(Source Node, Community Membership, Graph)
endif

In the reducer phase, the clustering process is being carried out by developing a dendrogram structure for the network. Dendogram structure is the hierarchical structure from where one can identify communities at a certain level of granularity. If the structure is partitioned at a certain level, the groups formed below the partition can be considered as individual communities. The input to the reducer phase is in the form of the key–value pair where community number detected in a mapper phase is treated as the key and the rest of the parameters such as walk length, node id, and a number of detected community together is treated as the value for the corresponding key. Dendogram structure is detected at two levels. One is at the inner level of each community and the other at a global level where each detected community is treated as a node structure. At the inner hierarchy, all of the nodes are detected in the same walk length in mapper phase belong to the same level in the dendrogram structure. Nodes that are assigned in the same community at a particular walk length are grouped into a community. To develop the dendogram structure at higher level hierarchy, we have proposed an efficient distance measure between each pair of communities that depend on the dissimilarity between the communities. It is calculated by measuring maximum dissimilarity that may be possible between any two nodes of different communities. It can be mathematically represented as follows:

$$distance(C_i, C_j) = max\{sim(i,j)\} \quad textwhere \ \forall i \in C_i, \forall j \in C_j \ and \ i \neq j, \tag{11}$$

where C_i and C_j are the two different communities and i and j are the two nodes belonging to C_i and C_j, respectively. Here, $sim(i,j)$ has been already calculated from Equation (10). The pair of communities is first identified with minimum distance for grouping at the same level in the dendogram structure. After the grouping, it is treated as a single community. The next pair of communities is then identified by following the same procedure. The process is continued until all of the same communities are involved in the structure. The steps followed for obtaining the dendogram structure is presented in Algorithm 2. The clustering process of detected communities is presented in Figure 2.

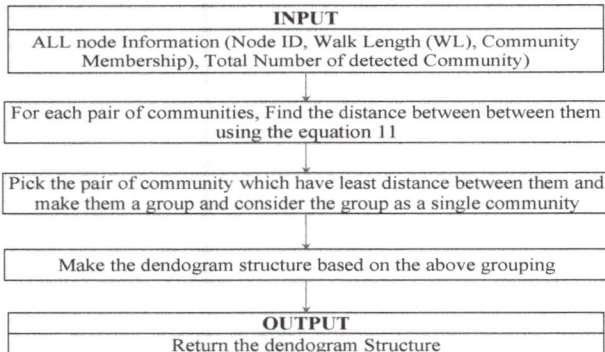

INPUT
ALL node Information (Node ID, Walk Length (WL), Community Membership), Total Number of detected Community)

For each pair of communities, Find the distance between between them using the equation 11

Pick the pair of community which have least distance between them and make them a group and consider the group as a single community

Make the dendogram structure based on the above grouping

OUTPUT
Return the dendogram Structure

Figure 2. The process of developing dendogram structure to identify communities.

Algorithm 2: Clustering of detected communities.

Reducer(Community_Number, Walk_Length, Node_Id, Parent_Id)

Input : Community ids for each node in the form of (Key,Value) pair.
Here Community_Number is the Key and (WL, Node_Id, Parent_Node) is the Value.
Output : Community Structure in the form of Dendrogram.
1. for every Community_Number sort Node_Id according to their Walk_Length
2.   for every Walk_Length in Community_Number
3.   Combine the nodes at the same level
4.   Increase the level for the next walk_Length
5.   end for
6.   Choose the next Community for further clustering
5. end for
2. for every pair of communities
3.   Find the distance by using Equation (11).
3.   Combine the communities with smallest distance at the same level
4.   Increase the level to choose next pair of communities
5. end for
6.
5. Return the Dendrogram structure for the Network.

---

6. Implementation

6.1. Metrics for Evaluation Performance

The following evaluation metrics have been considered for measuring performance of the proposed algorithm.

- **Normalized Mutual Information (NMI):** NMI is a suitable measure to compare the quality of different community partitions. It can be evaluated with the help of confusion matrix (CM), where each row corresponds to the community, present in the real partition and each column corresponds to the community, detected through the proposed algorithm. Confusion matrix has been obtained based on the number of communities and community memberships for each node, which is available as ground truth in the datasets [34,35]. Each element in the confusion matrix CM_{ij} represents the number of vertices in ith real community, which is also present in jth detected community. NMI of the detected partition may be formulated as:

$$NMI(X,Y) = \frac{-2\sum_{i=1}^{n_X}\sum_{j=1}^{n_Y} CM_{ij}log(\frac{CM_{ij}CM}{CM_iCM_j})}{\sum_{i=1}^{n_X} CM_i log(\frac{CM_i}{CM}) + \sum_{j=1}^{n_Y} CM_i log(\frac{CM_i}{CM})}, \quad (12)$$

where X and Y are the community partition structure corresponding to ground truth and detected structure, respectively. CM_i and CM_j indicate the communities in true and detected community partition, respectively.

- **Modularity (Q):** Modularity is a metric used to quantify the quality of community partition. This measure is proposed by Girvan and Newman [13]. It is defined as the difference between the number of edges existing inside the communities and the number of edges, which would have been present in a random assignment in the network with similar degree distribution. The expected number of edges between i and j with degree d_i and d_j, respectively, is $d_id_j/2$. Modularity value for a given partition $P = \{c_1, c_2, ...c_k\}$ in the graph $G = (V, E)$ is defined as follows [36]:

$$Q = \frac{1}{2E} \sum_{C_l \in P, (1 \leq l \leq k)} \sum_{i,j \in C_l} \left(A_{ij} - \frac{d_id_j}{2E}\right). \quad (13)$$

- **F-Measure:** F-measure is a metric used to find the accuracy of the proposed algorithm when the ground truth about the communities are available in the dataset. It is the harmonic mean of precision and recall, where precision and recall can be obtained from the confusion matrix obtained from the experiment. Confusion matrix for community detection has been described in

Table 1. In this work, all pairs of nodes are considered to get the value of a, b, c and d for each dataset considered, where

- a = number of pairs, in the same community in ground truth and assigned in same community after community detection. It is treated as True Positive (TP).
- b = number of pairs, belonging in different communities but assigned in same community after community detection. It is treated as False Positive (FP).
- c = number of pairs, belonging in same communities but assigned in different communities after community detection. It is treated as False Negative (FN).
- d = number of pairs, belonging to different communities and assigned in different communities after community detection. It is treated as True Negative (TN).

Precision can be defined as follows:

$$Precision = \frac{TP}{TP + FN} = \frac{a}{a + c} \tag{14}$$

Recall can be defined as follows:

$$Recall = \frac{TP}{TP + FP} = \frac{a}{a + b} \tag{15}$$

F-measure can be defined as follows:

$$F - measure = \frac{2 \times Precision \times Recall}{Precision + Recall}. \tag{16}$$

Table 1. Confusion matrix for community detection.

		Ground Truth	
		$C(v_i = C(v_j)$	$C(v_i) \neq C(v_j)$
Clustering result	$C(v_i) = C(v_j)$	$a(TP)$	$b(FP)$
	$C(v_i) \neq C(v_j$	$c(FN)$	$d(TN)$

- **Execution Time**: A major issue in community detection algorithms is to uncover communities in a reasonable amount of time. In this paper, performance of different algorithms has been measured in terms of execution time. Execution time includes only CPU running time without considering the external time factor. Execution time for all community detection algorithms has been measured in machines with the i7 processor with 3.4 GHz clock speed. Running times have been measured in units of seconds.

6.2. Datasets Used

Social network dataset is often represented in the form of the graph structure, where nodes in the graph represent social entities and edges represent the relationships among the entities. In this paper, the experiment has been carried out using both synthetic and real-world datasets. Details of the datasets are listed in Table 2.

Table 2. Datasets used for evaluation.

Datasets	Nodes	Edges	No. of Communities	Average Path Length	Clustering Coefficient
Synthetic data1	10,000	27,365	620	8.79	0.326
Synthetic data2	20,000	31,569	900	11.59	0.169
com-DBLP	317,080	1,049,866	13,477	7.23	0.6324
com-Amazon	334,863	925,872	75,149	28.67	0.3967
com-Youtube	1,134,890	2,987,624	8385	11.38	0.0808

6.2.1. Synthetic Dataset

The Lancichinetti–Fortunato–Radicchi (LFR) benchmark has been used for generating synthetic data for the social network. This benchmark is observed to be an established one for evaluating different community detection algorithms. Synthetic networks that resemble real-world social networks have been generated by tuning a set of parameters in an LFR benchmark. Parameters in the LFR benchmark include a number of nodes in the network, degree distribution, community size distribution, the maximum and average degree of node, etc. The degree distribution in the network follows the power-law in the LFR benchmark. The probability of having a node with degree k varies with the parameter γ as mentioned below:

$$P_k \ \alpha \ k^{-\gamma}. \tag{17}$$

Here, the value of γ is assigned to vary between 2 to 3 in order to resemble real-world social networks. μ is considered as another parameter in the LFR benchmark, which is also known as mixing parameter. A small value of μ indicates more sparsity between the planted communities in the network. The complexity of the network increases by scaling the μ value. In this paper, the complexity of the network has been increased by scaling mixing parameter from 0.2 to 0.5. β parameter in LFR has been considered for community size distribution in the network. β value often varies between 1 and 2. For each of the synthetic datasets, performance of algorithms has been measured by tuning mixing parameter from 0.2 to 0.5 by increasing 0.05 at each step. Thus, a total of fourteen data points have been generated from two artificial datasets listed in Table 2 and performance has been measured for these data points.

6.2.2. Real World Datasets

Real-world datasets are more complex and heterogeneous as compared to synthetic data. Revealing communities in real-world networks is a NP (non-deterministic polynomial)-hard problem. For measuring the performance of the proposed algorithm, the following real-world datasets have been taken into consideration:

- DBLP;
- Amazon;
- Youtube.

All of these datasets have been collected from the Stanford Large Network Dataset Collection (SNAP), which is publicly available for social network analysis [35].

7. Experimental Results

The experiment has been carried out on a cluster of five nodes, each with an i7 processor with 3.4 Ghz clock speed. The master node has a configuration with a 1 TB hard disk and 10 GB RAM. It also acts as a worker node. Each of the other four nodes acts as a slave or worker node. They all have a symmetric configuration with 1TB hard disk and 20 GB of RAM. A similarity measure based on the random walk has been used to identify the neighboring nodes for inclusion in the community. The threshold value for similarity measure has been considered to be 0.5. A number of synthetic datasets have been generated by tuning the parameters available in LFR benchmark. The proposed algorithm i.e., CDSW, has been implemented both on synthetic and real-world social network datasets. It has been compared with the following community detection algorithms, available in literature:

a   Infomap Community Detection (INF) [30];
b   Spin-Glass Community Detection (SG) [26];
c   Girvan Newman Community Detection (GN) [13];
d   Walktrap Community Detection (WT) [25].

NMI is one of the accepted measures for comparing detected clusters with ground truth partitions. NMI for each partition obtained from different algorithms has been evaluated. The higher the

NMI value, the better is the community partition. Figure 3a shows the box plot analysis for NMI of all data points. It is observed that the median of NMI for the proposed community detection algorithm using small world phenomenon (CDSW) is more than 0.8, which is better as compared to other algorithms. Minimum NMI for CDSW algorithm is higher than the minimum value for other algorithms. The WT algorithm performs better with respect to NMI, which is close to the performance of the proposed algorithm.

Figure 3b shows the box plot analysis of F-measure for different algorithms. F-measure is related to the accuracy of algorithms. It has been calculated using the confusion matrix for community partition. The structure of the confusion matrix has been presented in Table 1. From the Figure 3b, it is observed that the median F-measure of the CDSW algorithm is higher than 0.85. Although the maximum of F-measure values for the CDSW algorithm is not as good as those obtained for WT and GN algorithms, its average and maximum values have been observed to be higher than all other algorithms.

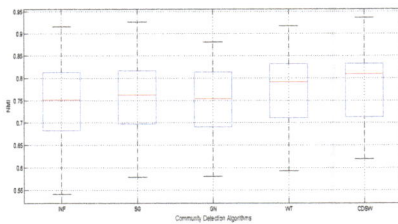

(**a**) box plot for *NMI* obtained in different algorithms

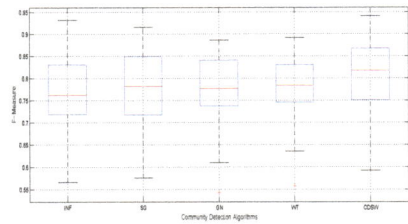

(**b**) box plot for F-measure of different algorithms

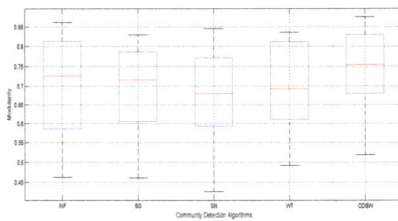

(**c**) box plot for modularity obtained in different algorithms

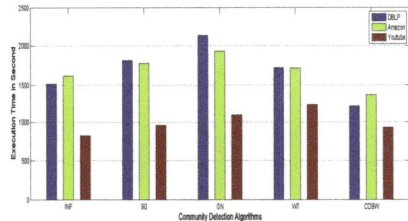

(**d**) execution time comparison in real-time datasets

Figure 3. Comparative study of different community detection algorithms.

Figure 3c shows the box plot analysis of modularity value for community partition generated from different algorithms. The higher the modularity value, the better is the community partition. Modularity value decreases when link density between the communities or the value of mixing parameter μ increases. Median modularity value for the proposed algorithm is observed to be better as compared to other community detection algorithms. Modularity values obtained from Spin-Glass and Infomap algorithms are relatively similar. From Figure 3c, it is observed that the CDSW algorithm provides better community structure as compared to other traditional algorithms.

In the social network, the number of entities and their relationships are observed to be increasing exponentially. Since community detection in the large-scale network in a reasonable amount of time is the focus of this study, an effort has been made in measuring execution times for different algorithms in community detection. Figure 3d shows the comparative study of execution time for different community detection algorithms. The structure of synthetic network changes when mixing parameters of benchmark increases. In this work, comparative analysis of execution time has been carried out separately for both synthetic and real-world datasets. The comparative analysis of execution time for synthetic network is presented in Figure 4a,b. The mean execution time of CDSW is observed to be less as compared to the values of other algorithms in both real-world and synthetic datasets.

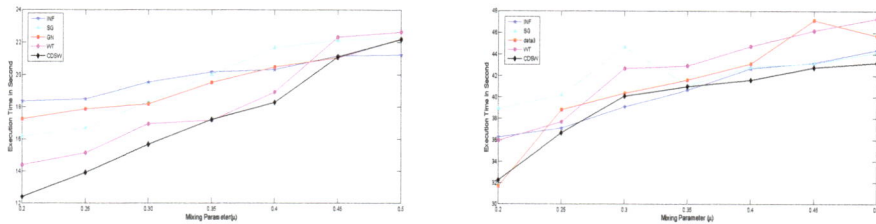

(a) execution time for synthetic dataset1

(b) execution time for synthetic dataset2

Figure 4. Comparative analysis of execution time (s) for synthetic datasets.

The evaluation metrics such as min-value, max-value, upper (25%), middle (50%) and lower (75%) quartile values have been listed in Table 3. *NMI* value provided by the CDSW algorithm at each point in the box plot is observed to be better. It may be observed that the Walktrap algorithm performs close to the proposed algorithm with respect to the *NMI* metric. Modularity metric indicates the quality of community partition.

Table 3. Box plot statistics for modularity, *NMI* and F-measure.

	Modularity					*NMI*					F-Measure				
	MIN	25%	50%	75%	MAX	MIN	25%	50%	75%	MAX	MIN	25%	50%	75%	MAX
INF	0.46	0.59	0.73	0.81	0.86	0.54	0.68	0.75	0.81	0.92	0.57	0.72	0.76	0.83	0.93
SG	0.46	0.6	0.72	0.79	0.83	0.58	0.7	0.76	0.82	0.93	0.58	0.72	0.78	0.85	0.92
GN	0.43	0.59	0.68	0.77	0.85	0.58	0.69	0.75	0.81	0.88	0.54	0.74	0.78	0.84	0.89
WT	0.49	0.61	0.69	0.81	0.84	0.59	0.71	0.79	0.83	0.92	0.56	0.75	0.78	0.83	0.89
CDSW	0.52	0.68	0.75	0.83	0.88	0.62	0.71	0.81	0.83	0.94	0.59	0.75	0.82	0.87	0.94

A *t*-test has been performed for static analysis of all mentioned community detection algorithms. Mean deviation and *p*-value have been extracted from the *t*-test. The mean deviation of evaluation metrics for all the algorithms are presented in Table 4. It may be noticed from Table 4 that the CDSW algorithm has positive deviation for all evaluation metrics, and it implies that it performs better with respect to all metrics. *p*-value in the *t*-test represents a significant difference of the CDSW algorithm with respect to other algorithms. Table 5 lists the statistics of *p*-value between all algorithms. An algorithm is said to be significantly different from others if its *p*-value is less than 0.05. It can be further noticed from Table 5, that *p*-value for the proposed algorithm in all of the evaluation metrics is less than 0.05. From this observation, it may be inferred that the CDSW algorithm is significantly different from all other algorithms with respect to metrics considered for comparison.

Table 4. Mean deviation in modularity, *NMI* and F-measure for different algorithms.

	Modularity					*NMI*					F-Measure				
	INF	SG	GN	WT	CDSW	INF	SG	GN	WT	CDSW	INF	SG	GN	WT	CDSW
INF	0.000	−0.010	−0.002	−0.028	−0.042	0.000	−0.010	−0.002	0.028	−0.042	0.000	−0.005	−0.006	−0.009	−0.036
SG	0.010	0.000	0.008	−0.018	−0.032	0.010	0.000	0.008	−0.018	−0.032	0.005	0.000	−0.001	−0.003	−0.030
GN	0.002	−0.008	0.000	−0.026	−0.040	0.002	−0.008	0.000	−0.026	−0.040	0.006	0.001	0.000	−0.003	−0.029
WT	0.028	0.018	0.026	0.000	−0.014	0.028	0.018	0.026	0.000	−0.014	0.009	0.003	0.003	0.000	−0.027
CDSW	0.042	0.032	0.040	0.014	0.000	0.042	0.032	0.040	0.014	0.000	0.036	0.030	0.029	0.027	0.000

Table 5. *p*-value obtained from *t*-test for modularity, *NMI* and F-measure.

	Modularity					NMI					F-Measure				
	INF	SG	GN	WT	CDSW	INF	SG	GN	WT	CDSW	INF	SG	GN	WT	CDSW
INF	-	0.241	0.761	0.001	0.000	-	0.241	0.761	0.001	0.000	-	0.324	0.442	0.234	0.000
SG	0.241	-	0.203	0.008	0.000	0.241	-	0.203	0.008	0.000	0.324	-	0.875	0.562	0.000
GN	0.761	0.203	-	0.000	0.000	0.761	0.203	-	0.000	0.000	0.442	0.875	-	0.515	0.000
WT	0.001	0.008	0.000	-	0.015	0.001	0.008	0.000	-	0.015	0.234	0.562	0.515	-	0.000
CDSW	0.000	0.000	0.000	0.015	-	0.000	0.000	0.000	0.015	-	0.000	0.000	0.000	0.000	-

8. Threat to Validity

- In the proposed work, threshold value for similarity index has been considered to be 0.5. This work may have a validation threat when the threshold value for similarity index is chosen to be higher than 0.8.
- Hadoop framework has been used for distributing computation in multiple nodes where data has been stored in the HDFS file system. All the data present in the HDFS file system are immutable in nature. The proposed algorithm may not perform well when the social network is dynamic in nature i.e., the structure of the network changes in course of time. In future work, issues regarding community detection in the dynamic network may be resolved.

9. Conclusions

The community detection problem is one of the challenging ones in social network analysis. In this study, communities have been detected in a distributed manner in order to have lesser computation time. Large-scale networks always follow small world phenomenon, which is based on the six degrees of separation principle. In this paper, similarity value between every pair of nodes has been obtained based on the random walk up to 3rd walk length. From the experimental analysis, it is observed that the proposed algorithm i.e., CDSW, provides better performance based on values of modularity, NMI and F-measure as compared to a few other community detection algorithms referred to in literature. From *t*-test analysis, the proposed algorithm is observed to be significantly different from other algorithms. It also provides better performance in terms of execution time, especially when large-scale networks are considered.

For the sake of simplicity, we have implemented all the algorithms in the undirected network. However, the proposed algorithm can be extended to the directed network by carefully choosing the source node in the network. As the network is a directed one, the source node can't be chosen randomly. The node with only the in-degree feature cannot be the source node for further processing. By imposing a constraint on choosing criteria for source node (i.e., source node must have at least one out-degree), the proposed algorithm can be extended to detect community for the directed network.

The proposed algorithm can be further extended to dynamic social networks, where a large number of nodes along with their relationships are added more frequently. In future, other distributed frameworks like Spark and Storm may be implemented in the community detection process, in order to have further improvement in execution time.

Acknowledgments: Thanks to the authorities of the NIT, Rourkela for availing the platform for doing this research study. Support also came from Covenant University Centre for Research and Innovation Development, Ota, Nigeria; and Research Cluster Fund of Faculty of Informatics, Kaunas University of Technology, Kaunas, Lithuania.

Author Contributions: All authors discussed the contents of the manuscript and contributed to its preparation. Santanu Kumar Rath supervised the research. Ranjan Kumar Behera contributed the idea, performed the numerical simulations. Sanjay Misra, Robertas Damaševičius and Rytis Maskeliūnas helped in the analysis of the framework developed and the writing of the manuscript.

Conflicts of Interest: The authors declare no conflict of interest.

References

1. Kossinets, G.; Watts, D.J. Empirical analysis of an evolving social network. *Science* **2006**, *311*, 88–90.
2. Carrington, P.J.; Scott, J.; Wasserman, S. *Models and Methods in Social Network Analysis*; Cambridge University Press: New York, NY, USA, 2005; Volume 28.
3. Freeman, L.C. Centrality in social networks conceptual clarification. *Soc. Netw.* **1978**, *1*, 215–239.
4. Bakshy, E.; Rosenn, I.; Marlow, C.; Adamic, L. The role of social networks in information diffusion. In Proceedings of the 21st International Conference on World Wide Web, Lyon, France, 16–20 April 2012; pp. 519–528.
5. Liben-Nowell, D.; Kleinberg, J. The link-prediction problem for social networks. *J. Assoc. Inf. Sci. Technol.* **2007**, *58*, 1019–1031.
6. Clauset, A.; Newman, M.E.J.; Moore, C. Finding community structure in very large networks. *Phys. Rev. E* **2004**, *70*, 066111.
7. Stephen, A.T.; Toubia, O. Explaining the power-law degree distribution in a social commerce network. *Soc. Netw.* **2009**, *31*, 262–270.
8. Newman, M.E.J.; Watts, D.J. Scaling and percolation in the small-world network model. *Phys. Rev. E* **1999**, *60*, 7332.
9. Borgatti, S.P.; Cross, R. A relational view of information seeking and learning in social networks. *Manag. Sci.* **2003**, *49*, 432–445.
10. Travers, J.; Milgram, S. The small world problem. *Phychol. Today* **1967**, *1*, 61–67.
11. Behera, R.K.; Rath, S.K. An efficient modularity based algorithm for community detection in social network. In Proceedings of the International Conference on Internet of Things and Applications (IOTA), Pune, India, 22–24 January 2016; pp. 162–167.
12. Shu, W.; Chuang, Y.-H. The perceived benefits of six-degree-separation social networks. *Int. Res.* **2011**, *21*, 26–45.
13. Newman, M.E.J. Detecting community structure in networks. *Eur. Phys. J. B-Condens. Matter Complex Syst.* **2004**, *38*, 321–330.
14. Behera, R.K.; Rath, S.K.; Jena, M. Spanning tree based community detection using min-max modularity. *Procedia Comput. Sci.* **2016**, *93*, 1070–1076.
15. Newman, M.E.J.; Watts, D.J.; Strogatz, S.H. Random graph models of social networks. *Proc. Natl. Acad. Sci. USA* **2002**, *99* (Suppl. 1), 2566–2572.
16. Kempe, D.; Kleinberg, J.; Tardos, É. Maximizing the spread of influence through a social network. In Proceedings of the Ninth ACM SIGKDD International Conference on Knowledge Discovery and Data Mining, Washington, DC, USA, 24–27 August 2003; pp. 137–146.
17. Leskovec, J.; Mcauley, J.J. Learning to discover social circles in ego networks. In Proceedings of the Advances in Neural Information Processing Systems, Lake Tahoe, NV, USA, 3–6 December 2012; pp. 539–547.
18. Newman, M.E.J. The structure and function of complex networks. *SIAM Rev.* **2003**, *45*, 167–256.
19. Chopade, P.; Zhan, J. Structural and functional analytics for community detection in large-scale complex networks. *J. Big Data* **2015**, *2*, 1–28.
20. Cook, D.J.; Holder, L.B. *Mining Graph Data*; John Wiley & Sons: Hoboken, NJ, USA, 2006.
21. Newman, M.E.J. Analysis of weighted networks. *Phys. Rev. E* **2004**, *70*, 61311–61319.
22. De Meo, P.; Ferrara, E.; Fiumara, G.; Provetti, A. Mixing local and global information for community detection in large networks. *J. Comput. Syst. Sci.* **2014**, *80*, 72–87.
23. Breiger, R.L.; Boorman, S.A.; Arabie, P. An algorithm for clustering relational data with applications to social network analysis and comparison with multidimensional scaling. *J. Math. Psychol.* **1975**, *12*, 328–383.
24. Gregory, S. Finding overlapping communities in networks by label propagation. *New J. Phys.* **2010**, *12*, 103018.
25. Pons, P.; Latapy, M. Computing communities in large networks using random walks. In *Computer and Information Sciences-ISCIS 2005*; Springer: Istanbul, Turkey, 2005; pp. 284–293.
26. Eaton, E.; Mansbach, R. A spin-glass model for semi-supervised community detection. In Proceedings of the Twenty-Sixth AAAI Conference on Artificial Intelligence, Toronto, Ontario, Canada, 22–26 July 2012; pp. 900–906.

27. Agreste, S.; De Meo, P.; Fiumara, G.; Piccione, G.; Piccolo, S.; Rosaci, D.; Sarné, G.M.L.; Vasilakos, A.V. An empirical comparison of algorithms to find communities in directed graphs and their application in web data analytics. *IEEE Trans. Big Data* **2017**, *3*, 289–306.
28. Sun, P.G.; Gao, L. A framework of mapping undirected to directed graphs for community detection. *Inf. Sci.* **2015**, *298*, 330–343.
29. Deng, X.; Zhai, J.; Lv, T.; Yin, L. Efficient vector influence clustering coefficient based directed community detection method. *IEEE Access* **2017**, *5*, 17106–17116.
30. Rosvall, M.; Bergstrom, C.T. Maps of random walks on complex networks reveal community structure. *Proc. Natl. Acad. Sci. USA* **2008**, *105*, 1118–1123.
31. Orman, G.K.; Labatut, V.; Cherifi, H. On accuracy of community structure discovery algorithms. *arXiv* **2011**, arXiv:1112.4134.
32. Zhang, P.; Wang, J.; Li, X.; Li, M.; Di, Z.; Fan, Y. Clustering coefficient and community structure of bipartite networks. *Phys. A Stat. Mech. Appl.* **2008**, *387*, 6869–6875.
33. Kemeny, J.G.; Snell, J.L. *Finite Markov Chains*; Springer: Princeton, NJ, USA, 1960; Volume 356.
34. McDaid, A.F.; Greene, D.; Hurley, N. Normalized mutual information to evaluate overlapping community finding algorithms. *arXiv* **2011**, arXiv:1110.2515.
35. Leskovec, J.; Krevl, A. {SNAP Datasets}:{Stanford} Large Network Dataset Collection. Available online: http://snap.stanford.edu/data (accessed on 31 August 2015).
36. Newman, M.E.J. Modularity and community structure in networks. *Proc. Natl. Acad. Sci. USA* **2006**, *103*, 8577–8582.

![applied sciences logo] *applied sciences*

MDPI

Article

A Parallel Approach for Frequent Subgraph Mining in a Single Large Graph Using Spark

Fengcai Qiao [1,*,†], Xin Zhang [1,†], Pei Li [1,†], Zhaoyun Ding [1,†], Shanshan Jia [2,†] and Hui Wang [1,†]

[1]   College of Engineering System, National University of Defense Technology, Changsha 410073, Hunan, China; ijunzhanggm@gmail.com (X.Z.); peili@nudt.edu.cn (P.L.); zyding@nudt.edu.cn (Z.D.); huiwang@nudt.edu.cn (H.W.)
[2]   Digital Media Center, Hunan Education Publishing House, Changsha 410073, Hunan, China; wandou_2007@163.com
*   Correspondence: fcqiao@nudt.edu.cn; Tel.: +86-0731-84574331
†   Current address: No.109, Deya Road, Changsha 410073, Hunan, China.

Received: 3 January 2018; Accepted: 31 January 2018; Published: 2 February 2018

Abstract: Frequent subgraph mining (FSM) plays an important role in graph mining, attracting a great deal of attention in many areas, such as bioinformatics, web data mining and social networks. In this paper, we propose SSIGRAM (Spark based Single Graph Mining), a Spark based parallel frequent subgraph mining algorithm in a single large graph. Aiming to approach the two computational challenges of FSM, we conduct the subgraph extension and support evaluation parallel across all the distributed cluster worker nodes. In addition, we also employ a heuristic search strategy and three novel optimizations: load balancing, pre-search pruning and top-down pruning in the support evaluation process, which significantly improve the performance. Extensive experiments with four different real-world datasets demonstrate that the proposed algorithm outperforms the existing GRAMI (Graph Mining) algorithm by an order of magnitude for all datasets and can work with a lower support threshold.

Keywords: frequent subgraph mining; parallel, algorithm; constraint satisfaction problem; Spark

---

1. Introduction

Many relationships among objects in a variety of applications such as chemical, bioinformatics, computer vision, social networks, text retrieval and web analysis can be represented in the form of graphs. Frequent subgraph mining (FSM) is a well-studied problem in the graph mining area which boosts many real-world application scenarios such as retail suggestion engines [1], protein–protein interaction networks [2], relationship prediction [3], intrusion detection [4], event prediction [5], text sentiment analysis [6], image classification [7], etc. For example, mining frequent subgraphs from a massive event interaction graph [8] can help to find recurring interaction patterns between people or organizations which may be of interest to social scientists.

There are two broad categories of frequent subgraph mining: (i) graph transaction-based FSM; and (ii) single graph-based FSM [9]. In graph transaction-based FSM, the input data comprise a collection of small-size or medium-size graphs called transactions, i.e., a graph database. In single graph-based FSM, the input data comprise one very large graph. The FSM task is to enumerate all subgraphs with support above the minimum support threshold. Graph transaction-based FSM uses transaction-based counting support while single graph-based FSM adopts occurrence-based counting. Mining frequent subgraphs in a single graph is more complicated and computationally demanding because multiple instances of identical subgraphs may overlap. In this paper, we focus on frequent subgraph mining in a single large graph.

The "bottleneck" for frequent subgraph mining algorithms on a single large graph is the computational complexity incurred by the two core operations: (i) efficient generation of all subgraphs with various size; and (ii) subgraph isomorphism evaluation (support evaluation), i.e., determining whether a graph is an exact match of another one. Let N and n be the number of vertexes of input graph G and subgraph S, respectively. Typically, the complexity of subgraph generation is $\mathcal{O}(2^{N^2})$ and support evaluation is $\mathcal{O}(N^n)$. Thus, the total complexity of an FSM algorithm is $\mathcal{O}(2^{N^2} \cdot N^n)$, which is exponential in terms of problem size. In recent years, numerous algorithms for single graph-based FSM have been proposed. Nevertheless, most of them are sequential algorithms that require much time to mine large datasets, including SiGraM (Single Graph Mining) [10], GERM (Graph Evolution Rule Miner) [11] and GRAMI (Graph Mining) [12]. Meanwhile, researchers have also used parallel and distributed computing techniques to accelerate the computation, in which two parallel computing frameworks are mainly used: Map-Reduce [13–18] and MPI (Message Passing Interface) [19]. The existing MapReduce implementations of parallel FSM algorithms are all based on Hadoop [20] and are designed for graph transaction and not for a single graph, often reaching IO (Input and Output) bottlenecks because they have to spend a lot of time moving the data/processes in and out of the disk during iteration of the algorithms. Besides, some of these algorithms cannot support mining via subgraph extension [14,15]. That is to say, users must provide the size of subgraph as input. In addition, although the MPI based methods, such as DistGraph [19], usually have a good performance, it is geared towards tightly interconnected HPC (High Performance Computing) machines, which are less available for most people. In addition, for MPI-based algorithms, it is hard to combine multiple machine learning or data mining algorithms into a single pipeline from distributed data storage to feature selection and training, which is common for machine learning. Fault tolerance is also left to the application developer.

In this paper, we propose a parallel frequent subgraph mining algorithm in a single large graph using Apache Spark framework, called SSIGRAM. The Spark [21] is an in-memory MapReduce-like general-purpose distributed computation platform which provides a high-level interface for users to build applications. Unlike previous MapReduce frameworks such as Hadoop, Spark mainly stores intermediate data in memory, effectively reducing the number of disk input/output operations. In addition, a benefits of the "ML (Machine Learning) Pipelines" design [22] in Spark is that we can not only mine frequent subgraphs efficiently but also easily combine the mining process seamlessly with other machine learning algorithms like classification, clustering or recommendation.

Aiming at the two computational challenges of FSM, we conduct the subgraph extension and support evaluation across all the distributed cluster worker nodes. For subgraph extension, our approach generates all subgraphs in parallel through FFSM-Join (Fast Frequent Subgraph Mining) and FFSM-Extend proposed by Huan et al. [23], which is an efficient solution for candidate subgraphs enumeration. When computing subgraph support, we adopt the constraint satisfaction problem (CSP) model proposed in [12] as the support evaluation method. The CSP support satisfies the downward closure property (DCP), also known as anti-monotonic (or Apriori property), which means that a subgraph g is frequent if and only if all of its subgraphs are frequent. As a result, we employ a breadth first search (BFS) strategy in our SSIGRAM algorithm. At each iteration, the generated subgraphs are distributed to every executor across the Spark cluster for solving the CSP. Then, the infrequent subgraphs are removed while the remaining frequent subgraphs are passed to the next iteration for candidate subgraph generation.

In practice, the support evaluation is more complicated than subgraph extension and will cost most time during the mining process. As a result, besides parallel mining, our SSIGRAM algorithm also employs a heuristic search strategy and three novel optimizations: load balancing, pre-search pruning and top-down pruning in the support counting process, which significantly improve the performance. Noteworthily, SSIGRAM can also be applied to directed graphs, weighted subgraph mining or uncertain graph mining with slight modifications introduced in [24,25]. In summary, our main contributions to the frequent subgraph mining in a single large graph are three-pronged:

- First, we propose SSIGRAM, a novel parallel frequent subgraph mining algorithm in a single large graph using Spark, which is different from the Hadoop MapReduce based and MPI based parallel algorithms. SSIGRAM can also easily combine with the bottom Hadoop distributed storage data and other machine learning algorithms.
- Second, we conduct in parallel subgraph extension and support counting, respectively, aiming at the two core steps with high computational complexity in frequent subgraph mining. In addition, we provide a heuristic search strategy and three optimizations for the support computing operation.
- Third, extensive experimental performance evaluations are conducted with four real-world graphs. The proposed SSIGRAM algorithm outperforms the GRAMI method by at least one order of magnitude with the same memory allocated.

The paper is organized as follows. The problem formalization is provided in Section 2. Our SSIGRAM algorithm and its optimizations are presented in Section 3. In Section 4, extensive experiments to evaluate the performance of the proposed algorithm are conducted and analyzed. The work is summarized and conclusions are drawn in Section 5.

2. Formalism

A graph $G = (V, E)$ is defined to be a set of vertexes (nodes) V which are interconnected by a set of edges (links) $E \subseteq V \times V$ [26]. A labelled graph also consists of a labeling function L besides V and E that assigns labels to V and E. Usually, the graphs used in FSM are assumed to be labelled simple graphs, i.e., un-weighted and un-directed labeled graphs with no loops and no multiple links between any two distinct nodes [27]. To simplify the presentation, our SSIGRAM is illustrated with an undirected graph with a single label for each node and edge. Nevertheless, as mentioned above, the SSIGRAM can also be extended to support either directed or weighted graphs. In the following, a number of widely used definitions used later in this paper are introduced.

Definition 1. *(Labelled Graph): A labelled graph can be represented as $G = (V, E, L_V, L_E, \varphi)$, where V is a set of vertexes, $E \subseteq V \times V$ a set of edges. L_V and L_E are sets of vertex and edge labels respectively. φ is a label function that defines the mappings $V \rightarrow L_V$ and $E \rightarrow L_E$.*

Definition 2. *(Subgraph): Given two graphs $G_1 = (V_1, E_1, L_{V_1}, L_{E_1}, \varphi_1)$ and $G_2 = (V_2, E_2, L_{V_2}, L_{E_2}, \varphi_2)$, G_1 is a subgraph of G_2, if and only if (i) $V_1 \subseteq V_2$, and $\forall v \in V_1$, $\varphi_1(v) = \varphi_2(v)$; (ii) $E_1 \subseteq E_2$, and $\forall (u, v) \in E_1$, $\varphi_1(u, v) = \varphi_2(u, v)$. G_2 is also called a supergraph of G_1.*

Definition 3. *(Subgraph Isomorphism): Let $G_1 = (V_1, E_1, L_{V_1}, L_{E_1}, \varphi_1)$ be a subgraph of G. G_1 is subgraph isomorphic to graph G, if and only if there exists another subgraph $G_2 = (V_2, E_2, L_{V_2}, L_{E_2}, \varphi_2) \subseteq G$ and a bijection $f : V_1 \rightarrow V_2$ satisfying: (i) $\forall u \in V_1$, $\varphi_1(u) = \varphi_2(f(u))$; (ii) $\forall (u, v) \in E_1 \iff (f(u), f(v)) \in E_2$; and (iii) $\forall (u, v) \in E_1$, $\varphi_1(u, v) = \varphi_2(f(u), f(v))$. G_2 is also called an embedding of G_1 in G.*

For example, Figure 1a illustrates a labelled graph of event interaction graph. Node labels represent the event actor's type (e.g., GOV: government) in CAMEO (Conflict And Mediation Event Observations) codes [28] and edge labels represent the event type [28] between the two actors. Figure 1b,c shows two subgraphs of Figure 1a. Figure 1b ($v_1 \overset{3}{-} v_2 \overset{9}{-} v_3$) has three isomorphisms with respect to graph (a): $u_2 \overset{3}{-} u_4 \overset{9}{-} u_5$, $u_6 \overset{3}{-} u_5 \overset{9}{-} u_4$ and $u_7 \overset{3}{-} u_8 \overset{9}{-} u_{10}$.

Definition 4. *(Frequent Subgraph in Single Graph): Given a labelled single graph G and a minimum support threshold τ, the frequent subgraph mining problem is defined as finding all subgraphs G_i in G,*

$$Sup(G_i, G) \geq \tau, \quad \forall G_i \in S,$$

where S denotes the set of subgraphs in G with support greater or equal to τ.

For a subgraph G_1 and input graph G, the straightforward way to compute the support of G_1 in graph G is to count all its isomorphisms of G_1 in G [29]. Unfortunately, such a method does not satisfy the downward closure property (DCP) since there are cases where a subgraph appears fewer times than its supergraph. For example, in Figure 1a, the single node graph REF (Refugee) appears three times, while its supergraph REF$\overset{3}{-}$GOV appears four times. Without the DCP, the search space cannot be pruned and the exhaustive search is unavoidable [30]. To address this issue, we employ the minimum image (MNI) based support which is anti-monotonic introduced in [31].

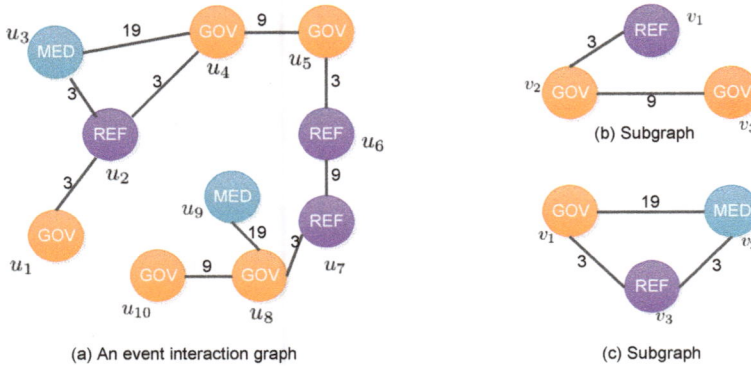

Figure 1. (**a**) An event interaction graph; nodes represent the event actors (labelled with their type code) and edges represent the event (labelled with event type code); and (**b,c**) two subgraphs of (**a**) (MED: Media, REF: Refugee, GOV: Government).

Definition 5. (MNI Support): *For a subgraph G_1, let $\sum(S) = \{\phi_1, \phi_2, ..., \phi_m\}$ denote the set of all isomorphisms/embeddings of G_1 in the input graph G. The MNI support of a subgraph G_1 in G is based on the number of unique nodes in the input graph G that a node of the subgraph $G_1 = (V_1, E_1)$ is mapped to, which is defined as:*

$$Sup_{MNI}(G_1, G) = \min_{v \in V_1}\{|\Phi(v)|\},$$

where $\Phi(v)$ is the set of unique mappings for each $v \in V_1$, denoted as

$$\Phi(v) = \bigcup_{i=1}^{|\Sigma(S)|} \phi_i(v).$$

Figure 2 illustrates the four isomorphisms of a subgraph $G_1 \equiv A\text{-}B\text{-}C\text{-}A$ in the input graph. For example, one of the isomorphisms is $\phi = \{u_1, u_4, u_6, u_8\}$, shown in the second column in Figure 2c. There are four isomorphisms for the subgraph G_1 in Figure 2b. Therefore, the set of unique mappings for the vertex v_1 is $\{u_1, u_2, u_8\}$. The number of unique mappings over all the subgraph vertices $\{v_1, v_2, v_3, v_4\}$ are 3, 3, 2 and 2, respectively. Thus, the MNI support of G_1 is $Sup_{MNI}(G_1, G) = \min\{3, 3, 2, 2\} = 2$.

Definition 6. (Adjacency Matrix): *The adjacency matrix of a graph G_i with n vertexes is defined as a $n \times n$ matrix M, in which every diagonal entry corresponds to a distinct vertex in G_i and is filled with the label of this vertex and every off-diagonal (for an undirected graph, the upper triangle is always a mirror of the lower* triangle) entry in the lower triangle part corresponds to a pair of vertices in G_i and is filled with the label of the edge between the two vertices and zero if there is no edge.

Definition 7. *(Maximal Proper Submatrix): For a $m \times m$ matrix A, a $n \times n$ matrix B is the maximal proper submatrix of A, iff B is obtained by removing the last nonzero entry from A. For example, the last non-zero entry of M_2 in Figure 3 is y in the bottom row.*

Definition 8. *(Canonical Adjacency Matrix): Let matrix M denote the adjacency matrix of graph G_i. Code(M) represents the code of M, which is defined as the sequence formed by concatenating lower triangular entries of M (including entries on the diagonal) from left to right and top to bottom, respectively. The canonical adjacency matrix (CAM) of graph G_i is the one that produces the maximal code, using lexicographic order [23]. Obviously, a CAM's maximal proper submatrix is also a CAM.*

Figure 3 shows three adjacency matrices M_1, M_2, M_3 for the graph G_1 on the left. After applying the standard lexicographic order, we have $Code(M_1) : aybyxb0y0c00y0d > Code(M_2) : aybyxb00yd0y00c > Code(M_3) : bxby0d0y0cyy00a$. In fact, M_1 is the canonical adjacency matrix of G_1.

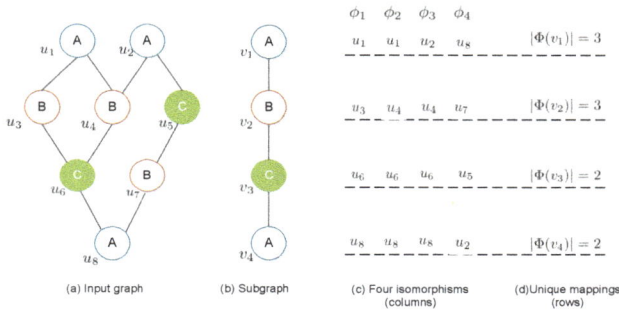

Figure 2. Minimum image (MNI) support of a subgraph $G_1 \equiv$ *A-B-C-A* in a single graph *G* with eight vertexes and three vertex labels.

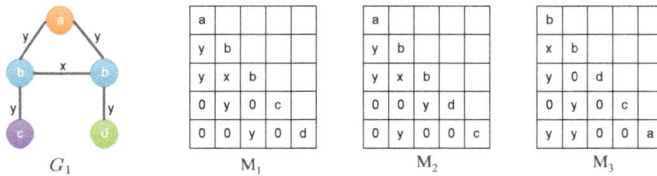

Figure 3. Three adjacency matrices M_1, M_2, and M_3 for the graph G_1.

Definition 9. *(Suboptimal CAM): Given a graph G, a suboptimal canonical adjacency matrix (suboptimal CAM) of G is an adjacency matrix M of G such that its maximal proper submatrix N is the CAM of the graph that N represents. [23]. A CAM is also a suboptimal CAM. A proper suboptimal CAM is denoted as a suboptimal CAM that is not the CAM of the graph it represents.*

3. The SSiGraM Approach

This section first elaborates upon the framework of the proposed algorithm, before describing the detailed procedure of parallel subgraph extension, parallel support evaluation and three optimization strategies.

3.1. Framework

Figure 4 illustrates the proposed framework of our SSIGRAM approach. It mainly contains two major components: parallel subgraph extension and parallel support evaluation. The green notations are main Spark RDD (a resilient distributed dataset (RDD) is the core abstraction of Spark, which is a fault-tolerant collection of elements that can be operated on in parallel) transformations or actions used during the algorithm pipeline, e.g., map, join, etc. We will discuss details of the framework below.

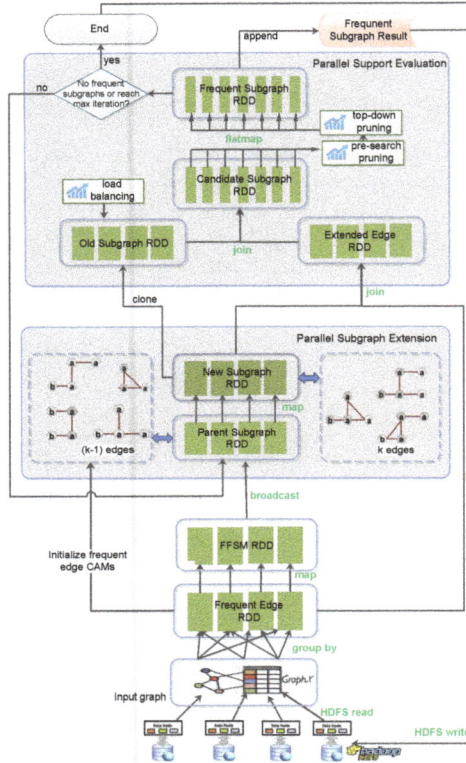

Figure 4. Framework of the SSIGRAM(Spark based Single Graph Mining) approach. Green notations on the right side are main Spark resilient distributed dataset (RDD) transformations or actions used during the algorithm pipeline. Abbreviations: CAM (Canonical Adjacency Matrix), HDFS (Hadoop Distributed File System), FFSM (Fast Frequent Subgraph Mining).

3.2. Parallel Subgraph Extension

Our approach employs a breadth first search (BFS) strategy that generates all subgraphs in parallel through FFSM-Join and FFSM-Extend proposed in [23]. Similarly, we organize all the suboptimal CAMs of subgraphs in a graph G into a rooted tree, that follows the rules: (i) The root of the tree is an empty matrix. (ii) Each node in the tree is a distinct subgraph of G, represented by its suboptimal CAM that is either a CAM or a proper suboptimal CAM. (iii) For a given non-root node (with suboptimal CAM M), its parent is the graph represented by the maximal proper submatrix of M. The completeness of the suboptimal CAM tree is guaranteed by the Theorem 1. For the formal proof, we refer to the appendix in [23].

Theorem 1. *For a graph G, let C_{k-1} and C_k be sets of the suboptimal CAMs of all the subgraphs with $(k-1)$ edges and k edges $(k \geq 2)$. Every member of set C_k can be enumerated unambiguously either by joining two members of set C_{k-1} or by extending a member in C_{k-1}.*

Algorithm 1 shows how the subgraph extension process is conducted in parallel. Actually, the extension process is implemented in parallel at the parent subgraph scale (Lines 6–15), which means that each group of subgraphs with the same parent will be sent to an executor for extension in the Spark cluster. The FFSM operator is provided by [32], which implements the FFSM-Join and FFSM-Extend. After extension, all of the extended results are collected to the driver node (Line 16) from the cluster. Because of the extension on more than one executors at the same time, the indexes of the new generated subgraphs from different executors may be duplicated. As a result, the subgraph indexes are reassigned at the end (Line 17).

To perform a parallel subgraph extension, Line 11 and Line 13 conduct the joining and extension of CAM across all Spark executors. The overall complexity is $\mathcal{O}(n^2 \cdot m)$ where n is the number of nodes in subgraph and m number of edges. A complete graph with n vertices consists of $n(n-1)/2$ edges. Thus, the final complexity is $\mathcal{O}(m^2)$.

Algorithm 1 Parallel Subgraph Extension.

Input: frequent subgraphs S_{k-1}, broadcasted FFSM operator $FFSMRDD$, Spark context sc
Output: new generated S_k that extend S_{k-1}
 1: **function** PARAGENSUBGRAPHS(S_{k-1},$FFSMRDD$,sc)
 2:     $S_k \leftarrow \varnothing$
 3:     $FFSMOperator \leftarrow FFSMRDD$
 4:     $PGraph_{S_{k-1}} \leftarrow S_{k-1}$, group by parents
 5:     $PGraphRDD_{S_{k-1}} \leftarrow sc.\textbf{parallelize}(PGraph_{S_{k-1}})$
 6:     $S_k \leftarrow PGraphRDD_{S_{k-1}}.\textbf{map}\{$
 7:         $S_t \leftarrow \varnothing$
 8:         $C_{list} \leftarrow$ child subgraphs of present parent
 9:         **for** s_i in C_{list} **do**
10:             **for** s_j in C_{list} **do**
11:                 $S_t \leftarrow S_t \cup FFSMOperator.join(s_i, s_j)$
12:             **end for**
13:             $S_t \leftarrow S_t \cup FFSMOperator.extend(s_i)$
14:         **end for**
15:         **return** S_t
16:     $\}.$**collect**
17:     reassign subgraph indexes in S_k
18:     **return** S_k
19: **end function**

3.3. Parallel Support Evaluation

Our SSIGRAM approach employs the CSP model [12] as the subgraph support evaluation strategy. The constraint satisfaction problem (CSP) is an efficient method for finding subgraph isomorphisms (Definition 3), which is illustrated as follows:

Definition 10. (CSP Model): *Let $G_1 = (V_1, E_1, L_{V_1}, L_{E_1}, \varphi_1)$ be a subgraph of a graph $G = (V, E, L_V, L_E, \varphi)$. Finding isomorphisms of G_1 in G is a CSP($\mathcal{X}, \mathcal{D}, \mathcal{C}$) where:*
1. $\mathcal{X}$ is an ordered set of variables which contains a variable x_v for each node $v \in V_1$.
2. $\mathcal{D}$ is the set of domains for each variable $x_v \in \mathcal{X}$. Each domain is a subset of V.
3. Set $\mathcal{C}$ contains the following constraint rules:

- $\forall x_v, x_{v'} \in \mathcal{X}, x_v \neq x_{v'}$.
- $\forall x_v \in \mathcal{X}$ and the corresponding $v \in V_1, \varphi(x_v) = \varphi_1(v)$.

- $\forall x_v, x_{v'} \in \mathcal{X}$ and the corresponding $v, v' \in V_1$, $\varphi(x_v, x_{v'}) = \varphi_1(v, v')$.

For example, the CSP model of a subgraph in Figure 1b under graph Figure 1a is:

$$\begin{pmatrix} \mathcal{X} : \{x_{v_1}, x_{v_2}, x_{v_3}\}, \\ \mathcal{D} : \{\{u_1, u_2, ..., u_{10}\}, \{u_1, u_2, ..., u_{10}\}, \{u_1, u_2, ..., u_{10}\}\}, \\ \mathcal{C} : \{x_{v_1} \neq x_{v_2} \neq x_{v_3}, \\ \varphi(v_1) = \text{REF}, \varphi(v_2) = \text{GOV}, \varphi(v_3) = \text{GOV}, \\ \varphi(v_1, v_2) = 3, \varphi(v_2, v_3) = 9\} \end{pmatrix}$$

Theorem 2 [12] describes the relation between subgraph isomorphism and the CSP model. Intuitively, the CSP model is similar to a template, in which each variable in $\mathcal{X}$ is a slot. A *solution* is a correct slot fitting which assigns a different node of G to each node of G_1, such that the labels of the corresponding nodes and edges match. For instance, a solution to the CSP of the above example is the assignment $(x_{v_1}, x_{v_2}, x_{v_3}) = (u_2, u_4, u_5)$. If there exists a solution that assigns a node u to variable x_v, then this assignment is valid. $x_{v_1} = u_2$ is a valid assignment while $x_{v_1} = u_1$ is invalid in this example.

Theorem 2. *A solution of the subgraph G_1 to graph G CSP corresponds to a subgraph isomorphism of G_1 to G.*

Theorem 3. *Let $(\mathcal{X}, \mathcal{D}, \mathcal{C})$ be the subgraph CSP of G_1 under graph G. The MNI support of G_1 in G satisfying:*

$$Sup_{MNI}(G_1, G) \geq \tau \iff \forall x_v \in \mathcal{X}, ASS_{valid}(x_v) \geq \tau.$$

where $ASS_{valid}(x_v)$ is the total count of valid assignments of variable x_v.

According to Theorem 3 [12], we can consider the CSP of subgraph G_1 to graph G and check the count of valid assignments of each variable. If there exist τ or more valid assignments for every variable, in other words, at least τ nodes in each domain $D_1, ..., D_n$ for the corresponding variables $x_{v_1}, ..., x_{v_n}$, then subgraph G_1 is frequent under the MNI support. Thus, the main idea of the heuristic search strategy is elaborated as: if any variable domain remains with less than τ candidates during the search process, then the subgraph cannot be frequent.

3.4. Optimizing Support Evaluation

After subgraph extension, all the new generated subgraphs are sent to the next procedure for support evaluation. As mentioned in the Introduction, support evaluation is an NP-hard problem which takes $\mathcal{O}(N^n)$ time. The complexity is exponential if we brutally search all the valid assignments.

Owing to the iterative and incremental design of RDD and the join transformation in Spark, we save the CSP domain data of every generated subgraph. As the two green labels *join* shown in Figure 4, the first join operation combines the new generated subgraphs and frequent edges to get the extended edges, while the second *join* combines new generated subgraphs and extended edges to generate the search space, i.e., the CSP domain data. In addition, to speed up the support evaluation process, we also propose three optimizations, namely, load balancing, pre-search pruning and top-down pruning, the execution order of which is illustrated on the headpiece of Figure 4.

3.4.1. Load Balancing

The support evaluation process is implemented in parallel at subgraph scale, which means that each subgraph will be sent to an executor in the Spark cluster for support evaluation. The search space is highly dependent on the subgraph's CSP domain size. Nevertheless, new subgraphs may have different domain sizes which result in the phenomenon that some executors may finish searching fast while others are very slow. The final execution time of the whole cluster depends upon the last finished executor.

To overcome this unbalance, generally, the subgraphs distributed to various executors must have roughly the same domain sizes. Algorithm 2 illustrates the detailed process. Because the domain of the present subgraph is incrementally generated from the parent subgraph's domain of last iteration, we save the domain sizes of all subgraphs in each iteration. Then, according to the saved domain sizes of parent subgraphs, new generated subgraphs are re-ordered and partitioned to different executors (Lines 6–9).

Let n be the number of nodes of subgraph S, i.e., the domain size. Load balancing can be done in $\mathcal{O}(n)$ time.

Algorithm 2 Load Balancing

Input: RDD of subgraph set $SRDD$, Spark cluster parallelism n, Spark context sc
Output: $SRDD_{balance}$: balanced RDD of subgraph set
1: **function** LOADBALANCING($SRDD$,n,sc)
2:     $S_{balance} \leftarrow \varnothing$
3:     $S \leftarrow SRDD$
4:     sort S in descending domain size order
5:     initialize n empty partitions
6:     **for** s in S **do**
7:         $S_{balance}(p) \leftarrow$ smallest partition
8:         add s to partition $S_{balance}(p)$
9:     **end for**
10:     $SRDD_{rp} \leftarrow sc.parallelize(S_{balance}, n)$
11:     **return** $SRDD_{balance}$
12: **end function**

3.4.2. Pre-Search Pruning

Because the input single large graph we consider is an undirected labeled graph, if a node and its neighbors have the same node label and edges between them also have the same label, it will bring redundant search space. This phenomenon can be common in graphs, especially when the graphs have few node labels and edge labels. For example, in Figure 5, G is the input graph and G_1 a subgraph. The CSP search space of G_1 is illustrated at the bottom. The assignments in dashed lines are added to the search space when iteratively building the CSP domain data of G_1 whereas they are redundant space violating the first rule in Definition 10. Here, u_1 is assigned twice to v_1 and v_3 ($\{u_1, u_2, u_1\}$, $\{u_1, u_3, u_1\}$, $\{u_1, u_4, u_1\}$). If this redundant search space is pruned before calculating the actual support, the search speed will be much accelerated.

Let N and n be the number of nodes of input graph G and subgraph S. Pre-search pruning will search for redundant space for every node of S between its neighbors in G, the complexity of which is $\mathcal{O}(n \cdot N^3)$.

3.4.3. Top-Down Pruning

Either FFSM-Join operation or FFSM-Extend operation add an edge to the parent subgraph at a time when generating new subgraphs and constructing the suboptimal CAM tree. Therefore, as the parent subgraph at upside of suboptimal CAM tree is a substructure of its child subgraph, those assignments that were pruned from the domains of the parent, can also not be valid assignments for any of its children [12]. For instance, Figure 6a shows a part of a subgraph generation tree, which is constructed from G_1 which is extended to G_2 and G_3 and last, G_4 via G_3. The marked nodes in different colors represent the pruned assignments from the top to bottom. Invalid assignments from parent subgraphs are pruned from all their child subgraphs. Thus, the search space is reduced a lot. Take subgraph G_4 in Figure 6 as an example, when considering variable x_{v_1}, the search space has a size of $3 \times 2 \times 3 \times 2 = 36$ combinations, while without top-down pruning the respective search space size is $5 \times 3 \times 5 \times 4 = 300$ combinations.

Top-down pruning iterates for every node in S and for every value in each domain. Thus, the overall complexity is $\mathcal{O}(m \cdot N)$.

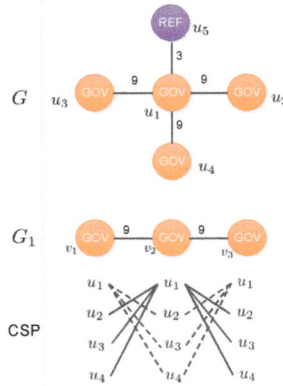

Figure 5. Constraint satisfaction problem (CSP) search space of the subgraph G_1 with the input graph G.

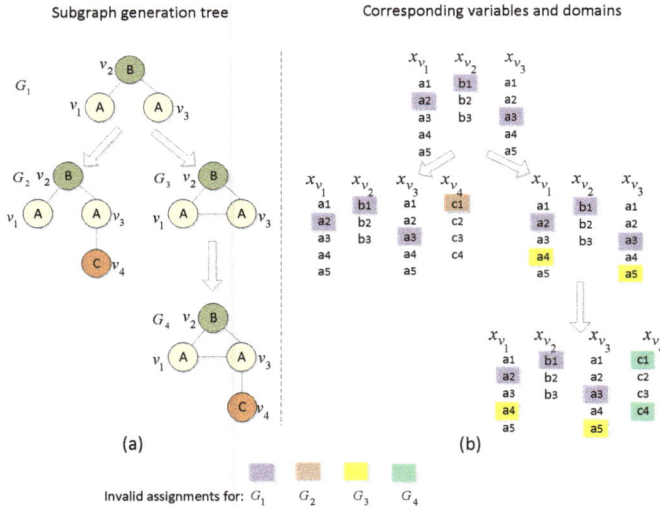

Figure 6. (a) The subgraph generation tree; and (b) the corresponding variables and domains. Marked nodes represent the pruned assignments from top to bottom.

After introducing pre-search pruning and top-down pruning, we give the pseudocode IsFREQUENT of heuristically checking whether a subgraph s is frequent in Algorithm 3. Pre-search pruning is conducted at Line 4 to Line 7 while top-down pruning Line 11 to Line 23.

3.5. The SSIGRAM Algorithm

Finally, we show the detail pipeline of the SSIGRAM approach in Algorithm 4. SSIGRAM starts by loading the input graph using Spark GraphX (Line 2). Then all frequent edges are identified at

Line 4. For each iteration, parallel subgraph extension is conducted at Line 11 and parallel support evaluation Line 14 to Line 22 in which load balancing, pre-search pruning and top-down pruning are conducted. The complexity of SSiGram is $\mathcal{O}(n \cdot N \cdot (N^2 + n) + n^4 + n)$ based on the above processes.

Algorithm 3 Support Evaluation.

Input: A subgraph s, domain data D_s, threshold τ
Output: true if s is a frequent subgraph of G, false otherwise
 1: **function** IsFrequent(s, D_s, τ)
 2:    **for** variable v in D_s **do**
 3:        get neighbors of v: $N(v) \leftarrow v$
 4:        **if** $N(v).size > 1$ **then**
 5:        **for** element u of $D_s(v)$ **do**
 6:           remove redundant u
 7:        **end for**
 8:    **end for**
 9:    **if** the size of any domain in D_s is less than τ **then**
10:    **return** *false*
11:    **for** variable v in D_s **do**
12:        *count* $\leftarrow 0$
13:        **for** element u of $D_s(v)$ **do**
14:           **if** u is already marked **then**
15:           *count++*
16:           **else if** a *solution* that assigns u to v exists **then**
17:           Mark corresponding values in D_s
18:           *count++*
19:           **else** Remove u from v's domain in D_s
20:           **if** *count* $= \tau$ **then**
21:           Move to next variable v
22:        **end for**
23:    **end for**
24:    **return** *true*
25: **end function**

---

4. Experimental Evaluation

In this section, the performance of the proposed algorithm SSiGram is evaluated using four real-world datasets with different sizes from different domains. Firstly, the experimental setup is introduced. The performance of SSiGram is then evaluated.

4.1. Experimental Setup

Dataset: We experiment on four real graph datasets, whose main characteristics are summarized in Table 1.

Table 1. Datasets and their characteristics.

Dataset	#Node	#L(Node)	#Edge	#L(Edge)	Density
DBLP	151,574	7	191,840	17	Medium
Aviation	101,185	6173	133,087	41	Sparse
GDELT	1,515,712	14,816	2,832,692	8	Sparse
Twitter	11,316,811	40	85,331,846	1	Dense

Algorithm 4 The SSIGRAM Algorithm.

Input: A graph G, support threshold τ and Spark context sc, parallelism n
Output: All subgraphs S of G such that $Sup_{MNI}(S, G) \geq \tau$
1: result $S \leftarrow \varnothing$, intermediate subgrah set $S_{rdd} \leftarrow \varnothing$
2: load $Graph_{rdd}$ from G using $sc.GraphX$
3: detect connected areas of $Graph_{rdd}$
4: $FreEdge_{rdd} \leftarrow Graph_{rdd}$
5: $FFSM \leftarrow FreEdge_{rdd}$
6: $FFSM_{rdd} = sc.\textbf{broadcast}(FFSM)$
7: initial frequent subgraphs $S_1 \leftarrow FreEdge_{rdd}$
8: $S_{rdd} \leftarrow S_1$
9: **while** $S_{rdd}.count > 0$ **do**
10:     $S_{k-1} \leftarrow S_{rdd}$
11:     $S_k \leftarrow \textbf{PARAGENSUBGRAPHS}(S_{k-1}, FFSM_{rdd}, sc)$
12:     $NewSub_{rdd} \leftarrow S_k$
13:     $OldSub_{rdd} \leftarrow NewSub_{rdd}.clone$
14:     $ExtEdge_{rdd} \leftarrow NewSub_{rdd}.\textbf{join}(FreEdge_{rdd})$
15:     $OldSub_{rdd} \leftarrow \textbf{LOADBALANCING}(OldSub_{rdd}, n, sc)$
16:     $CandSub_{rdd} \leftarrow OldSub_{rdd}.\textbf{join}(ExtEdge_{rdd})$
17:     $S_{rdd} \leftarrow CandSub_{rdd}.\textbf{flatmap}\{$
18:         let current subgraph be s
19:         $D_s \leftarrow s$
20:         **if** $\textbf{ISFREQUENT}(s, D_s, \tau)$ **then**
21:             **return** s
22:     $\}.\textbf{collect}$
23:     $S \leftarrow S \cup S_{rdd}$
24: **end while**
25: **return** S

DBLP (http://dblp.uni-trier.de/db/). The DBLP (DataBase systems and Logic Programming) bibliographic dataset models the co-authorship information and consists of 150 K nodes and nearly 200 K edges. Vertices represent authors and are labeled with the author's field of interest. Edges represent collaboration between two authors and are labeled with the number of co-authored papers.

Aviation (http://ailab.wsu.edu/subdue/). This dataset contains a list of event records extracted from the aviation safety database. The events are transformed to a graph which consists of 100 K nodes and 133 K edges. The nodes represent event ids and attribute values. Edges represent attribute names and the "near_to" relationship between two events.

GDELT (https://bigquery.cloud.google.com/table/gdelt-bq:full.events?pli=1). This dataset is constructed from part of the raw events exported from the GDELT (Global Data on Events, Location and Tone) dataset. It consists of 1.5 M nodes and 2.8 M edges. Similar to the Aviation dataset, nodes represent events and attribute values (and are labeled with event types and actual attribute values). Edges represents attribute name and the "relate_to" relationship between two events.

Twitter (http://socialcomputing.asu.edu/datasets/Twitter). This graph models the social news of Twitter and consists of 11M nodes and 85 M edges. Each node represents a Twitter user and each edge represents an interaction between two users. The original graph does not have labels, so we randomly added 40 labels to the nodes, the randomization of which follows a Gaussian distribution. In detail, the mean value was set to 50 and the std-deviance 15. The generated vertex labels less than 0 were all set to 1.

Comparison Method: We compare the proposed SSIGRAM algorithm with the GRAMI [12] and we use the *GRAMI_UNDIRECTED_SUBGRAPHS* version of GraMi provided by the authors.

Running Environment: All the experiments with SSIGRAM are conducted on Apache Spark (version 1.6.1) deployed on Apache Hadoop YARN (version 2.7.1). The total executors is set to 20 with 6 GB memory and 1 core running at 2.4 GHz for each executor. The memory of driver program is also 6 GB and max results 2 GB. Thus, the total memory allocated from YARN is 128 GB. For the sake of fairness, GRAMI is conducted on a Linux (Ubuntu 14.04) machine running at 2.4 GHz with 128 GB RAM.

Performance Metrics: The support threshold τ is the key evaluation metric as it determines whether a subgraph is frequent. Decreasing τ results in an exponential increase in the number of possible candidates and thus exponential decrease in the performance of the mining algorithms. For a given time budget, an efficient algorithm should be able to solve mining problems for low τ values. When τ is given, efficiency is determined by the running time. In addition, we also give the total subgraphs each algorithm identified under each τ value, proving the correctness of the SSIGRAM algorithm.

4.2. Experimental Results

Performance: At the top part of Figure 7, we show the performance comparison between SSIGRAM and GRAMI on DBLP, Aviation, GDELT and Twitter datasets. The number of results grows exponentially when the support threshold τ decreases. Thus, the running time of all algorithms also grows exponentially. Our results indicate that SSIGRAM outperforms GRAMI by an order of magnitude for all datasets. For bigger dataset GDELT and lower τ (9600), GRAMI ran out of memory and was not able to produce a result. For smaller dataset Aviation and bigger τ (2200, 2300 and 2400), GRAMI is faster because the resource scheduling of Hadoop YARN in SSIGRAM will cost 10 to 20 s. Actually, in this circumstance, there is no need to use parallel mining algorithms since GRAMI can give results within a few seconds. For the Twitter dataset, SSIGRAM is about five times faster than GRAMI because of the existence of nodes with a big degree. When a subgraph involves such a node, SSIGRAM will not go to the next iteration until the executor finishes calculating the support of this subgraph.

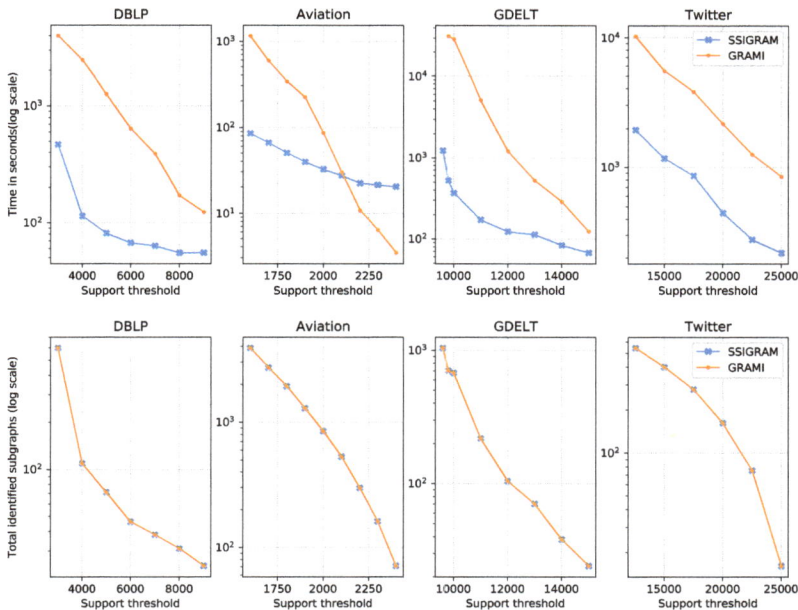

Figure 7. Performance of SSIGRAM and GRAMI on the four different datasets. Abbreviations: DBLP (DataBase systems and Logic Programming), GDELT (Global Data on Events, Location and Tone).

The bottom part of Figure 7 illustrates the total numbers of identified frequent subgraphs on each dataset. The identical numbers of frequent subgraphs of SSiGRAM and GRAMI elaborate the correctness of our SSiGRAM algorithm.

Optimization: Figure 8 demonstrates the effect of the three optimizations discussed above on the DBLP and GDELT datasets. For both datasets, the SSiGRAM with all optimizations (denoted by All opts. in Figure 8) definitely performs best. For the DBLP dataset, when $\tau > 3500$, load balancing is the most effective optimization while as τ becomes bigger, pre-search pruning becomes the most effective. For the GDELT dataset, the pre-pruning is always the most effective optimization. When no optimization is involved (denoted by No opt. in Figure 8), the algorithm performs worst. Actually, the effect of each optimization strategy varies with input graphs and different thresholds.

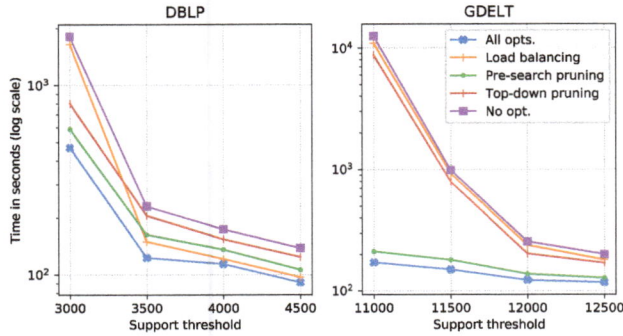

Figure 8. The effect of each optimization. **All opts.**: All optimization enabled; **Load balancing:** Only load balancing enabled; **Pre-pruning:** Only pre-search pruning enabled; **Top-down prune:** Only top-down pruning enabled; **No Opt.:** No optimization strategies involved.

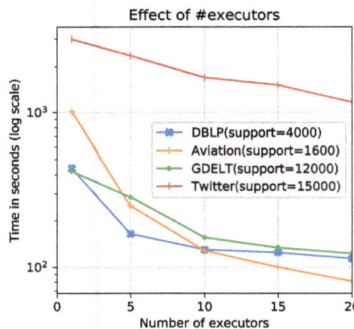

Figure 9. The effect of the number of executors on each dataset.

Parallelism: Finally, to evaluate the effect of the number of executors, we fix the supports of each dataset and vary the *num-executor* parameter of the Spark configuration file. According to the principle of the same allocated memory from YARN, we set *num-executors* to 20, 15, 10, 5 and 1 with *executor-memory* being 6 GB, 8 GB, 12 GB, 24 GB and 120 GB respectively. The results shown in Figure 9 lead to three major observations. First, compared with GRAMI's performance shown in Figure 7, the proposed algorithm outperforms the GRAMI algorithm even when only one executor was used. This is because the complexity of SSGRAMI is less than that of GRAMI ($\mathcal{O}(n \cdot N^{n-1})$), especially when n is large. Second, the runtime decreases with the increment of parallelism for each dataset overall. Third, when the *num-executor* is bigger than 10, the performance improvement is less obvious because the final performance will be dependent on a few time-consuming subgraphs. Thus, most executors will wait until these subgraphs are finished. More executors cannot avoid this phenomenon.

5. Conclusions

In this paper, we propose SSIGRAM, a novel parallel frequent subgraph mining algorithm in a single large graph using Spark, which conducts in parallel subgraph extension and support counting, respectively, focusing on the two core steps with high computational complexity in frequent subgraph mining. In addition, we also provide a heuristic search strategy and three optimizations for the support computing operation. Finally, extensive experimental performance evaluations are conducted with four graph datasets, showing the effectiveness of the proposed SSIGRAM algorithm.

Currently, the parallel execution is conducted on the scale of every generated subgraph. When a subgraph involves a node with very big degree, SSIGRAM will not go to next iteration until the executor finishes calculating the support of this subgraph. In the future, we plan to design a strategy that decomposes the evaluation task for this type of subgraph to all executors, accelerating the search speed further.

6. Discussion

The proposed algorithm in this paper is applied to FSM on a single large certain graph data, in which each edge definitely exists. However, uncertain graphs are also common and have practical importance in the real world, e.g., the telecommunication or electrical networks. In the uncertain graph model [33], each edge of a graph is associated with a probability to quantify the likelihood that this edge exists in the graph. Usually the existence of edges is assumed to be independent.

There are also two types of FSM on uncertain graphs: transaction based and single graph based. Most existing work on FSM on uncertain graphs is developed on transaction settings, i.e., multiple small/medium uncertain graphs. FSM on uncertain graph transactions under expected semantics considers a subgraph frequent if its expected support is greater than the threshold. Representative algorithms include Mining Uncertain Subgraph patterns (MUSE) [34], Weighted MUSE (WMUSE) [35], Uncertain Graph Index(UGRAP) [36] and Mining Uncertain Subgraph patterns under Probabilistic semantics (MUSE-P) [37]. They are proposed under expected semantics or the probabilistic semantics. Here, we mainly discuss the measurement of uncertainty and applications of techniques proposed in this paper considering single uncertain graph.

The measurement of uncertainty is important when considering an uncertain graph. Combining labelled graph in Definition 1 in this paper, an uncertain graph is a tuple $G^u = (G, P)$, where G is the *backbone* labelled graph, and $P : E \rightarrow (0, 1]$ is a probability function that assigns each edge e with an existence probability, denoted by $P(e), e \in E$. An uncertain graph G^u implies $2^{|E|}$ possible graphs in total, each of which is a structure G^u may exist as. The existence probability of G^i can be computed by the joint probability distribution:

$$P(G^u \Rightarrow G^i) = \prod_{e \in E_{Gi}} P(e) \prod_{e \in E_G/E_{Gi}} (1 - P(e)).$$

Generally speaking, FSM on single uncertain graph can also be divided into two phases: subgraph extension and support evaluation. The subgraph extension phase is the same as that for FSM on the backbone graph G. Thus, techniques used in this paper, such as canonical adjacency matrix for representing subgraphs and the parallel extension for extending subgraphs, can be used in the single uncertain graph.

The biggest difference lies in the support evaluation phase. The support of a subgraph g in an uncertain graph G^u is measured by *expected support*. A straightforward procedure to compute the expected support is generating all implied graphs, computing and aggregating the support of the subgraph in every implied graph, and last deriving the expected support, which can be accomplished by the CSP model used in this paper. Formally, the expected support is a probability distribution over the support in implied graphs:

$$eSup(g, G^u) = \sum_{i=1}^{2^{|E|}} P(G^u \Rightarrow G^i) \cdot Sup(g, G^i),$$

where G^i is an implied graph of G^u. The support measure Sup can be the MNI support introduced in Definition 5, which is computed efficiently. Thus, given an uncertain graph $G^u = (G, P)$ and an expected support threshold τ, FSM on an uncertain graph finds all subgraphs g whose expected support is no less than the threshold, i.e., $G = \{g | eSup(g, G^u) \geq \tau \wedge g \subseteq G\}$.

Furthermore, let $P_j(g, G^u)$ denote the aggregated probability that the support of g in an implied graph is no less than j:

$$P_j(g, G^u) = \sum_{G^i \in \Delta_j(g)} P(G^u \Rightarrow G^i),$$

where $\Delta_j(g) = \{G^i | Sup g, G^i \geq j\}$. The expected support can be reformulated as:

$$eSup(g, G^u) = \sum_{j=1}^{M_s} P_j(g, G^u),$$

where M_s is the maximum support of g among all implied graphs of G^u. For the detailed proof, we refer to [25]. However, it is #P-hard to compute $eSup(g, G^u)$ because of the huge number of implied graphs ($2^{|E|}$), which means that it is rather time consuming to draw exact frequent subgraph results even using the parallel evaluation with Spark platform proposed in this paper. Approximate evaluation with an error tolerance to allow some false positive frequent subgraphs is a common method. Some special optimization techniques other than optimizations in this paper must also be designed. Therefore, the modifications of expected support and some potential optimizations are still problems to be further studied to make the proposed algorithm be fit to mine frequent subgraphs on single uncertain graph.

Acknowledgments: This work was supported by (i) Hunan Natural Science Foundation of China(2017JJ336): Research on Individual Influence Prediction Based on Dynamic Time Series in Social Networks; and (ii) The Subject of Teaching Reform in Hunan: A Research on Education Model on Internet plus Assignments.

Author Contributions: Fengcai Qiao and Hui Wang conceived and designed the experiments; Fengcai Qiao performed the experiments; Xin Zhang and Pei Li analyzed the data; Zhaoyun Ding contributed reagents/materials/analysis tools; Fengcai Qiao and Shanshan Jia wrote the paper.

Conflicts of Interest: The authors declare no conflict of interest.

References

1. Chakrabarti, D.; Faloutsos, C. Graph mining: Laws, generators, and algorithms. *ACM Comput. Surv. (CSUR)* **2006**, *38*, 2.
2. Yan, X.; Han, J. gSpan: Graph-based substructure pattern mining. In Proceedings of the 2002 IEEE International Conference on Data Mining (ICDM 2002), Maebashi City, Japan, 9–12 December 2002; pp. 721–724.
3. Liu, Y.; Xu, S.; Duan, L. Relationship Emergence Prediction in Heterogeneous Networks through Dynamic Frequent Subgraph Mining. In Proceedings of the 23rd ACM International Conference on Information and Knowledge Management, Shanghai, China, 3–7 November 2014; ACM: New York, NY, USA, 2014; pp. 1649–1658.
4. Herrera-Semenets, V.; Gago-Alonso, A. A novel rule generator for intrusion detection based on frequent subgraph mining. *Ingeniare Rev. Chil. Ing.* **2017**, *25*, 226–234.
5. Qiao, F.; Wang, H. Computational Approach to Detecting and Predicting Occupy Protest Events. In Proceedings of the 2015 International Conference on Identification, Information, and Knowledge in the Internet of Things (IIKI), Beijing, China, 22–23 October 2015; pp. 94–97.
6. Pak, A.; Paroubek, P. Extracting Sentiment Patterns from Syntactic Graphs. In *Social Media Mining and Social Network Analysis: Emerging Research*; IGI Global: Hershey, PA, USA, 2013; pp. 1–18.

7. Choi, C.; Lee, Y.; Yoon, S.E. Discriminative subgraphs for discovering family photos. *Comput. Vis. Media* **2016**, *2*, 257–266.

8. Keneshloo, Y.; Cadena, J.; Korkmaz, G.; Ramakrishnan, N. Detecting and forecasting domestic political crises: A graph-based approach. In Proceedings of the 2014 ACM Conference on Web Science, Bloomington, IN, USA, 23–26 June 2014; ACM: New York, NY, USA, 2014; pp. 192–196.

9. Jiang, C.; Coenen, F.; Zito, M. A survey of frequent subgraph mining algorithms. *Knowl. Eng. Rev.* **2013**, *28*, 75–105.

10. Kuramochi, M.; Karypis, G. Finding Frequent Patterns in a Large Sparse Graph. *Data Min. Knowl. Discov.* **2005**, *11*, 243–271.

11. Berlingerio, M.; Bonchi, F.; Bringmann, B.; Gionis, A. Mining graph evolution rules. In *Machine Learning and Knowledge Discovery in Databases*; MIT Press: Cambridge, MA, USA, 2009; pp. 115–130.

12. Elseidy, M.; Abdelhamid, E.; Skiadopoulos, S.; Kalnis, P. GraMi: Frequent Subgraph and Pattern Mining in a Single Large Graph. *Proc. VLDB Endow.* **2014**, *7*, 517–528.

13. Wang, K.; Xie, X.; Jin, H.; Yuan, P.; Lu, F.; Ke, X. Frequent Subgraph Mining in Graph Databases Based on MapReduce. In *Advances in Services Computing, Proceedings of the 10th Asia-Pacific Services Computing Conference (APSCC 2016), Zhangjiajie, China, 16–18 November 2016*; Springer: Berlin, Germany, 2016; pp. 464–476.

14. Liu, Y.; Jiang, X.; Chen, H.; Ma, J.; Zhang, X. Mapreduce-based pattern finding algorithm applied in motif detection for prescription compatibility network. In *Advanced Parallel Processing Technologies*; Springer International Publishing AG: Cham, Switzerland, 2009; pp. 341–355.

15. Shahrivari, S.; Jalili, S. Distributed discovery of frequent subgraphs of a network using MapReduce. *Computing* **2015**, *97*, 1101–1120.

16. Aridhi, S.; d'Orazio, L.; Maddouri, M.; Nguifo, E.M. Density-based data partitioning strategy to approximate large-scale subgraph mining. *Inf. Syst.* **2015**, *48*, 213–223.

17. Hill, S.; Srichandan, B.; Sunderraman, R. An iterative mapreduce approach to frequent subgraph mining in biological datasets. In Proceedings of the ACM Conference on Bioinformatics, Computational Biology and Biomedicine, Orlando, FL, USA, 7–10 October 2012; ACM: New York, NY, USA, 2012; pp. 661–666.

18. Luo, Y.; Guan, J.; Zhou, S. Towards Efficient Subgraph Search in Cloud Computing Environments. In *Database Systems for Adanced Applications*; Springer International Publishing AG: Cham, Switzerland, 2011; pp. 2–13.

19. Talukder, N.; Zaki, M.J. A distributed approach for graph mining in massive networks. *Data Min. Knowl. Discov.* **2016**, *30*, 1024–1052.

20. White, T. *Hadoop: The Definitive Guide*, 1st ed.; O'Reilly Media, Inc.: Newton, MA, USA, 2009.

21. Zaharia, M.; Chowdhury, M.; Franklin, M.J.; Shenker, S.; Stoica, I. Spark: Cluster Computing with Working Sets. In Proceedings of the 2nd USENIX Conference on Hot Topics in Cloud Computing (HotCloud'10), Boston, MA, USA, 22–25 June 2010; USENIX Association: Berkeley, CA, USA, 2010; p. 10.

22. Meng, X.; Bradley, J.; Yavuz, B.; Sparks, E.; Venkataraman, S.; Liu, D.; Freeman, J.; Tsai, D.; Amde, M.; Owen, S.; et al. Mllib: Machine learning in apache spark. *J. Mach. Learn. Res.* **2016**, *17*, 1235–1241.

23. Huan, J.; Wang, W.; Prins, J. Efficient mining of frequent subgraphs in the presence of isomorphism. In Proceedings of the Third IEEE International Conference on Data Mining (ICDM 2003), Melbourne, FL, USA, 22–22 November 2003; pp. 549–552.

24. Jiang, C. Frequent Subgraph Mining Algorithms on Weighted Graphs. Ph.D. Thesis, University of Liverpool, Liverpool, UK, 2011.

25. Chen, Y.; Zhao, X.; Lin, X.; Wang, Y. Towards frequent subgraph mining on single large uncertain graphs. In Proceedings of the 2015 IEEE International Conference on Data Mining (ICDM), Atlantic City, NJ, USA, 14–17 November 2015; pp. 41–50.

26. Gibbons, A. *Algorithmic Graph Theory*; Cambridge University Press: Cambridge, UK, 1985.

27. West, D.B. *Introduction to Graph Theory*; Prentice Hall Upper Saddle River: Bergen County, NJ, USA; 2001; Volume 2.

28. Gerner, D.J.; Schrodt, P.A.; Yilmaz, O.; Abu-Jabr, R. *Conflict and Mediation Event Observations (CAMEO): A New Event Data Framework for the Analysis of Foreign Policy Interactions*; International Studies Association: New Orleans, LA, USA, 2002.

29. He, H.; Singh, A.K. Graphs-at-a-time: Query language and access methods for graph databases. In Proceedings of the 2008 ACM SIGMOD International Conference on Management of Data, Vancouver, BC, Canada, 9–12 June 2008; ACM: New York, NY, USA, 2008; pp. 405–418.

30. Fiedler, M.; Borgelt, C. Subgraph support in a single large graph. In Proceedings of the Seventh IEEE International Conference on Data Mining Workshops (ICDM Workshops 2007), Omaha, NE, USA, 28–31 October 2007; pp. 399–404.
31. Bringmann, B.; Nijssen, S. What is Frequent in a Single Graph? In Proceedings of the 12th Pacific-Asia Conference on Advances in Knowledge Discovery and Data Mining (PAKDD'08), Osaka, Japan, 20–23 May 2008; Springer: Berlin/Heidelberg, Germany, 2008; pp. 858–863.
32. Wörlein, M.; Meinl, T.; Fischer, I.; Philippsen, M. A quantitative comparison of the subgraph miners MoFa, gSpan, FFSM, and Gaston. In *Knowledge Discovery in Database: PKDD 2005*; Jorge, A., Torgo, L., Brazdil, P., Camacho, R., Gama, J., Eds.; Lecture Notes in Computer Science; Springer: Berlin, Germany, 2005; Volume 3721, pp. 392–403.
33. Jin, R.; Liu, L.; Aggarwal, C.C. Discovering highly reliable subgraphs in uncertain graphs. In Proceedings of the 17th ACM SIGKDD International Conference on Knowledge Discovery and Data Mining, San Diego, CA, USA, 21–24 August 2011; ACM: New York, NY, USA, 2011; pp. 992–1000.
34. Zou, Z.; Li, J.; Gao, H.; Zhang, S. Frequent subgraph pattern mining on uncertain graph data. In Proceedings of the 18th ACM Conference on Information and Knowledge Management, Hong Kong, China, 2–6 November 2009; ACM: New York, NY, USA, 2009; pp. 583–592.
35. Jamil, S.; Khan, A.; Halim, Z.; Baig, A.R. Weighted muse for frequent sub-graph pattern finding in uncertain dblp data. In Proceedings of the 2011 International Conference on Internet Technology and Applications (iTAP), Wuhan, China, 16–18 August 2011; pp. 1–6.
36. Papapetrou, O.; Ioannou, E.; Skoutas, D. Efficient discovery of frequent subgraph patterns in uncertain graph databases. In Proceedings of the 14th International Conference on Extending Database Technology, Uppsala, Sweden, 21–24 March 2011; ACM: New York, NY, USA, 2011; pp. 355–366.
37. Zou, Z.; Gao, H.; Li, J. Discovering frequent subgraphs over uncertain graph databases under probabilistic semantics. In Proceedings of the 16th ACM SIGKDD International Conference on Knowledge Discovery and Data Mining, Washington, DC, USA, 25–28 July 2010; ACM: New York, NY, USA, 2010; pp. 633–642.

![applied sciences logo] *applied sciences*

MDPI

Article

Predicting Human Behaviour with Recurrent Neural Networks

Aitor Almeida [†,*] and Gorka Azkune [†]

DeustoTech-Deusto Foundation, University of Deusto, Av. Universidades 24, 48007 Bilbao, Spain;
gorka.azkune@deusto.es
* Correspondence: aitor.almeida@deusto.es
† These authors contributed equally to this work.

Received: 15 December 2017; Accepted: 9 February 2018; Published: 20 February 2018

Abstract: As the average age of the urban population increases, cities must adapt to improve the quality of life of their citizens. The City4Age H2020 project is working on the early detection of the risks related to mild cognitive impairment and frailty and on providing meaningful interventions that prevent these risks. As part of the risk detection process, we have developed a multilevel conceptual model that describes the user behaviour using *actions*, *activities*, and *intra-* and *inter-activity behaviour*. Using this conceptual model, we have created a deep learning architecture based on long short-term memory networks (LSTMs) that models the inter-activity behaviour. The presented architecture offers a probabilistic model that allows us to predict the user's next actions and to identify anomalous user behaviours.

Keywords: long short-term memory networks; behavior modelling; intelligent environments; activity recognition

1. Introduction

As a result of the growth of the urban population worldwide [1], cities are consolidating their position as one of the central structures in human organizations. This concentration of resources and services around cities offers new opportunities to be exploited. Smart cities [2,3] are emerging as a paradigm to take advantage of these opportunities to improve their citizens' lives. Smart cities use the sensing architecture deployed in the city to provide new and disruptive city-wide services both to the citizens and the policy-makers. The large quantity of data available allows for improving the decision-making process, transforming the whole city into an intelligent environment at the service of its inhabitants. One of the target groups for these improved services is the "young–old" category, which comprises people aged from 60 to 69 [4] and that are starting to develop the ailments of old age. The early detection of frailty and mild cognitive impairments (MCI) is an important step to treat those problems. In order to do so, ambient assisted living (AAL) [5] solutions must transition from the homes to the cities.

City4Age is a H2020 research and innovation project with the aim of enabling age-friendly cities. The project aims to create an innovative framework of ICT tools and services that can be deployed by European cities in order to enhance the early detection of risk related to frailty and MCI, as well as to provide personalized interventions that can help the elderly population to improve their daily life by promoting positive behaviour changes. As part of the tools created for the framework, we have developed a series of algorithms for activity recognition and behaviour modelling. The recognized activities and the behaviour variations are then used to ascertain the frailty and MCI risks levels of the users. In the past, we have worked on creating single-user activity recognition algorithms for home environments [6–8], but in the case of City4Age, we have created algorithms that take into account both the large-scale scenarios of the project and the variation in the behaviours of the users.

In this paper, we present a conceptual classification of the user behaviour in intelligent environments according to the different levels of granularity used to describe it (Section 3). Using this classification, we also describe the algorithm developed to automatically model the inter-activity behaviour (Section 3). This algorithm uses long short-term memory networks (LSTMs) [9] to create a probabilistic model of user behaviour that can predict the next user action and detect unexpected behaviours. Our evaluation (Section 4) analyses different architectures for the creation of a statistical model of user behaviour.

2. Related Work

There are two main monitoring approaches for automatic human behaviour and activity evaluation, namely, vision- and sensor-based monitoring. For a review of vision-based approaches, [10] can be consulted. When approaching human behaviour and activity evaluation recognition in intelligent environments, sensor-based behaviour and activity evaluation are the most widely used solutions [11], as vision-based approaches tend to generate privacy concerns among the users [12]. Sensor-based approaches are based on the use of emerging sensor network technologies for behaviour and activity monitoring. The generated sensor data from sensor-based monitoring are mainly time series of state changes and/or various parameter values that are usually processed through data fusion, probabilistic or statistical analysis methods and formal knowledge technologies for activity recognition. There are two main approaches for sensor-based behaviour and activity recognition in the literature: data- and knowledge-driven approaches.

The idea behind data-driven approaches is to use data mining and machine learning techniques to learn behaviour and activity models. These are usually presented as a supervised learning approach, for which different techniques have been used to learn behaviours and activities from collected sensor data. Data-driven approaches need big datasets of labelled activities to train different kinds of classifiers. The learning techniques used in the literature are broad, from simple naive Bayes classifiers [13] to hidden Markov models [14], dynamic Bayesian networks [15], support vector machines [16] and online (or incremental) classifiers [17]. The knowledge-driven approaches try to use the existing domain knowledge in order to avoid using labelled datasets for the training. Rashidi and Cook [18] tried to overcome the problem of depending on manually labelled activity datasets by extracting activity clusters using unsupervised learning techniques. These clusters are used to train a boosted hidden Markov model, which is shown to be able to recognise several activities. Chen et al. [19] used logic-based approaches for activity recognition, on the basis of previous knowledge from the domain experts. Others have adopted ontology-based approaches, which allow for a commonly agreed explicit representation of activity definitions independent of algorithmic choices, thus facilitating portability, interoperability and reusability [20–22]. User behaviour in intelligent environments builds on user activities to describe the conduct of the user. Modelling user behaviour entails an abstraction layer over activity recognition. Behaviour models describe how specific users perform activities and what activities comprise their daily living. User behaviour prediction is an important task in intelligent environments. It allows us to anticipate user needs and to detect variations in behaviour that can be related to health risks. In the Mavhome project [23], the authors created algorithms to predict the users' mobility patterns and their device usage. Their algorithms, based mainly on sequence matching, compression and Markov models [24], allowed the intelligent environments to adapt to the user needs. Other authors have used prediction methods to recognize the user activities in smart environments [16]. An analysis of the importance of prediction in intelligent environments can be found in [25]. Prediction has also been used in intelligent environments for the control of artificial illumination [26] using neuro-fuzzy systems or for the control of climate parameters on the basis of user behaviour [27]. A more in-depth analysis of the use of user behaviour prediction for comfort management in intelligent environments can be found in [28]. As explained in the following section, we identify two types of behaviours: intra-activity behaviour (which describes how the user performs activities) and inter-activity behaviour (which describes the actions and activity sequences that compose the user's daily life). To model and predict the inter-activity behaviour, we use action sequences, as this allows us to have a fine-grained description of the user conduct while abstracting the model from specific sensor technology.

3. Actions, Activities and Behaviours

User behaviours are comprised of a large collection of defining elements, making them a complex structure. In order to properly describe it, we have defined a series of concepts on the basis of those proposed in [29]: *actions*, *activities* and *behaviours*. Actions describe the simplest conscious movements, while behaviours describe the most complex conduct (see Figure 1). We have extended the model proposed in [29] dividing the behaviours into two different types, *intra-activity behaviours* and *inter-activity behaviours*. This allows us to better model different aspects of the user's behaviour.

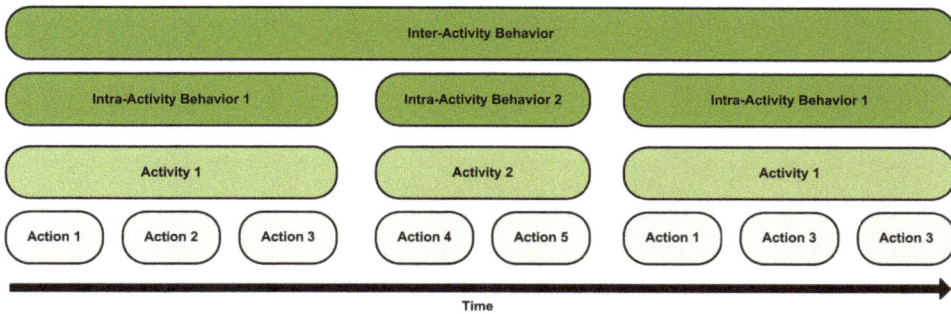

Figure 1. Elements of the user behaviour.

The different elements of the user behaviour are the following:

1. Actions are temporally short and conscious muscular movements made by the users (e.g., taking a cup, opening the fridge, etc.).
2. Activities are temporally longer but finite and are composed of several actions (e.g., preparing dinner, taking a shower, watching a movie, etc.).
3. Behaviours describe how the user performs these activities at different times. We have identified two types of behaviours. The intra-activity behaviours describe how a single activity is performed by a user at different times (e.g., while the user is preparing dinner, sometimes they may gather all the ingredients before starting, while on other occasions, the user may take them as they are needed). The inter-activity behaviours describe how the user chains different activities (e.g., on Mondays after having breakfast, the user leaves the house to go to work, but in the weekends they go to the main room).

The algorithm we present in this paper models the inter-activity behaviour, using actions to describe it. One of the characteristics of our algorithm is that it works on the action-space instead of the sensor-space. To be able to work with a more flexible representation of the information in the intelligent environments, we map the raw sensor data to actions [19]. Other authors have proved that actions are a suitable approach to model behaviours [30,31]. Using actions for behaviour recognition has also been tackled in the domain of plan and goal recognition. Hoey et al. [32] use the same definition of actions (short and conscious muscular movements) to analyse the hand-washing process of patients with dementia. In their case, the mapping is from a video to a set of actions. Krüger et al. [33] use actions to model activities using computational state-space models. In their case, the mapping is from inertial measurements to actions, and the actions are grouped in different classes. Although different types of sensors (video by Hoey et al., inertial measurements by Krüger et al., and binary sensors in our evaluation) and mapping methods are used, the same concept is present, working on the action-space instead of the sensor-space. The mapping procedure changes according to the sensors used.

The advantage of working on the action-space is that different sensor types may detect the same action type, simplifying and reducing the hypothesis space. This is even more important when using semantic embeddings to represent these actions in the model (as we explain in Section 3.1), as the reduced number of actions produces more significant embedding representations.

3.1. Semantic Embeddings for Action Representation

Traditionally in activity recognition or behaviour modelling tasks, the inputs (when using actions) have been represented as IDs, strings or one-hot vectors. The problem with this type of representation is that it does not contain any information about the action meaning. Using a one-hot vector alone, it is not possible to compute how similar two actions are, and this information is not available for the model that will use the actions. A similar problem occurs in the area of natural language processing (NLP) with the representation of words. The solution to this is to use embeddings [34] to represent the words, and we have used the same approach for the actions.

While one-hot vectors are sparse and the features of the model increase with the action dictionary size, embeddings are dense and more computationally efficient, with the number of features being constant, regardless of the number of action types. Most significantly for our model, embeddings provide semantic meaning to the representation of the actions. Each action is represented as a point in a multidimensional plane, which place them at a distance of the other actions, thus providing relations of similitude and significance between them. It is important to note that the system proposed in this paper does not compute explicitly the distances between the actions as an input for the model. Figure 2 shows an example of action embeddings. Given a sequence of actions $S_{act} = [a_1, a_2, ..., a_{l_a}]$, where l_a is the sequence length and $a_i \in \Re^{d_a}$ indicates the action vector of the ith action in the sequence, we let $Context(a_i) = [a_{i-n}, ..., a_{i-1}, a_{i+1}, ..., a_{i+n}]$ be the context of a_i, where $2n$ is the length of the context window. We let $p(a_i|Context(a_i))$ be the probability of the ith action in the sequence for action a_i. The target of the model used to create the embeddings is to optimize the log maximum likelihood estimation (logMLE):

$$L_a(MLE) = \sum_{a_i \in S} \log p(a_i|Context(a_i)) \tag{1}$$

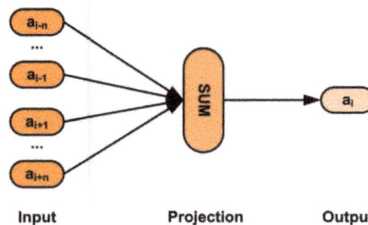

Figure 2. Representation of actions as embeddings.

In our model, we use the Word2Vec algorithm proposed by Mikolov et al. [34] to calculate the embeddings, a widely used embedding model inspired originally by the neural network language model developed by Bengio et al. [35]. Word2Vec has been used successfully in recent years in very varied tasks, such as knowledge path extraction [36], sentiment analysis [37] or sentence classification [38]. This combination of positive results and an easy to perform unsupervised learning process are the reason why we have decided to use the Word2Vec algorithm. In our model, we use the Word2Vec implementation in Gensim to calculate the embedding values for each action in the dataset. Gensim is one of the most popular Python vector-space modelling libraries. We represent each action with a vector of 50 float values, because of the small number of action instances compared with the number of words that are usually used in NLP tasks. Instead of providing the values directly to our model, we have included an embedding layer as the input to the model. In this layer, we store the procedural information on how to transform an action ID to its embedding. Adding this layer allows us to train it with the rest of the model and, in this way, fine-tune the embedding values to the current task, improving the general accuracy of the model.

3.2. LSTM-Based Network for Behaviour Modelling

In order to create a probabilistic model for behaviour prediction, we have used a deep neural network architecture (https://github.com/aitoralmeida/c4a_behavior_recognition) (see Figure 3) based on recurrent neural networks, specifically on LSTMs [9]. LSTMs are universal, in the sense that they can theoretically compute anything that a computer can, given enough network units. These types of networks are particularly well suited for modelling problems in which temporal relations are relevant and the time gaps between the events are of unknown size. LSTMs have also been shown to be prepared for sequential data structures [39]. In inter-activity behaviour modelling, the prediction of the activity label for an action depends on the actions registered before. The recurrent memory management of LSTMs allows us to model the problem considering those sequential dependencies.

While the LSTMs are the central element of the proposed architecture, they can be divided into three different parts: the input module, the sequence modelling module and the predictive module. The input module receives raw sensor data and maps it to actions using previously defined equivalences [6]. These actions are then fed to the embedding layer. As discussed in Section 3.1, embeddings provide a more expressive and dense model for the actions. The embedding layer receives the action IDs and transforms them into embeddings with semantic meaning. This layer is configured as trainable, that is, able to learn during the training process. The layer weights are initialized using the values obtained by using the Word2Vec algorithm, as explained in the previous section. The action embeddings obtained in the input module are then processed by the sequence modelling module. As discussed before, LSTMs are the type of network layer most suitable to do this. The layer has a size of 512 network units.

Finally, after the LSTM layer, we have the predictive module, which uses the sequence models created by the LSTMs to predict the next action. This module is composed by densely connected layers with different types of activations. This is a standard design in deep neural models. These densely connected layers are those that learn to predict the next actions on the basis of the models extracted by the previous layers. First we use two blocks of densely connected layers with rectified linear unit (ReLU) activations [40]. ReLU activations are one of the most widely used activations for intermediate densely connected layers in both classification and prediction problems. Each of these layers has a size of 1024 network units. After the ReLU activation, we use dropout regularization [41] with a value of 0.8. Dropout regularization prevents the complex co-adaptations of the fully connected layers by ignoring randomly selected neurons during the training process. This prevents overfitting during the training process. Finally, we use a third fully connected layer with a softmax activation function to obtain the next action predictions. As we want to select the most probable actions for a given sequence, softmax activation is the natural choice for the output layer.

In Section 4, we describe and analyse the different architectures and configurations that we have evaluated, which include combinations of multiple LSTMs and fully connected layers.

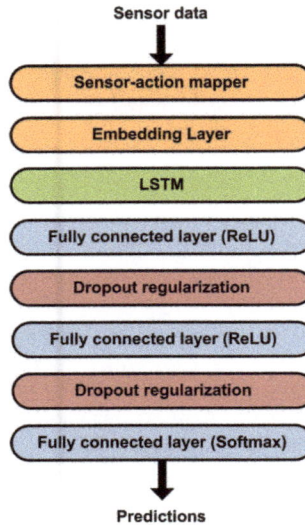

Figure 3. Long short-term memory network (LSTM) for behaviour modelling.

4. Evaluation

4.1. Experimental Setup

In order to validate our algorithm, we have used the dataset published by Kasteren et al. [14]. The dataset has been selected because it is widely used in the activity recognition and intelligent environment literature. This allows other researchers working in both areas to better compare the results of this paper with their own work. The dataset is the result of monitoring a 26-year-old man in a three-room apartment where 14 binary sensors were installed. These sensors were installed in locations such as doors, cupboards, refrigerators, freezers or toilets. Sensor data for 28 days was collected for a total of 2120 sensor events and 245 activity instances. The annotated activities were the following: "LeaveHouse", "UseToilet", "TakeShower", "GoToBed", "Prepare Breakfast", "Prepare Dinner" and "Get Drink". In this specific case, the sensors were mapped one to one to actions, resulting in the following set of actions: "UseDishwasher", "OpenPansCupboard", "ToiletFlush", "UseHallBedroomDoor", "OpenPlatesCupboard", "OpenCupsCupboard", "OpenFridge", "UseMicrowave", "UseHallBathroomDoor", "UseWashingmachine", "UseHallToiletDoor", "OpenFreezer", "OpenGroceriesCupboard" and "UseFrontdoor".

For the training process, the dataset was split into a training set (80% of the dataset) and a validation set (20% of the dataset) of continuous days. These sets were composed by the raw sensor data provided by Kasteren et al. In order to make the training process more streamlined, we apply the sensor to action mappings offline. This allows us to train the deep neural model faster while still having the raw sensor data as the input. To do the training, we use n actions as the input (as described in the sequence-length experiments) to predict the next action (see Section 4.2 for a description of how accuracy is evaluated). That is, the training examples are the sequences of actions, and the label is the next action that follows that sequence, being a supervised learning problem. The proposed architectures have been implemented using Keras and were executed using TensorFlow as the back-end. Each of the experiments was trained for 1000 epochs, with a batch size of 128, using categorical cross-entropy as the loss function and Adam [42] as the optimizer. After the 1000 epochs, we selected the best model using the validation accuracy as the fitness metric. The action embeddings were calculated using the full training set extracted from the Kasteren dataset and using the Word2Vec [34] algorithm, and the embedding layer was configured as trainable.

To validate the results of the architecture, we performed three types of experiments:

- Architecture experiments: we evaluated different architectures, varying the number of LSTMs and fully connected dense layers.
- Sequence length experiments: we evaluated the effects of altering the input action sequence length.
- Time experiments: we evaluated the effects of taking into account the timestamps of the input actions.

For the architecture experiments (see Table 1), we tried different dropout values (with a dropout regularization after each fully connected layer with a ReLU activation), different numbers of LSTM layers, different types of LSTM layers (normal and bidirectional [43]), different numbers of fully connected layers and different sizes of fully connected layers. We also compared using embeddings for the representation of the actions versus the more traditional approach of using one-hot vectors, in order to ascertain the improvements that the embeddings provide.

Table 1. Architecture experiments. *Dropout* is the value of the dropout regularizations. *LSTM No.* is the number of long short-term memory network (LSTM) layers in the architecture. *LSTM size* is the size of the LSTM layers. *Dense No.* is the number of fully connected layers with a rectified linear unit (ReLU) activation (all the architectures have a final fully connected layer with a softmax activation). *Dense size* is the size of the fully connected layers with a ReLU activation. *Sequence length* is the length of the input action sequence. *Coding* is the codifying strategy used for the actions, either embeddings or a one-hot vector.

ID	Dropout	LSTM #	LSTM Size	Dense #	Dense Size	Sequence Length	Coding
A1	0.4	1 (Standard)	512	1	1024	5	Embedding
A2	0.8	1 (Standard)	512	1	1024	5	Embedding
A3	0.8	1 (Standard)	512	2	1024	5	Embedding
A4	0.8	2 (Standard)	512	2	1024	5	Embedding
A5	0.2	1 (Standard)	512	5	50	5	Embedding
A6	0.8	1 (Bidirectional)	512	2	1024	5	Embedding
A7	0.8	1 (Standard)	512	2	1024	5	One-hot vector
A8	0.8	1 (Standard)	512	1	1024	5	One-hot vector
A9	0.8	2 (Standard)	512	2	1024	5	One-hot vector

For the sequence-length experiments (see Table 2), we varied the length of the input action sequence in a network, but maintained the rest of the values to a dropout regularization of 0.8, one LSTM layer with a size of 512, two fully connected layers with ReLU activation with a size of 1024 and one final fully connected layer with softmax activation. This was the same configuration used in experiment A3.

Table 2. Sequence-length experiments. Each experiment followed the same configuration used in A3 (see Table 1).

ID	Sequence Length
S1	3
S2	1
S3	4
S4	6
S5	10
S6	30

Finally for the time experiments (see Figure 4), we tried different ways of taking into account the timestamps of the actions in the input sequence. We analysed three different options. In the first configuration (T1), we used two parallel LSTM layers, one for the action embeddings and the other for the timestamps, concatenating the results of both layers before the fully connected layers (late fusion strategy [44]). In the second configuration (T2), we concatenated the action embeddings and the timestamps before a single LSTM layer (early fusion strategy [44]). In the third configuration (T3), the embeddings were connected to an LSTM layer, whose output was concatenated with the timestamps and sent to another LSTM layer (slow fusion strategy [44]). All the configurations used a dropout regularization of 0.8, a LSTM layer size of 512, two fully connected layers with ReLU activation with a size of 1024 and one final fully connected layer with softmax activation, as in experiment A3. In all three architectures, the timestamps were represented as a two-position vector ($[x_t, y_t]$), representing the time in a circle with a radius of 1 and centred at the (0,0) coordinate. The x_t and y_t values were calculated respectively as the cosine and sine of the angle ang_{sec}:

$$ang_{sec} = \frac{secs_{timestamp} * 2 * \pi}{secs_{day}} \tag{2}$$

where $secs_{timestamp}$ is the timestamp expressed as seconds since midnight and $secs_{day}$ are the total seconds in a day. Using ang_{sec}, x_t is expressed as

$$x_t = \cos(ang_{sec}) \tag{3}$$

and y_t is expressed as

$$y_t = \sin(ang_{sec}) \tag{4}$$

4.2. Metrics

To properly validate the predicting capabilities of the proposed architectures, we evaluated how they perform using the top-k accuracy. The top-k accuracy is a standard metric in different prediction and modelling tasks. The top-k accuracy (acc_at_k) is defined as

$$acc\_at\_k = \frac{1}{N} \sum_{i=1}^{N} 1[a_i \in C_i^k] \tag{5}$$

where a_i is the expected action and C_i^k is the set of the top k predicted actions; $1[.] \rightarrow \{0,1\}$ represents the scoring function; when the condition in the first part is true, the function value is 1; otherwise, the value is 0. In our case, if the ground-truth action is in the set of k predicted actions, the function value is 1. To evaluate our models, we provide the accuracy for k values of 1, 2, 3, 4 and 5.

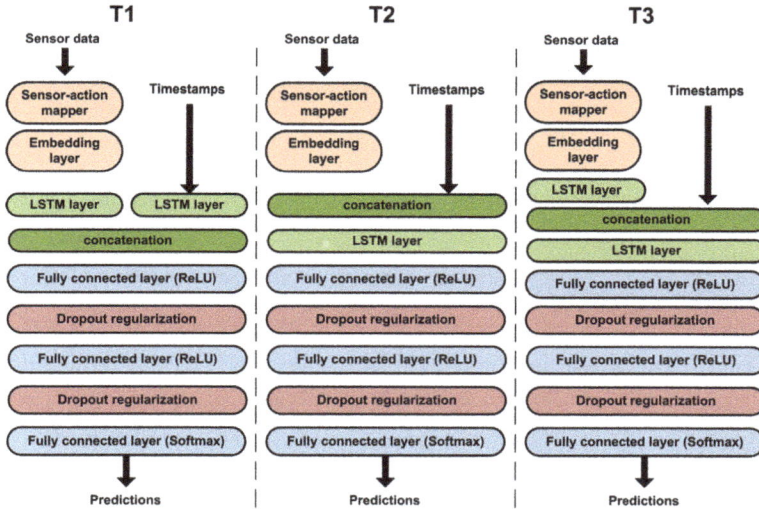

Figure 4. Architectures used in the time experiments. Each of the experiments used a different fusion strategy: T1 used late fusion, T2 used early fusion and T3 used slow fusion.

4.3. Results and Discussion

Tables 3–5 show the results for each type of experiment. In the case of the architecture experiments (Table 3), adding more LSTM layers (A4) or more smaller, densely connected layers (A5) reduced the overall accuracy of the model no matter the number of predictions. The usage of bidirectional LSTMs (A6) was also detrimental. Generally, higher dropout regularization values (A2 and A3) produced better results. Overall, the A3 configuration offered the best results across the board. As can be seen in the table, the use of embeddings obtained better results when compared to the same configurations using one-hot vectors (A3 vs. A7, A2 vs. A8 and A4 vs. A9). A more semantically rich representation of the actions allows the same architecture to better model the user behaviour. The advantage of using embeddings is that the training process to obtain them is completely unsupervised, not requiring a significant overheat when comparing them against other representation strategies such as one-hot vectors.

For the sequence-length experiments (Table 4), the best results were achieved using sequences with a length of 4 (S3) or 5 (A3) actions. In this specific case, the optimal sequence length was closely related with each deployment, being determined by the average action length of the activities in each scenario. This value should be adjusted in each specific case in order to achieve the best results, and it cannot be generalized.

Finally for the time experiments (Table 5), the evaluation shows that none of the proposed options to take into account the timestamps (T1, T2 and T3) improved the results. When comparing these with an architecture with a similar configuration and no timestamps (A3), the results were clearly worse. Our initial intuition was that taking into account the timestamps would help to model the temporal patterns in the actions, but the results show that this information is not so relevant for the behaviour modelling task. We expect that the temporal data will be much more relevant in the activity recognition task, particularly when discriminating between activity patterns with similar actions that happen at different periods (preparing breakfast vs. preparing dinner).

Table 3. Architecture experiments: Accuracy for different number of predictions.

ID	acc_at_1	acc_at_2	acc_at_3	acc_at_4	acc_at_5
A1	0.4487	**0.6367**	0.7094	**0.7948**	0.8461
A2	0.4530	0.6239	**0.7222**	0.7692	0.8504
A3	**0.4744**	0.6282	0.7179	0.7905	**0.8589**
A4	0.4444	0.5940	0.6965	0.7735	0.8247
A5	0.4402	0.5982	0.7136	0.7820	0.8418
A6	0.4487	0.6068	0.7136	0.7905	0.8376
A7	0.4572	0.6153	0.7094	0.7820	0.8376
A8	0.4529	0.5811	0.7051	0.7735	0.8376
A9	0.4102	0.5940	0.7008	0.7777	0.8247

Table 4. Sequence-length experiments: Accuracy for different number of predictions.

ID	acc_at_1	acc_at_2	acc_at_3	acc_at_4	acc_at_5
A3	**0.4744**	0.6282	0.7179	0.7905	**0.8589**
S1	0.4553	0.5957	0.7021	0.8	0.8553
S2	0.4255	0.6255	0.7021	**0.8085**	0.8382
S3	0.4658	**0.6452**	**0.7264**	0.7948	0.8504
S4	0.4700	0.6196	0.6965	0.7692	0.8461
S5	0.4592	0.6351	0.7210	0.7896	0.8369
S6	0.4192	0.5589	0.6593	0.7554	0.8122

Table 5. Time experiments: Accuracy for different number of predictions. The A3 experiment depicted in the first row did not use any time information.

ID	acc_at_1	acc_at_2	acc_at_3	acc_at_4	acc_at_5
A3	**0.4744**	**0.6282**	**0.7179**	**0.7905**	**0.8589**
T1	0.4487	0.6239	0.7094	0.7692	0.8076
T2	0.4487	0.6111	0.7008	0.7692	0.8247
T3	0.3846	0.5940	0.7051	0.7564	0.8076

A summary of the best results can be seen in Figure 5.

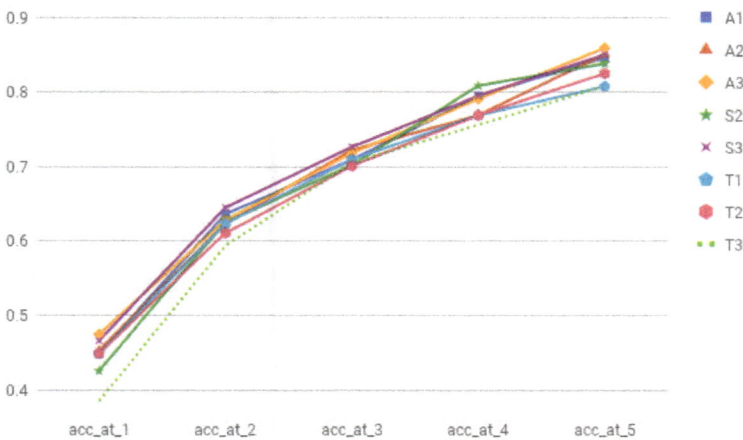

Figure 5. Best results in all the experiments.

5. Conclusions and Future Work

In this paper, we have proposed a multilevel conceptual model that describes the user behaviour using actions, activities, intra-activity behaviour and inter-activity behaviour Using this conceptual model, we have presented a deep learning architecture based on LSTMs that models inter-activity behaviour. Our architecture offers a probabilistic model that allows us to predict the user's next actions and to identify anomalous user behaviours. We have evaluated several architectures, analysing how each one of them behaves for a different number of action predictions. The proposed behaviour model is being used in the City4Age H2020 project to detect the risks related to the frailty and MCI in young elders.

As future work, we plan to continue studying how the temporal and spatial information could be integrated into the architecture to improve the statistical model. We would like to explore other deep learning architectures, for example, using convolutional neural networks (CNNs) instead of LSTMs. Using CNNs, we intend to model the different action n-grams that occur in the inter-activity behaviour. This will allow us to compare different sequence modelling approaches using deep neural models. We also plan to develop architectures that will cover other aspects of the proposed multilevel conceptual model, starting with the activity recognition task. We plan to use the insights gained with the deep learning architectures proposed in this paper in order to create an activity recognition algorithm.

Acknowledgments: This work has been supported by the European Commission under the City4Age Project Grant Agreement (No. 689731). We gratefully acknowledge the support of the NVIDIA Corporation with the donation of the Titan X Pascal GPU used for this research.

Author Contributions: Aitor Almeida and Gorka Azkune conceived, designed and performed the experiments; analyzed the data; and wrote the paper.

Conflicts of Interest: The authors declare no conflict of interest.

References

1. United Nations. *World Urbanization Prospects: The 2014 Revision, Highlights*; Department of Economic and Social Affairs, Population Division, United Nations: New York, NY, USA, 2014.
2. Caragliu, A.; Del Bo, C.; Nijkamp, P. Smart cities in Europe. *J. Urban Technol.* **2011**, *18*, 65–82.
3. Shapiro, J.M. Smart cities: Quality of life, productivity, and the growth effects of human capital. *Rev. Econ. Stat.* **2006**, *88*, 324–335.
4. Forman, D.E.; Berman, A.D.; McCabe, C.H.; Baim, D.S.; Wei, J.Y. PTCA in the Elderly: The "Young-Old" versus the "Old-Old". *J. Am. Geriatr. Soc.* **1992**, *40*, 19–22.
5. Rashidi, P.; Mihailidis, A. A survey on ambient-assisted living tools for older adults. *IEEE J. Biomed. Health Inform.* **2013**, *17*, 579–590.
6. Azkune, G.; Almeida, A.; L López-de-Ipiña, D.; Chen, L. Extending knowledge-driven activity models through data-driven learning techniques. *Expert Syst. Appl.* **2015**, *42*, 3115–3128.
7. Azkune, G.; Almeida, A.; López-de-Ipiña, D.; Chen, L. Combining users' activity survey and simulators to evaluate human activity recognition systems. *Sensors* **2015**, *15*, 8192–8213.
8. Bilbao, A.; Almeida, A.; López-de-Ipiña, D. Promotion of active ageing combining sensor and social network data. *J. Biomed. Inform.* **2016**, *64*, 108–115.
9. Hochreiter, S.; Schmidhuber, J. Long short-term memory. *Neural Comput.* **1997**, *9*, 1735–1780.
10. Weinland, D.; Ronfard, R.; Boyer, E. A survey of vision-based methods for action representation, segmentation and recognition. *Comput. Vis. Image Underst.* **2011**, *115*, 224–241.
11. Chen, L.; Hoey, J.; Nugent, C.D.; Cook, D.J.; Yu, Z. Sensor-based activity recognition. *IEEE Trans. Syst. Man Cybern. Part C (Appl. Rev.)* **2012**, *42*, 790–808.
12. Yilmaz, A.; Javed, O.; Shah, M. Object tracking: A survey. *ACM Comput. Surv. (CSUR)* **2006**, *38*, 13, doi:10.1145/1177352.1177355.
13. Bao, L.; Intille, S.S. Activity recognition from user-annotated acceleration data. In Proceedings of the International Conference on Pervasive Computing, Linz and Vienna, Austria, 21–23 April 2004; Springer: Berlin/Heidelberg, Germany, 2004; pp. 1–17.

14. Van Kasteren, T.; Noulas, A.; Englebienne, G.; Kröse, B. Accurate activity recognition in a home setting. In Proceedings of the 10th International Conference on Ubiquitous Computing, Seoul, Korea, 21–24 September 2008; ACM: New York, NY, USA, 2008; pp. 1–9.

15. Oliver, N.; Garg, A.; Horvitz, E. Layered representations for learning and inferring office activity from multiple sensory channels. *Comput. Vis. Image Underst.* **2004**, *96*, 163–180.

16. Fatima, I.; Fahim, M.; Lee, Y.K.; Lee, S. A unified framework for activity recognition-based behavior analysis and action prediction in smart homes. *Sensors* **2013**, *13*, 2682–2699.

17. Ordóñez, F.J.; Iglesias, J.A.; De Toledo, P.; Ledezma, A.; Sanchis, A. Online activity recognition using evolving classifiers. *Expert Syst. Appl.* **2013**, *40*, 1248–1255.

18. Rashidi, P.; Cook, D.J. COM: A method for mining and monitoring human activity patterns in home-based health monitoring systems. *ACM Trans. Intell. Syst. Technol. (TIST)* **2013**, *4*, 64, doi:10.1145/2508037.2508045.

19. Chen, L.; Nugent, C.D.; Mulvenna, M.; Finlay, D.; Hong, X.; Poland, M. A logical framework for behaviour reasoning and assistance in a smart home. *Int. J. Assist. Robot. Mechatron.* **2008**, *9*, 20–34.

20. Riboni, D.; Bettini, C. COSAR: Hybrid reasoning for context-aware activity recognition. *Pers. Ubiquitous Comput.* **2011**, *15*, 271–289.

21. Chen, L.; Nugent, C.D.; Wang, H. A knowledge-driven approach to activity recognition in smart homes. *IEEE Trans. Knowl. Data Eng.* **2012**, *24*, 961–974.

22. Aloulou, H.; Mokhtari, M.; Tiberghien, T.; Biswas, J.; Yap, P. An adaptable and flexible framework for assistive living of cognitively impaired people. *IEEE J. Biomed. Health Inform.* **2014**, *18*, 353–360.

23. Das, S.K.; Cook, D.J.; Battacharya, A.; Heierman, E.O.; Lin, T.Y. The role of prediction algorithms in the MavHome smart home architecture. *IEEE Wirel. Commun.* **2002**, *9*, 77–84.

24. Cook, D.J.; Youngblood, M.; Heierman, E.O.; Gopalratnam, K.; Rao, S.; Litvin, A.; Khawaja, F. MavHome: An agent-based smart home. In Proceedings of the First IEEE International Conference on Pervasive Computing and Communications, (PerCom 2003), Fort Worth, TX, USA, 26 March 2003; IEEE: Piscataway, NJ, USA, 2003; pp. 521–524.

25. Cook, D.J.; Das, S.K. How smart are our environments? An updated look at the state of the art. *Pervasive Mobile Comput.* **2007**, *3*, 53–73.

26. Kurian, C.P.; Kuriachan, S.; Bhat, J.; Aithal, R.S. An adaptive neuro-fuzzy model for the prediction and control of light in integrated lighting schemes. *Light. Res. Technol.* **2005**, *37*, 343–351.

27. Morel, N.; Bauer, M.; El-Khoury, M.; Krauss, J. Neurobat, a predictive and adaptive heating control system using artificial neural networks. *Int. J. Sol. Energy* **2001**, *21*, 161–201.

28. Dounis, A.I.; Caraiscos, C. Advanced control systems engineering for energy and comfort management in a building environment—A review. *Renew. Sustain. Energy Rev.* **2009**, *13*, 1246–1261.

29. Chaaraoui, A.A.; Climent-Pérez, P.; Flórez-Revuelta, F. A review on vision techniques applied to human behaviour analysis for ambient-assisted living. *Expert Syst. Appl.* **2012**, *39*, 10873–10888.

30. Schank, R.C. *Dynamic Memory: A Theory of Reminding and Learning in Computers and People*; Cambridge University Press: Cambridge, UK, 1983.

31. Schank, R.C.; Abelson, R.P. *Scripts, Plans, Goals and Understanding, an Inquiry into Human Knowledge Structures*; Lawrence Erlbaum Associates: Mahwah, NJ, USA, 1977.

32. Hoey, J.; Poupart, P.; von Bertoldi, A.; Craig, T.; Boutilier, C.; Mihailidis, A. Automated handwashing assistance for persons with dementia using video and a partially observable Markov decision process. *Comput. Vis. Image Underst.* **2010**, *114*, 503–519.

33. Krüger, F.; Nyolt, M.; Yordanova, K.; Hein, A.; Kirste, T. Computational state space models for activity and intention recognition. A feasibility study. *PLoS ONE* **2014**, *9*, e109381, doi:10.1371/journal.pone.0109381.

34. Mikolov, T.; Sutskever, I.; Chen, K.; Corrado, G.S.; Dean, J. Distributed representations of words and phrases and their compositionality. In *Advances in Neural Information Processing Systems*; The MIT Press: Cambridge, MA, USA, 2013; pp. 3111–3119.

35. Bengio, Y.; Ducharme, R.; Vincent, P.; Jauvin, C. A neural probabilistic language model. *J. Mach. Learn. Res.* **2003**, *3*, 1137–1155.

36. Wang, Z.; Zhang, J.; Feng, J.; Chen, Z. Knowledge Graph Embedding by Translating on Hyperplanes. In Proceedings of the Twenty-Eighth AAAI Conference on Artificial Intelligence, Hilton, QC, Canada, 27–31 July 2014; pp. 1112–1119.

37. Dos Santos, C.N.; Gatti, M. Deep Convolutional Neural Networks for Sentiment Analysis of Short Texts. In Proceedings of the COLING, the 25th International Conference on Computational Linguistics: Technical Papers, Dublin, Ireland, 23–29 August 2014; pp. 69–78.

38. Kim, Y. Convolutional Neural Networks for Sentence Classification. 2014, arXiv:1408.5882. arXiv.org e-Print archive. Available online: https://arxiv.org/abs/1408.5882 (accessed on 15 December 2017).

39. Lipton, Z.C.; Berkowitz, J.; Elkan, C. A Critical Review of Recurrent Neural Networks for Sequence Learning. 2015, arXiv:1506.00019. arXiv.org e-Print archive. Available online: https://arxiv.org/abs/1506.00019 (accessed on 15 December 2017).

40. Glorot, X.; Bordes, A.; Bengio, Y. Deep Sparse Rectifier Neural Networks. In Proceedings of the Fourteenth International Conference on Artificial Intelligence and Statistics, PMLR, Ft. Lauderdale, FL, USA, 11–13 April 2011; Volume 15, No. 106, p. 275.

41. Srivastava, N.; Hinton, G.E.; Krizhevsky, A.; Sutskever, I.; Salakhutdinov, R. Dropout: A simple way to prevent neural networks from overfitting. *J. Mach. Learn. Res.* **2014**, *15*, 1929–1958.

42. Kingma, D.; Ba, J. Adam: A method for stochastic optimization. In Proceedings of the 3rd International Conference for Learning Representations, San Diego, CA, USA, 7–9 May 2015.

43. Schuster, M.; Paliwal, K.K. Bidirectional recurrent neural networks. *IEEE Trans. Signal Process.* **1997**, *45*, 2673–2681.

44. Karpathy, A.; Toderici, G.; Shetty, S.; Leung, T.; Sukthankar, R.; Fei-Fei, L. Large-scale video classification with convolutional neural networks. In Proceedings of the IEEE Conference on Computer Vision and Pattern Recognition, Columbus, OH, USA, 23–28 June 2014; pp. 1725–173.

![applied sciences logo] *applied* *sciences*

MDPI

Article

Path Planning Strategy for Vehicle Navigation Based on User Habits

Pengzhan Chen, Xiaoyan Zhang, Xiaoyue Chen * and Mengchao Liu

School of Electrical Engineering and Automation, East China Jiaotong University, Nanchang 330013, China;
cyxcpz@163.com (P.C.); xyzhangbinggo@yeah.net (X.Z.); m15797916990@163.com (M.L.)
* Correspondence: wdmxdefeng@126.com; Tel.: +86-157-9791-9086

Received: 6 February 2018; Accepted: 6 March 2018; Published: 9 March 2018

Abstract: Vehicle navigation is widely used in path planning of self driving travel, and it plays an increasing important role in people's daily trips. Therefore, path planning algorithms have attracted substantial attention. However, most path planning methods are based on public data, aiming at different driver groups rather than a specific user. Hence, this study proposes a personalized path decision algorithm that is based on user habits. First, the categories of driving characteristics are obtained through the investigation of public users, and the clustering results corresponding to the category space are obtained by log fuzzy C-means clustering algorithm (LFCM) based on the driving information contained in the log trajectories. Then, the road performance personalized quantization algorithm evaluation is proposed to evaluate roads from the user's field of vision. Finally, adaptive ant colony algorithm is improved and used to validate the path planning based on the road performance personalized values. Results show that the algorithm can meet the personalized requirements of the user path selection in the path decision.

Keywords: individualization; dynamic path planning; driving habits; personalized performance evaluation

1. Introduction

With the rapid development of the modern automobile industry, self-driving groups are steadily increasing and road traffic pressure is increasing. These changes do not match users' demand for comfortable travel. In addition to the expansion of existing road networks, attempts have been made to avoid the clustering effect caused by the planning of the integration of personalized travel path for users. For example, the full use of existing resources in public transportation can ease traffic pressure and improve user experience in personalized travel. Therefore, studying the personalized paths of users is of great social and economic significance.

Mature navigation systems and path planning algorithms mainly focus on the fastest path [1–4], the shortest path [5–8] or the most comfortable path [4,9] between specified starting and target points. In recent years, path optimization algorithms of multi-objective optimization [10–14] and single-objective optimization [15,16], as well as a few traffic impact factors, have deepened [17,18] and popular routes planning [19,20]. Moreover, significant attention has been directed toward personalized path guidance systems. Campigotto P [20] introduces the favorite route recommendation (FAVOUR) approach to provide a personalized, situation-aware route based on the information updating through Bayesian learning, which was obtained from initial configuration files (home location, work place, mobility options, etc.). A personalized fuzzy path planning algorithm based on the fuzzy sorting of the center of gravity was proposed by Nadi [21], in which a optimization route according to user standard types was formulated through analyzing the uncertainty of user preferences through the expression of fuzzy linguistic preference relations. A personalized recommendation route algorithm based on large trajectory data was proposed by Dai J [22] where both drivers' driving preferences and multiple travel costs were considered to recommend personalized routes to individual drivers, however, the shortcoming was that the trajectory data were based on public drivers. A trip router planning method with individualized preferences was proposed by Letchner J [23] which presents a set of methods for including driver preferences and time-variant traffic condition estimates during route planning. Zhu X [24] put forward a personalized and time-sensitive route recommendation system, in which user preferences and time information were acquired through location-based social network where users' location and access information was shared by different users. Qiong Long [25] proposed a dynamic route guidance method for drivers' personalized needs where a qualitative and personalized evaluation was conducted on road networks with consideration of preferences according to the premise of obtaining user preferences, and then Dijkstra algorithm was used to find the optimal path.

The aforementioned results lay a good theoretical foundation; however, there still exist three shortcomings: (1) the personalized path planning algorithm is aimed at different driver groups but not at a specific user; (2) driver feature data are based on public trajectories or public information, which leads to the lack of pertinence; and (3) personalized performance values fail to be obtained in the evaluation of road traffic network performance.

Hence, to the best of our knowledge, this paper explores the possibility of providing a higher level of personalized path planning using trajectory information. Specifically, three major contributions have been made and structured as follows. First, a log C-means clustering algorithm (LFCM) is introduced in Section 2.1 to lead the characteristic mining and clustering process of a specific user who provides the log trajectories. Second, an individualized quantitative evaluation system of road performance is built in Section 2.2 to obtain specific performance value in the view of the owner of the log tracks. Third, adaptive ant colony algorithm is improved and utilized to validate the path planning based on the personalized road values in Section 2.3 to find the optimal individualized path.

2. Materials and Methods

2.1. Mining and Clustering of Users' Driving Habits

2.1.1. Establishment of Driving Style Space

Driving style space should be analyzed before driver style clustering to ensure the rationality and reliability of the results of the LFCM. A driver can choose from a wide selection of varied details, generally, including time-saving [1–3], economic [26,27] and comfort [4,9]. For ease of discussion, we record the set of three indicators above as a style space. For most users, the style features are not limited to a single element of the style space, but a coexistence of multiple coupling indexes, whose index weight varies in each individual.

2.1.2. Acquisition of Individualized Driving Styles

Based on the fuzzy C-means algorithm [28], we integrate driving behavior characteristics and introduce the log fuzzy C-means algorithm (LFCM) to guide the clustering of driving characteristics in the style space. The LFCM algorithm is shown in Figure 1.

- Selection of style clustering center

(1) Mining of driving characteristics

Driving characteristics in log trajectories should be mined [29] before the core process of feature clustering. Here, space–time paths of time geography [30] are used to excavate the driving characteristics (Figure 2), in which the two-dimensional coordinates represent the historical space position of the vehicle, while the vertical coordinates indicate the time of the vehicle arriving at the corresponding space position.

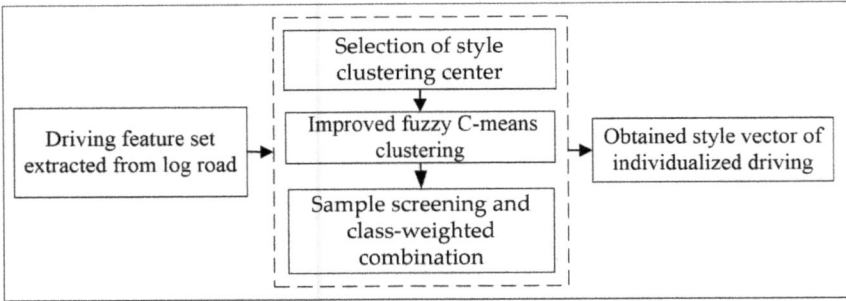

Figure 1. The log fuzzy C-means algorithm (LFCM) algorithm.

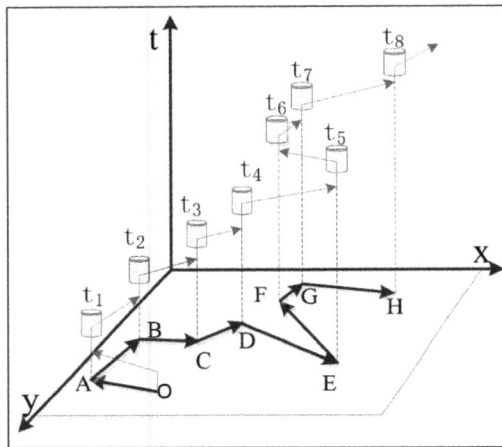

Figure 2. Vehicle space–time path.

The relationships among the speed of driving v_t at any time t, the acceleration of driving a_t, and the tangent slope of the inclined curve k_t is shown in Formulas (1) and (2).

$$v_t = \frac{1}{k_t} \tag{1}$$

$$a_t = \lim_{\tau \to t} \frac{\frac{1}{k_\tau} - \frac{1}{k_t}}{\tau - t} \tag{2}$$

The total square root acceleration of the total weight a_v in ISO2631 is used to approximate the comfort level of the human body; it is defined as follows:

$$a_v = \left(\lambda_x^2 a_x^2 + \lambda_y^2 a_y^2 + \lambda_z^2 a_z^2 \right)^{\frac{1}{2}} \tag{3}$$

where $a_x, a_y, a_z, \lambda_x, \lambda_y$, and λ_z, respectively, represent the square root acceleration and the direction factors in the forward, horizontal, and vertical directions of the vehicle. The comparison between the root acceleration of the total weight of different values and the comfort level of the human body is shown in Table 1.

The quality of the natural environment has a significant impact on the outcome of a user's path selection. The definition of the evaluation index is

$$I = \sum_{i=1}^{m} I_i = \sum_{i=1}^{m} \frac{C_i}{S_i} \tag{4}$$

where I_i, C_i, and S_i represent the environmental quality index, coverage rate, and evaluation standard of the natural environment factors i; m is the species number of i, including the road greening rate, river distribution, and air quality; and I denotes positive correlation with the comprehensive road environment.

Table 1. Corresponding relation between acceleration and comfort level.

Total Vibration $a_v/\text{m}\cdot\text{s}^{-2}$	Time Feature Weight ω_t
<0.315	Without malaise
0.315–0.63	Slightly uncomfortable
0.5–1	Quite uncomfortable
0.8–1.6	Uncomfortable
51.25–2.5	Extremely uncomfortable
>2	Terrible

(2) Selection of cluster center

To minimize the clustering convergence time and avoid the local optima [31,32] before the clustering of characteristics, we must fully utilize the human inductive ability to restrict and select the value of the initial cluster center. The following principles are adopted for selecting the time, economic value, and comfort index of the cluster center.

Time clustering center: $v \geq v_0, |a| \geq |a_0|$, where v_0, a_0 according depend on adjustments to a specific situation.

Economic cluster center: $|a| \leq |a_0|, c_0 \to c_{0\cdot\min}$, where c_0 represents the consumption rate for oil and facilities, and $c_{0\cdot\min}$ represents the minimum consumption rate for fuel and facilities.

Comfort clustering center: $|a_v| \leq |a_{v0}|, I \geq I_0$, where a_{v0} is routinely specified as 0.63 while I_0 is adjusted according to user requirements.

(3) Improved fuzzy C-means clustering

The minimum element, which contains the driving characteristics, is defined as the vector eigenvalue and is called the vector element. The Hausdorff distance [33–36] of each vector element to the time, economy, and comfort clustering centers is calculated, and the contribution of driving information to three different driving styles is measured. The fuzzy C-means algorithm is used to complete the clustering of the characteristics of each vector element. The concrete steps are as follows.

Step 1: A three-dimensional cluster coordinate system is established based on driving speed, consumption rate for fuel and facilities, total root mean square acceleration, and integrated environmental quality. The time, economic, and comfort information contained in each log vector element are expressed in the cluster coordinate system.

Step 2: According to the selection principle of clustering centers, the initial time, economic, and comfort clustering centers are ensured when constraint conditions are satisfied. In this way, the three cluster centers are moderately separated.

Step 3: The Hausdorff distance $d_{\mathrm{HD}}(f, o)$ of the log road vector element f to the time, economy, and comfort cluster centers O is calculated in Formula (5), where F is the finite set of vector elements, and O is the set of time, economy, and comfort clustering centers.

$$\begin{cases} d_{\mathrm{HD}}(f,o) = \min\big(h(F,O), h(O,F)\big) \\ h(F,O) = \max_{f \in F} \min_{o \in O} d(f,o) \\ h(O,F) = \max_{o \in O} \min_{f \in F} d(o,f) \end{cases} \tag{5}$$

Step 4: The membership degree of each vector element f to the three cluster centers is calculated in Formula (6).

$$R_{o \leftarrow f} = \frac{\mathrm{card} F / d_{\mathrm{HD}}^2(f,o)}{\sum\limits_{o \in O} \sum\limits_{f \in F} \big(1/d_{\mathrm{HD}}^2(f,o)\big)} \tag{6}$$

Step 5: The vector elements of the time, economy, and comfort cluster centers are updated according to the membership degree, as shown in Formula (7).

$$o' = \frac{\sum\limits_{f \in F} f \bullet R_{o \leftarrow f}^2}{\sum\limits_{f \in F} R_{o \leftarrow f}^2} \tag{7}$$

Step 6: If the conditions shown in Formula (8) are satisfied, the iteration stops. Otherwise, the process returns to Step 2, where O' and O represent the cluster center variables of the current time and the previous moment, respectively.

$$\max_{o \in O} \big\{ \|o' - o\| \big\} < \varepsilon \tag{8}$$

Step 7: Normalization is conducted according to Formula (9), where $j = \mathrm{t}, \mathrm{m}, \mathrm{c}$ represent the arbitrary sample time, economy, and comfort dimensions, respectively. $s_{i \cdot j}, s_{\min \cdot j}$, and $s_{\max \cdot j}$ represent the performance value, and the minimum and maximum performance values of sample i underdimension j, respectively. $s_{i \cdot j}^*$ is the new performance value of sample i under dimension j. after the normalization process.

$$s_{i \cdot j}^* = \frac{s_{i \cdot j} - s_{\min \cdot j}}{s_{\max \cdot j} - s_{\min \cdot j}} \tag{9}$$

The clustering results of the log road vector element features relative to the time (red), economic (green), and comfort (blue) cluster centers are shown in Figure 3.

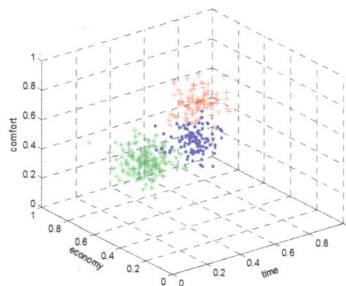

Figure 3. The result of LFCM.

- Sample screening and class-weighted combination

In the LFCM clustering, the validity of the statistics of the number of vectors in each cluster center directly affects the accuracy of driving style judgement. The effectiveness of each cluster center sample is as follows.

Step 1: The small effective ball radius r_0 is selected randomly (Figure 4a) to initialize the valid range of the sample. The corresponding plane projection of the valid ball is shown in Figure 4b.

Step 2: Taking the sample local density [33] and the Euclidean distance of the high-density point samples as the standard, the effective radius is adjusted positively according to Formula (10). The effective radius of clustering is finally determined as r. As shown in Figure 4c, the corresponding plane projection is shown in Figure 4d, where d_{io} and d_{ij} represent the Euclidean distance of sample i to the cluster center O and the high-density point sample j, respectively; d_c is the truncated distance, which is related to the average percentage of the total sample size and number of neighbor sample points; ρ_i and ρ_j denote the local sample density of i and j, respectively; δ_i is the shortest Euclidean distance between i and the high-density point sample j; and τ and ε are both related boundary parameters which are often substitution of empirical value.

$$\begin{cases} \rho_i = \sum_{j=1}^{n} \exp\left(-\frac{d_{ij}}{d_c^2}\right) \geq \tau \\ \delta_i = \min_{j:\rho_i > \rho_i}(d_{ij}) \leq \varepsilon \end{cases} \tag{10}$$

Step 3: The invalid samples of time, economic, and comfort cluster centers are deleted, as shown in Figure 4e. The corresponding plane projection is shown in Figure 4f. The number of samples within the effective radius r of each cluster center is counted as n_t, n_m, n_c.

Step 4: The distribution weight of time, economy, and comfort indexes is obtained through the normalization process shown in Formula (11).

$$\begin{cases} \omega_T = \frac{n_t}{n_t + n_e + n_c} \\ \omega_M = \frac{n_e}{n_t + n_e + n_c} \\ \omega_C = \frac{n_c}{n_t + n_e + n_c} \end{cases} \tag{11}$$

The user's driving style feature vector is expressed as $W = [\omega_T, \omega_M, \omega_C]$.

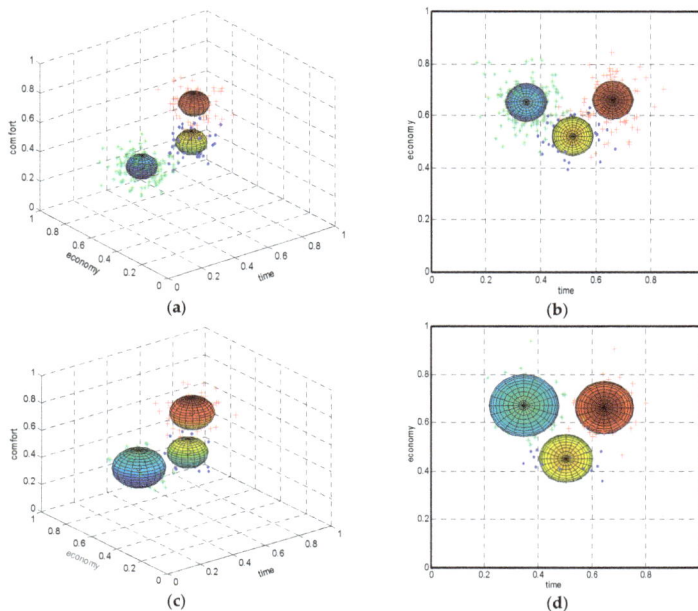

(a) (b) (c) (d)

Figure 4. *Cont.*

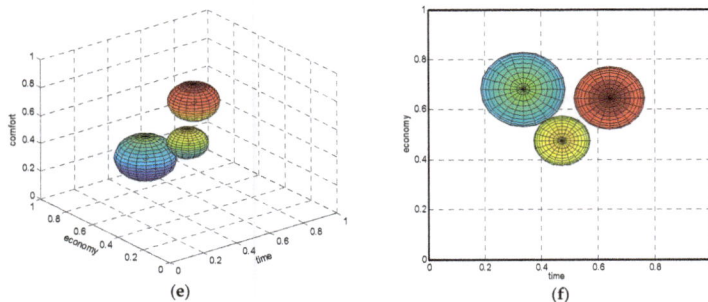

(e)

(f)

Figure 4. (**a**) Initialize the valid range of the sample; (**b**) The corresponding plane projection of the initialization process; (**c**) Determining the effective radius of clustering; (**d**) The corresponding plane projection of the effective radius of clustering; (**e**) Effective sample screening results; (**f**) The corresponding plane projection of the result of Effective sample screening process.

2.2. Individualized Quantitative Evaluation of Road Performance

2.2.1. Analysis of Influencing Factors of Roads

Road and its subsidiary factors [25,37,38], including peripheral environment, traffic laws and regulations, driver factors and other dynamic factors, whose varied details are listed in Table 2, can reflect road performance in different extents, and then affect user path selection results.

Table 2. Influencing factors of roads.

Factor Types		Influence Factor
Road factors	road self-factors	1. road alignment, road length, road intersection number, road level, road width, number of continuous turns, road evenness, isolation of green belts, and scribing isolation. 2. the ratio of green to credit, road traffic density, road evenness, number of crossings, average speed of main roads, blocking rate of intersections, average traffic delays and illegal occupancy Lane rates.
	Peripheral environmental factors	1. road greening and landscape organization 2. gas station, parking lot, traffic indicator, trunk road lighting rate, parking guidance and traffic guidance system setting 3. garbage field and operation station, pedestrian crossing facilities, to sign and rate of population density circulation area
Subsidiary factors	law and regulation factors	1. internship high speed limit 2. speed limit, limit height, weight limit, restriction entry 3. single line setting, single and double number limit line 4. influence, restriction, or prohibition of area and type of vehicle on the time of road separation
	Driver factors	1. restrictions on driving ability 2. drive-age limit
	Other dynamic factors	1. the fluctuation of the driver's mood 2. emergency (large-scale mass activities, major traffic accidents, temporary traffic control) occupation rate 3. weather, natural calamities, urban construction

2.2.2. Synchronous Standard Quantitative Evaluation of Multi-Index Road Performance

In this section, an accurate synchronous multi-index quantitative evaluation of road performance modeling system is established by maximizing the advantages of solving dimensions based on a T–S fuzzy neural network [39,40], whose architecture is shown in Figure 5.

The evaluation model using the "IF–THEN" form R^τ: IF f_{o1} is $\mu_{o1}^{i_1}$, f_{o2} is $\mu_{o2}^{i_2}$, $\cdots f_{ok}$ is $\mu_{ok}^{i_k}$ THEN $y_{o\tau} = p_{o0}^{i_1} + p_{o1}^{i_2} f_{o1} + p_{o2}^{i_3} f_{o2} + \cdots + p_{ok}^{i_k} f_{ok}$, where $o \in \{T, M, C\}$, $k \in \{|F_T|, |F_M|, |F_C|\}$ are the set of significant factors of time, economy, and comfort; $f_{oj} \in \{F_T, F_M, F_C\}$, $j = 1, 2, \cdots k$, μ_{oj}^i is the fuzzy set of index o; and p^i_{oj}, y_{oj} denote the corresponding parameter and output of the fuzzy rule, respectively.

The former network is used to calculate the applicability of fuzzy rules. For any f_{oj}, the membership degree of the fuzzy layer is first based on the Gaussian membership function shown in Formula (12), where a^i_j and b^i_j represent the function centers and widths, respectively; $i = 1, 2, \ldots, n$ where n is the fuzzy division number of f_{oj}. To satisfy the actual factors that affect the situation, n is assumed to be seven levels that represent the seven effects of - - -, - -, -, 0, +, + +, and +++, respectively.

$$\mu^i_{oj} = \exp\left(-\left(f_{oj} - a^i_j\right)^2 / b^i_j\right) \tag{12}$$

Then, we calculate the fitness of each rule by using a multiplicative operator in the fuzzy reasoning layer, as shown in Formula (13).

$$\alpha_{o\tau} = \mu^{i_1}_{o1} \cdot \mu^{i_2}_{o2} \cdot \mu^{i_3}_{o3} \cdots \mu^{i_k}_{ok} \left(\tau = 1, 2, \cdots, n^k\right) \tag{13}$$

Finally, to avoid model turbulence caused by the order of magnitude of significant road influencing factors, we perform normalization in the anti-fuzzification layer, as shown in Formula (14).

$$\overline{\alpha}_{o\tau} = \frac{\alpha_{o\tau}}{\sum\limits_{i=1}^{n^k} \alpha_{oi}} \tag{14}$$

The post-part network consists of three sub-networks of the same structure for outputting fuzzy rules of time, economy, and comfort indexes. The input layer needs to be supplemented with a constant term, that is, input parameter 1 of the 0th node, which is used to generate the constant term in the road performance level calculation result. The fuzzy inference layer calculates the consequent of each fuzzy rule, as shown in Formula (15).

$$y_{o\tau} = p^{i_1}_{o0} + p^{i_2}_{o1} f_{o1} + p^{i_3}_{o2} f_{o2} + \cdots + p^{i_k}_{ok} f_{ok} \tag{15}$$

The output of the network is the weighted sum of the fuzzy rules, as shown in Formula (16). The weighting coefficient is the applicability of the rules and the result of the road performance level after the output-clarified process.

$$\lambda_o = \overline{\alpha}_o \otimes y_o = \sum_{\tau=1}^{n^k} \overline{\alpha}_{o\tau} y_{o\tau} \tag{16}$$

According to the equilibrium characteristics of the weight distribution of a user's driving style, the overall characteristics of different path tendencies based on user choice is divided into three categories, as shown in the first four columns of Table 3, where c^1_3, c^2_3, c^3_3, respectively, represent the number of effective features in user eigenvectors.

Table 3. Driving style classification.

Style Type	ω_t	ω_m	ω_c	Personal-Value ω
	valid	—	—	$\omega_T \lambda_T$
c^1_3	—	valid	—	$\omega_M \lambda_M$
	—	—	valid	$\omega_c \lambda_c$
	valid	valid	—	$\omega_T \lambda_T + \omega_M \lambda_M$
c^2_3	valid	—	valid	$\omega_T \lambda_T + \omega_c \lambda_c$
	—	valid	valid	$\omega_M \lambda_M + \omega_c \lambda_c$
c^3_3	valid	valid	valid	$\omega_T \lambda_T + \omega_M \lambda_M + \omega_c \lambda_c$

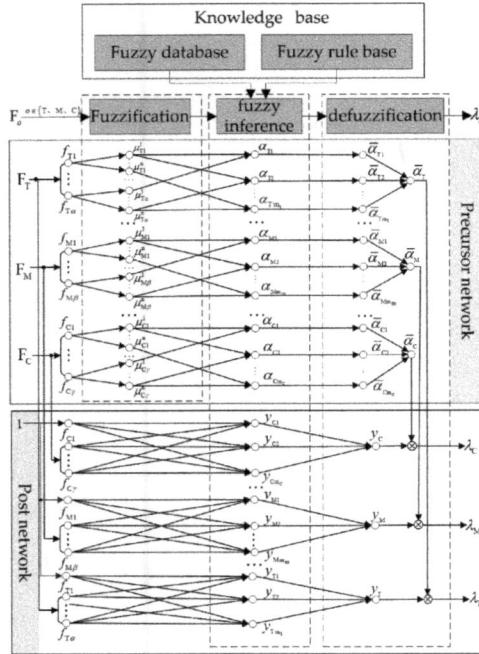

Figure 5. Performance synchronization evaluation model of road multidimensional standard.

In addition, to avoid excessive neutralization of the decision-making power of the high weight feature, and ensure the good consistency and universality of the quantitative and the actual evaluation results, an effective definition of the features is presented in Theorem 1.

Theorem 1. *For any vector element of the driving characteristics, if any of the following arbitrary conditions are satisfied, the element can be called an invalid feature; otherwise, it is an effective feature.*

(a) $\omega_0 > 0.35 \times \max W$

(b) $\overline{\omega}_o > 0.35\max W$ and $\omega_o \geq 0.5^* \overline{\omega}_o$, where $\overline{\omega}_o = W - \max W - \min W$

The personalized quantitative evaluation of roads is a weighted summation of the effective characteristics and the corresponding road standard quantification values. The calculation process for the personalized road performance from a user perspective is shown in the last column of Table 3.

To facilitate the dynamic optimization of the follow-up path, we standardize the deviation of the personalized road performance ω, adjust the original performance values linearly, and adjust the road area range of the traffic network to [0, 1], as shown in Formula (17).

$$\omega^* = \frac{\omega - \omega_{\min}}{\omega_{\max} - \omega_{\min}} \tag{17}$$

2.3. Dynamic Selection of Path

2.3.1. Description of the Problem of Individualized Optimal Path

To avoid the redundant consumption of roads and the interference of irrelevant paths, we introduce a linear transformation between road performance and road consumption. The use of road performance as an index is also avoided, as shown in Formula (18). For any two points in the road network, the optimal path [41] is solved based on the consumption value $\overline{\omega}(\overline{\omega} \in [0,1])$.

$$\overline{\omega} = \frac{1}{\omega^*} \tag{18}$$

The optimal path problem is described as follows. Let $G = \{V, L, \overline{W}\}$ be a road network digraph, where $V = \{1, 2, \cdots n\}$ is a set of branch points of the road network, $L \subseteq V \times V$ is a set of roads of G, and $\overline{W} = (\overline{\omega}_{ij})_{|L| \times |L|}$ is the set of consumption of roads. The path consumption of directed roads $(i,j) \in L$ between any connected branch points i and j is recorded as $\overline{\omega}_{ij}$, in which $i, j \in V; \overline{\omega}_{ij} \in [0,1]$. For $\forall A, B \in V$, the optimal path is the path of minimum cumulative consumption from point A to point B in the road network digraph.

2.3.2. Global Dynamic Selection of Path

The adaptive ant colony algorithm proposed in the literature [42–44] adaptively controls the proportion of pheromone concentration in the current optimal solution and updates the global pheromone concentration of the optimal path in real time by introducing a hyperbolic sine function as an adaptive dynamic factor $\sigma(\sigma \in (0,1))$. In this way, the path exploration results can be effectively implemented when the performance of each path is known. In view of the real-time changes in the characteristics of road traffic network status, we apply the adaptive ant colony algorithm to solve the personalized optimal path. We add dynamic updating links of road performance to simulate the road environment realistically. The flowchart of the improved adaptive ant colony algorithm and its contribution in this study are shown in Figure 6a,b (shadow), respectively.

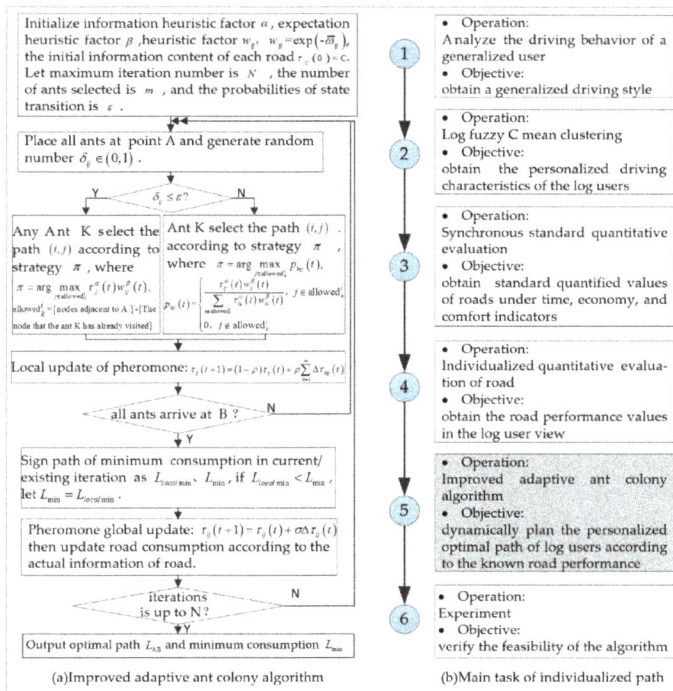

Figure 6. Improved adaptive ant colony algorithm and its contribution.

3. Results

3.1. Emulation

To verify the proposed theory and the effectiveness of the algorithm, we use a road network environment with a 5 × 5 grid simulated according to ideal settings from top to bottom and from left to right in the order of the road traffic environment simulation. The grid performance simulation of traffic assignment, the network structure, and the performance simulation of road vector are shown in Figure 7 and Table 4.

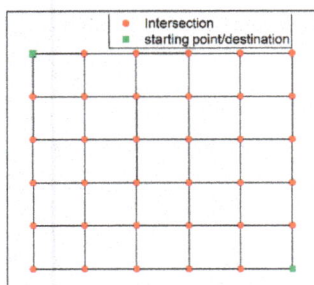

Figure 7. Analog road network.

Table 4. The road performance of the analog road network.

Road Number	Standard Road Performance (λ_t, λ_m, λ_c)
1–12	[0.77,0.50,0.45],[0.34,0.74,0.38],[0.56,0.69,0.72],[0.49,0.59,0.21] [0.73,0.69,0.66],[0.66,0.60,0.78],[0.47,0.41,0.79],[0.21,0.37,0.67] [0.69,0.40,0.46],[0.47,0.52,0.50],[0.57,0.64,0.33],[0.68,0.39,0.59]
13–24	[0.75,0.70,0.39],[0.64,0.54,0.78],[0.31,0.42,0.64],[0.44,0.62,0.45] [0.76,0.53,0.65],[0.75,0.47,0.36],[0.45,0.62,0.46],[0.74,0.57,0.76] [0.23,0.68,0.61],[0.41,0.77,0.33],[0.69,0.51,0.70],[0.21,0.73,0.58]
25–36	[0.28,0.30,0.28],[0.32,0.79,0.32],[0.32,0.36,0.56],[0.56,0.35,0.58] [0.36,0.73,0.42],[0.32,0.64,0.55],[0.21,0.28,0.47],[0.65,0.21,0.23] [0.47,0.74,0.22],[0.76,0.32,0.39],[0.48,0.38,0.21],[0.45,0.60,0.43]
37–48	[0.71,0.37,0.61],[0.52,0.48,0.26],[0.32,0.24,0.22],[0.60,0.79,0.57] [0.70,0.55,0.57],[0.21,0.45,0.21],[0.61,0.51,0.21],[0.43,0.40,0.31] [0.70,0.46,0.55],[0.50,0.34,0.23],[0.63,0.55,0.42],[0.46,0.66,0.58]
49–60	[0.38,0.52,0.63],[0.31,0.58,0.62],[0.32,0.33,0.25],[0.61,0.43,0.47] [0.38,0.67,0.47],[0.53,0.61,0.41],[0.29,0.48,0.29],[0.62,0.54,0.61] [0.43,0.68,0.62],[0.72,0.24,0.64],[0.71,0.56,0.49],[0.56,0.23,0.53]

We verify the reliability of the LFCM algorithm in driving style clustering under different conditions. Without loss of generality, the time-based, time–economic-based, and time–economic–comfort-based users are randomly selected as examples from category c_3^1, c_3^2, c_3^3, respectively (Table 4).

The clustering results are shown in Table 5, where "√" and "—" represent valid and invalid features, respectively. For ease of description, we present the type of time-based driving style as an example.

Step 1: Suppose a certain time-based driving style user A. Following his habits and the road performance vector shown in Table 4, 20 groups of different starting points and destination path combinations are selected as log samples.

Step 2: The smallest road vector elements containing users' driving information from the 20 log samples are obtained by the sampling operation. The fuzzy C-means algorithm is used on the tuple cluster of the log sample vector.

Step 3: The validity of the eigenvectors of the 20 groups of clustering results is determined.

Step 4: The 20 log samples are compared and counted to obtain the driving style and the presumed driving style.

The coincidence rate of the driving style clustering results with actual statistics is above 90% (Table 5), which can meet practical requirements.

Based on the clustering results, the performance of each road can be quantified in a personalized way according to the evaluation method in Table 3. Without loss of generality, the latest 20 sample log trajectories of a random time–economic-based user are selected as an example of personalized user quantification.

Step 1: The 20 groups of clustering data acquired from the user log samples are screened to eliminate false clustering data.

Step 2: The corresponding terms of the filtered clustering data are weighted and averaged to obtain the statistical driving feature vector: $W = [0.56,0.37,0.07]$.

Step 3: In accordance with Formula (19), the quantitative results are shown in Table 6.

$$\omega = \omega_T \lambda_T + \omega_M \lambda_M \tag{19}$$

According to the personalized path performance shown in Table 6, we reassign the road network performance and use the ant colony algorithm to achieve personalized dynamic route selection based on user driving habits, as shown in Figure 8, in which the blue route is a random track in the user log samples, the green line segment is the personalized path of the proposed algorithm, and the red part is the coincidence of the two paths.

Table 5. Driving style clustering verification.

Given-Style	Clustering Result ($\omega_t, \omega_m, \omega_t$)	Validity Determination	Correct Rate
T	[0.90,0.05,0.05],[0.75,0.20,0.05]	[√,—,—],[√,—,—]	100%
	[0.85,0.10,0.05],[0.70,0.10,0.20]	[√,—,—],[√,—,—]	
	[0.75,0.15,0.10],[0.75,0.10,0.15]	[√,—,—],[√,—,—]	
	[0.80,0.15,0.05],[0.70,0.15,0.15]	[√,—,—],[√,—,—]	
	[0.80,0.05,0.15],[0.80,0.10,0.10]	[√,—,—],[√,—,—]	
	[0.85,0.05,0.10],[0.65,0.25,0.10]	[√,—,—],[√,—,—]	
	[0.75,0.05,0.20],[0.70,0.20,0.10]	[√,—,—],[√,—,—]	
	[0.85,0.10,0.05],[0.90,0.05,0.05]	[√,—,—],[√,—,—]	
	[0.70,0.15,0.15],[0.85,0.05,0.10]	[√,—,—],[√,—,—]	
	[0.75,0.15,0.10],[0.80,0.05,0.15]	[√,—,—],[√,—,—]	
T+M	[0.50,0.45,0.05],[0.70,0.25,0.05]	[√,√,—],[√,√,—]	95%
	[0.60,0.35,0.05],[0.35,0.60,0.05]	[√,√,—],[√,√,—]	
	[0.55,0.40,0.05],[0.65,0.30,0.05]	[√,√,—],[√,√,—]	
	[0.40,0.55,0.05],[0.50,0.40,0.10]	[√,√,—],[√,√,—]	
	[0.55,0.35,0.10],[0.45,0.45,0.10]	[√,√,—],[√,√,—]	
	[0.70,0.25,0.05],[0.65,0.25,0.10]	[√,√,—],[√,√,—]	
	[0.50,0.40,0.10],[0.55,0.40,0.05]	[√,√,—],[√,√,—]	
	[0.55,0.30,0.15],[0.50,0.40,0.10]	[√,√,√],[√,√,—]	
	[0.65,0.30,0.05],[0.70,0.25,0.05]	[√,√,—],[√,√,—]	
	[0.50,0.40,0.10],[0.55,0.35,0.10]	[√,√,—],[√,√,—]	
T+M+C	[0.45,0.30,0.25],[0.55,0.20,0.25]	[√,√,√],[√,√,√]	90%
	[0.55,0.20,0.25],[0.45,0.25,0.30]	[√,√,√],[√,√,√]	
	[0.40,0.35,0.25],**[0.55,0.35,0.10]**	[√,√,√],[√,√,—]	
	[0.40,0.35,0.25],[0.45,0.35,0.20]	[√,√,√],[√,√,√]	
	[0.55,0.20,0.25],[0.50,0.25,0.25]	[√,√,√],[√,√,√]	
	[0.30,0.25,0.45],[0.50,0.20,0.30]	[√,√,√],[√,√,√]	
	[0.35,0.35,0.30],[0.35,0.35,0.30]	[√,√,√],[√,√,√]	
	[0.50,0.25,0.25],[0.50,0.30,0.20]	[√,√,√],[√,√,√]	
	[0.40,0.45,0.15],[0.40,0.40,0.20]	[√,√,—],[√,√,√]	
	[0.35,0.30,0.35],[0.55,0.25,0.20]	[√,√,√],[√,√,√]	

Figure 8 shows that the log sample trajectory from the starting point to the end point is 6-17-23-29-40-46-47-53-59-60 while the integrated user habits in the personalized planning path are presented as 6-17-23-29-40-46-52-58-59-60. The number of anastomosed sections is 9, the number of different sections is 2, and the absolute anastomosis rate is 82%. According to the depth analysis, the sample trajectory sections reach 47 and 53, which correspond to the road consumption values of 1.8 and 2.2, respectively. The different personalized path sections 52 and 58, which correspond to the road consumption values are both 2.0. Thus, the cumulative values of the road consumption of the two tracks match at a rate of 100%. Similarly, the remaining valid log samples for the user are verified one by one, and the results are weighted to all (the invalid sample weight is 0; otherwise, it is 1). The statistical results show that the absolute anastomosis rate is 89.5%, and the relative anastomosis rate is 98%.

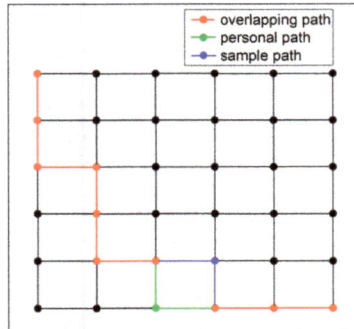

Figure 8. Comparison of simulation effect.

Table 6. Personalized performance quantification.

Road Number	Standard Road Performance Vector (λ_t, λ_m, λ_c)	Personal-Value ω
1–12	[0.77,0.50,0.45],[0.34,0.74,0.38],[0.56,0.69,0.72]	0.62,0.46,0.57
	[0.49,0.59,0.21],[0.73,0.69,0.66],[0.66,0.60,0.78]	0.49,0.67,0.59
	[0.47,0.41,0.79],[0.21,0.37,0.67],[0.69,0.40,0.46]	0.42,0.26,0.54
	[0.47,0.52,0.50],[0.57,0.64,0.33],[0.68,0.39,0.59]	0.45,0.55,0.502,
13–24	[0.75,0.70,0.39],[0.64,0.54,0.78],[0.31,0.42,0.64]	0.68,0.56,0.33
	[0.44,0.62,0.45],[0.76,0.53,0.65],[0.75,0.47,0.36]	0.48,0.62,0.59
	[0.45,0.62,0.46],[0.74,0.57,0.76],[0.23,0.68,0.61]	0.48,0.62,0.38
	[0.41,0.77,0.33],[0.69,0.51,0.70],[0.21,0.73,0.58]	0.52,0.58,0.38
25–36	[0.28,0.30,0.28],[0.32,0.79,0.32],[0.32,0.36,0.56]	0.27,0.47,0.31
	[0.56,0.35,0.58],[0.36,0.73,0.42],[0.32,0.64,0.55]	0.44,0.47,0.42
	[0.21,0.28,0.47],[0.65,0.21,0.23],[0.47,0.74,0.22]	0.22,0.44,0.53
	[0.76,0.32,0.39],[0.48,0.38,0.21],[0.45,0.60,0.43]	0.54,0.41,0.47
37–48	[0.71,0.37,0.61],[0.52,0.48,0.26],[0.32,0.24,0.22]	0.53,0.47,0.27
	[0.60,0.79,0.57],[0.70,0.55,0.57],[0.21,0.45,0.21]	0.63,0.60,0.29
	[0.61,0.51,0.21],[0.43,0.40,0.31],[0.70,0.46,0.55]	0.53,0.39,0.56
	[0.50,0.34,0.23],[0.63,0.55,0.42],[0.46,0.66,0.58]	0.41,0.55,0.50
49–60	[0.38,0.52,0.63],[0.31,0.58,0.62],[0.32,0.33,0.25]	0.41,0.39,0.30
	[0.61,0.43,0.47],[0.38,0.67,0.47],[0.53,0.61,0.41]	0.50,0.46,0.52
	[0.29,0.48,0.29],[0.62,0.54,0.61],[0.43,0.68,0.62]	0.34,0.55,0.49
	[0.72,0.24,0.64],[0.71,0.56,0.49],[0.56,0.23,0.53]	0.49,0.61,0.40

3.2. Real Routine Verification

We verify the practical performance of the proposed algorithm from two levels of the effectiveness of the index performance synchronization evaluation model and the effectiveness of the personalized planning.

3.2.1. Validation of Multi-Index Synchronization Performance Evaluation Model

(1)   Establishment of road performance standards

To test the accuracy of the road performance quantification system, we first refer to prior examples and existing *city road engineering design standards, city road network planning index systems, Changsha city traffic planning*, *Green Plan*, and other Changsha cases. Then, we integrate the examples and the actual characteristics of road networks, as well as the results of the Baidu map app and field survey. Subsequently, we develop quantitative descriptions of the selected road network density, road saturation, road quality, environmental quality, speed and economic consumption of six representative variables of road performance to establish road performance standards. In view of the actual complexity of traffic networks, we refer to the path in the city planning classification results and the grade of expansion according to the requirement of this research. The actual performance of road networks that is classified and described is shown in Table 7, along with the actual performance of each road grade.

Table 7. The performance of road network.

Road Classification	Road Density km/km²	Road Saturation a.u.	Road Quality a.u.	Environmental Quality a.u.	Drive Speed km/h	Economic Consumption yuan/km
G-Expressway	0.42	0.40	0.97	>0.40	60–100	0.72
Elevated express	0.06	0.48	0.97	>0.35	60–80	0.35
Central main road	1.52	0.82	0.97	0.25–0.30	40–50	0.58
Peripheral main road	1.31	0.76	0.97	0.25–0.40	40–60	0.51
Central sub Road	1.75	1.05	0.64	0.25–0.35	30–40	0.63
Peripheral sub road	1.60	0.80	0.64	0.30–0.40	30–50	0.52
Access Road	3.00	0.60	0.49	>0.20	20–40	0.55
landscape road	0.15	0.55	0.64	>0.60	30–50	0.50

To compare path performances easily, we perform the normalization operation after the data association process to unify the quantitative standard. In view of the high correlation between road time performance and user driving speed, we describe the time performance of the road in interval form. The performance standards of time, economy, and comfort obtained by normalization are shown in Table 8, where T, M, and C represent the time, economic, and comfort values, respectively.

Table 8. Performance standard of road time, economy and comfort of each road level.

Road Level	Road Classification	T	M	C
1	G-Expressway	0.6–1.0	0.12	0.77
2	Elevated Express	0.6–0.8	0.71	0.57
3	Central main road	0.4–0.5	0.26	0.48
4	Peripheral main road	0.4–0.6	0.35	0.52
5	Central sub Road	0.3–0.4	0.20	0.40
6	Peripheral sub road	0.3–0.5	0.34	0.44
7	Access Road	0.2–0.4	0.29	0.29
8	landscape road	0.3–0.5	0.37	0.61

(2)   Evaluation model verification

To facilitate the classification, marking and quantification of roads, we establish a simplified road network model based on path guidance in Changsha. We also conduct the process of road quantification and path planning based on the model. In view of the complexity and generality of road networks, we randomly select 10 road samples from the Changsha road network to demonstrate the time, economy, and comfort performance and verify its effectiveness. The results of sorting according to type are shown in Table 9.

Table 9 shows the standard errors σ of the samples in the road landscape of Binjiang road and Third Ring Road, with the minimum errors being 0.014 and 0.062, respectively. The corresponding error vectors for time, economy, and comfort are $[-0.05, 0.03, -0.09]$. The errors are within the acceptable value of 0.10. The average standard error of the sample road is 0.036, which meets the quantitative requirement of actual road performance.

Table 9. The time, economy, and comfort performance of iconic road.

Road Name	Level	T	M	C	Error Vector	σ
third ring road	1	0.75	0.15	0.68	[−0.05,0.03,−0.09]	0.062
Wan Jiali viaduct	2	0.67	0.65	0.59	[−0.03,−0.07,0.01]	0.044
Shaoshan South Road	3	0.42	0.25	0.43	[−0.03,−0.01,−0.05]	0.034
West Second Ring	3	0.46	0.31	0.45	[0.01,0.05,−0.03]	0.034
ThreeFenglin Road	4	0.53	0.33	0.54	[0.03,−0.02,0.02]	0.024
The middle of the people Road	5	0.30	0.19	0.41	[−0.05,−0.01,0.01]	0.030
Binhu West Road	6	0.43	0.32	0.40	[0.03,−0.02,−0.04]	0.031
The road of literature and art	7	0.34	0.27	0.32	[0.04,−0.02,0.03]	0.031
Binjiang landscape road	8	0.42	0.38	0.60	[0.02,0.01,−0.01]	0.014
Xiangjiang Middle Road	8	0.48	0.33	0.63	[0.08,−0.04,0.02]	0.053

3.2.2. Validation of the Effectiveness of Personalized Planning

Based on the basic evaluation model for the reliable evaluation of the results of the synchronization performance index, three drives (A, B and C) in two categories, who own their own log trajectory records, are randomly selected from the seven kinds of small classifications of driving users (Table 3) to the performance of the path planning algorithm. Specifically, inter-class experiments based on different type users A and B are designed to verify the universality of the algorithm. Meanwhile, intra-class validation experiment based on different type users B and C, who are in same type but possess different driving feature vectors are added to further verify the personalization level of algorithm. The validation verification framework is design as in Table 10. To facilitate the comparison of path planning results, this paper is based on the background of route guidance in Changsha City which is same as that of Reference [25], as well as the user types.

Table 10. Example validation framework.

Validation Category	Driver	Style	Feature Vector
Inter class validation	A	Economy-comfort	[0.05,0.32,0.63]
	B	Time-economic	[0.35,0.57,0.08]
Intra class validation	B	Time-economic	[0.35,0.57,0.08]
	C		[0.56,0.37,0.07]

- Inter-class validation

(1) Example 1

The class design verification in Table 10 shows that the economy + comfort user A ([0.05,0.32,0.63]) travels with only economy and comfort in mind and thus ignores the time factor. For this user, the economic attention level is 2, the influence coefficient is small at 0.32, the level of the concern for comfort is 1, and the influence coefficient is up to 0.63. The individual values of road performance depend only on the economy and comfort factors.

Based on the driving characteristics of user A, the case is weight coupled with the time, economy, and comfort performance values obtained by the evaluation model. Moreover, the individual road performance values in field A are obtained. Then, the best path is searched according to the road's personalized cost. In view of the large complexity of road networks, this example only enumerates the personalized performance and effective length of the road related to the path search results, as shown in Table 11, where the effective length of the road is the length of the path through the planning path. In Table 11, the personalized performance of each road in the personalized path that integrates user A's habits is higher than the personalized performance of the roads in the reference path, which lays the foundation for the integration of personalized path performance advantages of user A driving habits.

Table 11. Personalized performance of road based on A.

Path Name	Related Road	Level	T	M	C	Personal-Value	Effective Length/km
Individualization Route based on A	Yun Qi Road	4	0.49	0.35	0.53	0.45	3.77
	South Second Ring	4	0.49	0.35	0.53	0.45	1.89
	Xiaoxiang Middle Road	8	0.48	0.33	0.63	0.50	6.40
	Orange Chau Bridge	8	0.58	0.50	0.77	0.65	1.20
	Xiangjiang Middle Road	8	0.48	0.33	0.63	0.50	3.57
	31 Avenue	3	0.48	0.28	0.43	0.36	6.71
	Wan Jiali viaduct	2	0.67	0.65	0.59	0.58	2.49
	WanjialiNorth Road	6	0.37	0.32	0.47	0.40	2.54
	Xianghu West Road	7	0.28	0.30	0.34	0.31	1.13
literature Reference resources Route	Yun Qi Road	4	0.49	0.35	0.53	0.45	3.77
	South Second Ring	4	0.49	0.35	0.53	0.45	1.89
	Xiaoxiang Middle Road	8	0.48	0.33	0.63	0.50	7.87
	Yingpan Road Tunnel	3	0.50	0.35	0.48	0.41	1.35
	Xiangjiang Middle Road	8	0.48	0.33	0.63	0.50	2.43
	Xiangjiang North Road	8	0.48	0.40	0.63	0.50	2.45
	Fucheng Road	7	0.37	0.36	0.33	0.32	0.72
	Hibiscus North Road	5	0.30	0.25	0.40	0.33	0.17
	Fuyuan West Road	6	0.44	0.37	0.44	0.41	2.83
	The Fuyuanmiddle road	6	0.44	0.37	0.44	0.41	3.01
	Wan Jiali viaduct	2	0.67	0.65	0.59	0.58	0.43
	WanjialiNorth Road	6	0.37	0.32	0.47	0.40	2.54
	Xianghu West Road	7	0.28	0.30	0.34	0.31	1.13
log trajectory	Yun Qi Road	4	0.49	0.35	0.53	0.45	3.77
	South Second Ring	4	0.49	0.35	0.53	0.45	1.89
	Xiaoxiang Middle Road	8	0.48	0.33	0.63	0.50	6.40
	Orange Chau Bridge	8	0.58	0.50	0.77	0.65	1.20
	Xiangjiang Middle Road	8	0.48	0.33	0.63	0.50	3.57
	31 Avenue	3	0.48	0.28	0.43	0.36	6.71
	Wan Jiali viaduct	2	0.67	0.65	0.59	0.58	2.49
	WanJialiNorth Road	6	0.37	0.32	0.47	0.40	0.74
	Special road	6	0.43	0.32	0.41	0.36	1.71
	Xi Xia Road	6	0.49	0.36	0.47	0.41	1.53

The optimal path model and actual planning path for user A are shown in Figure 9a,b, respectively. In these figures, blue denotes the optimal planning path for users of the same type [20], red denotes the optimal path to merge user A's driving habits, and green denotes the log track of user A.

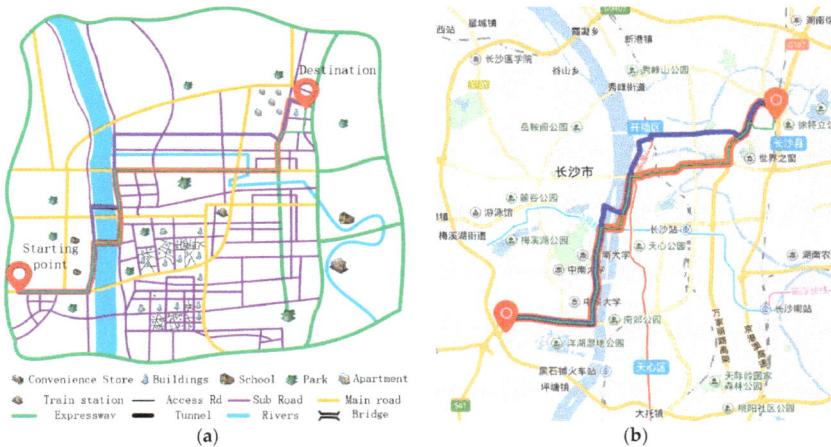

Figure 9. (**a**) The optimal path model for user A; (**b**) The optimal actual planning path for user A. Where, the green represents log trajectory, the blue represents literature Reference resources Route, and the red represents the Individualization Route based on A. Detailed performance of the relevant roads refer to Table 11.

Figure 9 shows that in the landscape area, the blue and red paths are dominated by the surrounding landscape of Juzizhou Road, which showcases fresh air, beautiful scenery and low noise. The blue line represents the Yingpan road tunnel, and the red path shows the Orange Island Bridge across the area, overlooking the Orange Island scenery. The latter is in line with user A's comfort requirements. In addition, the red route chooses the path from the perspective of user A, considering the comfort factors, and at the same time, it also focuses on the economic factors. In the non-sightseeing section, a small number of branches are allowed to reduce the detour distance to make up for the additional economic losses caused by the bypass of the two ends of the Orange Island Bridge. According to Table 11, the economic and comfort performance of the road and the effective length calculation indicate that the cumulative comfort consumption rates of the red planning path and blue reference path are 56.62 and 59.06, respectively, and the total economic losses are 88.28 and 88.19, respectively. In comparison with the reference path, the user personalization path that integrates user A's habits increases comfort performance while decreasing economic performance slightly. The comfort performance increment of user A is contrary to the economic performance increment, and comparing the performance of the red and blue paths is difficult.

Hence, in this work, the increase in comfort loss is $\Delta C_c = -2.44$, economic loss increment is $\Delta C_m = 0.09$ and is dimensionless, as shown in Formula (20). Between c_{total} and m_{total}, t represents the time consumption of the red path and the total amount of economic consumption.

$$
\begin{cases}
\Delta C_c' = \dfrac{|\Delta C_c|}{c_{total}} \\
\Delta C_m' = \dfrac{|\Delta C_m|}{m_{total}}
\end{cases}
\tag{20}
$$

The calculation shows that $\Delta C_c'$ and $\Delta C_m'$ are satisfied in Formula (21).

$$
\frac{\Delta C_m'}{0.32} < \frac{\Delta C_c'}{0.63}
\tag{21}
$$

Therefore, for user A, the increased comfort performance of the red path is greater than the economic performance reduction, the red path is more consistent with user A's driving habits.

(2) Example 2

Time + economic user B ([0.35,0.57,0.08]) and user A ([0.05,0.32,0.63]) show significant differences in driving characteristics. Specifically, the former places high importance in time and economic factors while ignoring the influence of comfort; that is, for time level 2, the influence coefficient is 0.35, and for time level 1, the influence coefficient is 0.57.

Reference example A makes personalized path planning from user B's perspective, and the path planning process is no longer duplicated. The individual performance and effective length of the related roads are shown in Table 12. Table 11 shows that the Yun Qi Road, South Second Ring, Wan Jiali viaduct, Wanjiali North Road, and Xianghu West Road appear in the path planning of users A and B; however, the road presents a considerable difference in terms of the personalized performance value and its effective length. Hence, verifying the performance of the proposed personalized path varies from person to person.

Table 12. Personalized performance of road based on B.

Path Name	Related Road	Level	T	M	C	Personal-Value	Effective Length/km
	Yun Qi Road	4	0.49	0.35	0.53	0.37	3.77
	South Second Ring	4	0.49	0.35	0.53	0.37	8.77
Individualization	Labor East Road	5	0.35	0.30	0.39	0.29	1.37
Route based on B	Wan Jiali viaduct	2	0.67	0.65	0.59	0.61	11.36
	Wanjiali North Road	6	0.37	0.32	0.47	0.31	2.54
	Xianghu West Road	7	0.28	0.30	0.34	0.27	1.13

Table 12. *Cont.*

Path Name	Related Road	Level	T	M	C	Personal-Value	Effective Length/km
Literature Reference Route	Yun Qi Road	4	0.49	0.35	0.53	0.45	3.77
	South Second Ring	4	0.49	0.35	0.53	0.40	8.77
	East Second Ring Road	3	0.37	0.22	0.38	0.25	8.78
	31 Avenue	3	0.48	0.28	0.43	0.33	1.95
	Wan Jiali viaduct	2	0.67	0.65	0.59	0.58	2.50
	Wan JialiNorth Road	6	0.37	0.32	0.47	0.31	2.54
	Xianghu West Road	7	0.28	0.30	0.34	0.27	1.13
Log trajectory	just as same as the personalized path that combines B's habits						

The optimal path model based on user B and the actual path, as shown in Figure 10a,b, respectively, remain unchanged in terms of the descriptions of the colors corresponding to the planning paths.

Figure 10 shows that the blue and red paths are driven mainly by the main road or viaduct, and they shorten the detour distance as much as possible. In comparison with the blue reference path, the red path representing the personalized path of user A can effectively avoid the East Second Ring Road, Bayi Road, South Road, Furong Road, Wuyi Road, and the bottleneck of a congested road. The road is relatively unimpeded. Moreover, by replacing Wan Jiali viaduct with going around the city at high speed, guaranteeing the time performance, it avoids the extra cost of passing through the road and the distance from the bypass, and improves the economic performance of the path.

According to the calculation results related to the time and economic performance and the effective road length in Table 12, the time consumption rates of the red planning path and blue reference path are 57.362 and 68.016, respectively, and the economic losses are 69.576 and 98.252, respectively. In comparison with the reference path, the personalized path that integrates user A's habits shows improved comfort and economic performance. This result is consistent with the driving characteristics of user A, focusing on time and economic performance.

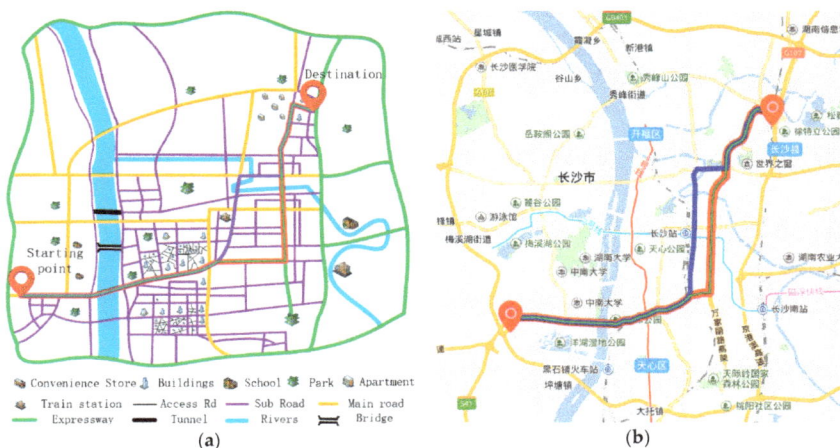

Figure 10. (**a**) The optimal path model for user B; (**b**) The optimal actual planning path for user B. Where, the green represents log trajectory, the blue represents literature Reference resources Route, and the red represents the Individualization Route based on B. Detailed performance of the relevant roads refer to Table 12.

- Intra-class validation

(3) Example 3

This example is the personalized path planning for time + economic user C ([0.56,0.37,0.07]), and user B's planning path constitutes the intra-class verification of personalized path performance. Users C and B ([0.35,0.57,0.08]) share similar features, belong to the same time + economic type, pay attention to time and economic factors, while ignore the factors of comfort effect. However, there are two different levels of attention in time and economy. C users are more inclined to time, with a level of 1, a coefficient of 0.56, a second economic factor, a level of 2, and a coefficient of 0.37.

Reference example A makes personalized path planning from user C's perspective. The personalized performance and effective length of the related roads are shown in Table 13.

Table 12 shows that the Yun Qi Road, South Second Ring, Wan Jiali viaduct, Wanjiali North Road, and Xianghu West Road also appear in the same type of path planning of users B and C. Path planning is of different types for users A and B. The same type of road users has different personalized performance, particularly for users B and C. With large effective length differences, the amplitude becomes increasingly small. The similarity obviously improves path planning. Path planning at a similar degree increases with the increase of similarity between user feature vectors.

The optimal path model based on user C's driving habits and the actual planning path, as shown in Figure 11a,b, remain unchanged in terms of the descriptions of the colors corresponding to the planning paths.

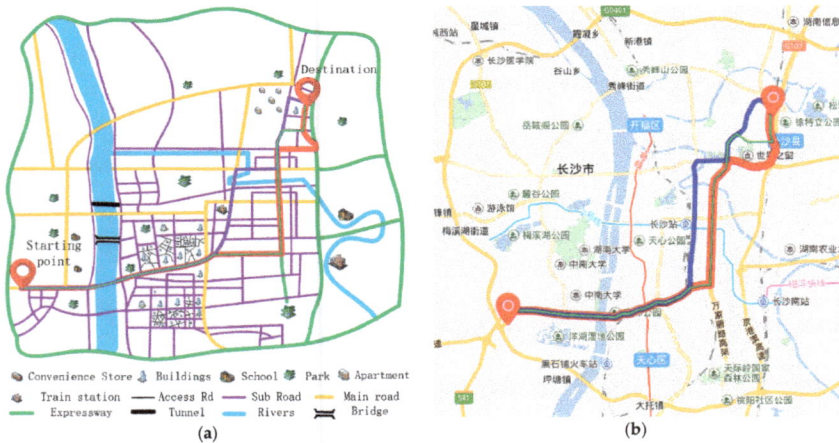

Figure 11. (**a**) The optimal path model for user C; (**b**) The optimal actual planning path for user C. Where, the green represents log trajectory, the blue represents literature Reference resources Route, and the red represents the Individualization Route based on B. Detailed performance of the relevant roads refer to Table 13.

Table 13. Personalized performance of road based on C.

Route Type	Related Road	Level	T	M	C	Personal-Value	Effective Length/km
	Yun-Qi Road	4	0.49	0.35	0.53	0.40	3.77
	South Second Ring	4	0.49	0.35	0.53	0.40	8.77
Individualization	Labor East Road	5	0.35	0.20	0.39	0.27	1.37
Route based on C	Wan Jiali	2	0.67	0.65	0.59	0.62	8.82
	31 Avenue	3	0.48	0.28	0.43	0.37	2.17
	Xi Xia Road	6	0.49	0.36	0.47	0.41	4.93

Table 13. *Cont.*

Route Type	Related Road	Level	T	M	C	Personal-Value	Effective Length/km
literature Reference Route	Yun Qi Road	4	0.49	0.35	0.53	0.45	3.77
	South Second Ring	4	0.49	0.35	0.53	0.40	8.77
	East Second Ring Road	3	0.37	0.22	0.38	0.29	8.78
	31 Avenue	3	0.48	0.28	0.43	0.37	1.95
	Wan Jiali viaduct	2	0.67	0.65	0.59	0.58	2.50
	Wan jiali North	6	0.37	0.32	0.47	0.33	2.54
	Xianghu West Road	7	0.28	0.30	0.34	0.27	1.13
log trajectory	Yun Qi Road	4	0.49	0.35	0.53	0.40	3.77
	South Second Ring	4	0.49	0.35	0.53	0.40	8.77
	Labor East Road	3	0.35	0.20	0.39	0.27	1.37
	Wan Jiali viaduct	2	0.67	0.65	0.59	0.58	10.94
	Fuyuan East Road	6	0.45	0.35	0.41	0.38	1.22
	Kaiyuan West Road	6	0.45	0.35	0.41	0.38	1.29
	Xi Xia Road	6	0.49	0.36	0.47	0.41	2.60

By comparing Figures 10 and 11, we can completely see that the blue reference path for user C in Figure 11 is consistent with that of user B in Figure 10, while the red planning path for user C in Figure 11 is only consistent with that of user B in Figure 10 at the beginning of the path planning, which avoids the East Second Ring Road and other regional centers and makes full use of time and economic advantages of Wanjiali viaduct; moreover, the inconsistent section steers clear of the Xianghu Road in densely populated areas, replacing it with Xi Xia Road that exhibits pedestrian sparsity, small confluence vehicles, and less traffic. According to the calculation results of time and economic performance and the effective road length shown in Table 13, the cumulative time consumption rates of the red planning path and blue reference path are 57.25 and 68.02, respectively, and the total economic losses are 77.69 and 98.25, respectively. In comparison with the reference path, the personalized path of user A shows significantly improved time and economic performance, as shown in Table 14.

Table 14. Inter class contrast experiment.

Comparison Object	ΔC_t	ΔC_m	$\Delta C_t'$	$\Delta C_m'$	Result
Bibliographic reference path	−10.76	−20.56	—	—	Time performance ↑ Econommic performance ↑
Personalization path of fusion Based on B's habits	−0.11	8.11	−0.002	0.104	$\frac{\Delta C_t'}{0.56} < \frac{\Delta C_m'}{0.37}$

To embody the private custom advantage of the personalized planning path proposed in this paper, the user C is used as the specific service object in this verification link, except for the example of the reference path (blue in Figure 11b) experiments, adding the same type of personalized path (red in Figure 10b) contrast link.

Obviously, the contrast experiment on the personalized planning path that integrates user B's habits does not meet the increment in time and economic consumption. Therefore, for the performance qualitative comparison link of Example 1, the dimensionless and performance comparison results are shown in Table 14.

Table 14 shows that in comparison with the reference path, the personalized planning path of user C shows greatly improved time and economic performance. This result is in line with the characteristics of the user related to time + economic performance. In comparison with the individual path planning of user B, the personalized path of user C shows improved time and economic performance; however, the benefits outweigh the lack of economic performance. This result is in line with user C's greater attention time than economic performance levels.

- Calculation of degree of anastomosis

To establish the evaluation criteria for different planning paths and evaluate the performance of planning paths under different algorithms quantitatively, we introduce the path coincidence rate to characterize the matching degree between personalized paths. User driving habits Λ are satisfied in Formula (22).

$$\Lambda = \frac{l_{\text{same}}}{l_{\text{total}}} \tag{22}$$

Type l_{same} is the path length for the coincidence of planning path and log trajectory, l_{total} is the planning path length, and $l_{\text{same}} \leq l_{\text{total}}$. The results of the above example are listed as follows: comfort + economic user A, time + economic user B, time + economic user C personalized path/literature reference path and user log trajectory matching the result shown in Table 15.

Table 15. Verification of path anastomosis.

Customer Type	Path Type	l_{same}/km	l_{total}/km	Λ/a.u.
A	Bibliographic reference path	15.69	30.01	52.3%
	Personalization path based on A's habits	26.77	30.01	89.2%
B	Bibliographic reference path	18.71	28.94	64.7%
	Personalization path based on B's habits	28.94	28.94	100%
C	Bibliographic reference path	14.59	29.96	48.7%
	Personalization path based on C's habits	25.33	29.96	84.5%

From Table 15, the coincidence degree levels of the personalized planning paths that integrates users A, B, and C are 89.2%, 100%, and 84.5%, respectively, which are different from the corresponding user log paths. The coincidence degrees of the planning paths are 52.3%, 64.7%, and 48.7%. In comparison with the reference path in the literature, the coincidence rate of the personalized path that integrates user habits and users' log trajectory significantly increases by 36.9%, 35.3%, and 35.8%.

4. Conclusions

We propose a personalized path decision algorithm that is based on user habits, implementing the navigation service from a previous similar trip group to a specific individual, which can essentially improve the personalization of the existing path planning algorithms. the results and shortcomings of the existing optimization algorithms based on a single optimization target or a single or a few special traffic factors are analyzed. We use the log road track that integrates user's habits to mine driving characteristics, and use LFCM to achieve user's driving style clustering, and get users' driving habits. Then, we use T-S based road multi index performance synchronous evaluation model to quantify the road's time, economy and comfort performance. At the same time, we combine user's driving style and feature to get the user's road performance. Finally, ant colony algorithm is used to search the shortest path consumption path. The algorithm is based on the user's own log trajectory and uses feature mining and clustering techniques to get the user's habits and implements the navigated service objects from the previous travel groups to the travel individual and customizes personalized path navigation system for users to meet their own driving habits. The experimental results show that the absolute anastomosis rate and the relative anastomosis rate of the method in the simulated traffic network are 82% and 100%, respectively. In the actual traffic environment, three comparative examples (i.e., A, B, and C) of user's personalized path and the corresponding user log path show consistent rates, reaching 89.2%, 100%, and 84.5%, respectively; in comparison with the individual path in the literature [20], the path coincidence rate significantly increases by 36.9%, 35.3%, and 35.8%, respectively. By improving the personalization level of existing path planning, the travel experience of self-driving users is greatly improved.

Acknowledgments: This project is partially supported by the National Natural Science Foundation of China (Grant nos. 61663011), the Provincial Natural Science Foundation of Jiangxi (Grant nos. 20161BAB212053), the postdoctoral fund of Jiangxi Province(2015KY19), and the National Natural Science Foundation of China (Grant nos. 51609088).

Author Contributions: Pengzhan Chen contributed to the conception of the reported research. Xiaoyan Zhang contributed significantly to the design and conduct of the experiments and the analysis of results, as well as contributed to the writing of the manuscript. Xiaoyue Chen helped design the experiments and perform the analysis with constructive discussions. Mengchao Liu helped deal with data and plot the graphs.

Conflicts of Interest: The authors declare no conflict of interest.

References

1. Ma, J.; Sun, G. Mutation Ant Colony Algorithm of Milk-Run Vehicle Routing Problem with Fastest Completion Time Based on Dynamic Optimization. *Discret. Dyn. Nat. Soc.* **2013**, *2013*, 418436. [CrossRef]
2. Guo, J.; Wu, Y.; Zhang, X.; Zhang, L.; Chen, W.; Cao, Z.; Zhang, L.; Guo, H. Finding the 'faster' path in vehicle routing. *IET Intell. Transp. Syst.* **2017**, *11*, 685–694. [CrossRef]
3. Tang, K.; Qian, M.; Duan, L. Choosing the fastest route for urban distribution based on big data of vehicle travel time. In Proceedings of the International Conference on Service Systems and Service Management, Dalian, China, 16–18 June 2017; pp. 1–4.
4. Li, Z.; Kolmanovsky, I.V.; Atkins, E.M.; Lu, J.; Filev, D.P.; Bai, Y. Road Disturbance Estimation and Cloud-Aided Comfort-Based Route Planning. *IEEE Trans. Cybern.* **2016**, *47*, 3879–3891. [CrossRef] [PubMed]
5. Faizian, P.; Mollah, M.A.; Yuan, X.; Pakin, S.; Lang, M. Random Regular Graph and Generalized De Bruijn Graph with k-shortest Path Routing. *IEEE Trans. Parallel Distrib. Syst.* **2018**, *29*, 144–155. [CrossRef]
6. Idri, A.; Oukarfi, M.; Boulmakoul, A.; Zeitouni, K.; Marsi, A. A distributed approach for shortest path algorithm in dynamic multimodal transportation networks. *Trans. Res. Procedia* **2018**. [CrossRef]
7. Zhang, Y.; Song, S.; Shen, Z.J.M.; Wu, C. Robust Shortest Path Problem with Distributional Uncertainty. *IEEE Trans. Intell. Transp. Syst.* **2017**. [CrossRef]
8. Zhang, D.; Yang, D.; Wang, Y.; Tan, K.-L.; Cao, J.; Shen, H.T. Distributed shortest path query processing on dynamic road networks. *VLDB J.* **2017**, *26*, 399–419. [CrossRef]
9. Mark, T.; Griffin, M.J. Motion sickness in public road transport: The effect of driver, route and vehicle. *Ergonomics* **1999**, *42*, 1646–1664.
10. Yao, Y.; Peng, Z.; Xiao, B.; Guan, J. An efficient learning-based approach to multi-objective route planning in a smart city. In Proceedings of the IEEE International Conference on Communications, Paris, France, 21–25 May 2017.
11. Siddiqi, U.F.; Shiraishi, Y.; Sait, S.M. Multi-objective optimal path selection in electric vehicles. *Artif. Life Robot.* **2012**, *17*, 113–122. [CrossRef]
12. Ahmed, F.; Deb, K. Multi-objective optimal path planning using elitist non-dominated sorting genetic algorithms. *Soft Comput.* **2013**, *17*, 1283–1299. [CrossRef]
13. Bazgan, C.; Jamain, F.; Vanderpooten, D. Approximate Pareto sets of minimal size for multi-objective optimization problems. *Oper. Res. Lett.* **2015**, *43*, 1–6. [CrossRef]
14. Kriegel, H.P.; Renz, M.; Schubert, M. Route skyline queries: A multi-preference path planning approach. In Proceedings of the IEEE International Conference on Data Engineering, Long Beach, CA, USA, 1–6 March 2010; pp. 261–272.
15. Song, X.; Cao, H.; Huang, J. Influencing Factors Research on Vehicle Path Planning Based on Elastic Bands for Collision Avoidance. *SAE Int. J. Passeng. Cars Electron. Electr. Syst.* **2012**, *5*, 625–637. [CrossRef]
16. Yook, D.; Heaslip, K. The effect of crowding on public transit user travel behavior in a large-scale public transportation system through modeling daily variations. *Transp. Plan. Technol.* **2015**, *38*, 935–953. [CrossRef]
17. Feng, P. The Research of Dynamic Path Planning Based on Improving Fuzzy Genetic Algorithm in the Vehicle Navigation. *Adv. Mater. Res.* **2012**, *424–425*, 73–76. [CrossRef]
18. Jiang, H.; Yuan, R.; Wang, T.; Wang, R. Research on Green Path Planning Model for Road Logistics Vehicle. In Proceedings of the International Conference of Logistics Engineering and Management, Chengdu, China, 8–10 October 2010; pp. 3082–3088.
19. Chen, Z.; Shen, H.T.; Zhou, X. Discovering popular routes from trajectories. In Proceedings of the IEEE, International Conference on Data Engineering, Hannover, Germany, 11–16 April 2011; IEEE Computer Society: Washington, DC, USA, 2011; pp. 900–911.
20. Campigotto, P.; Rudloff, C.; Leodolter, M.; Bauer, D. Personalized and Situation-Aware Multimodal Route Recommendations: The FAVOUR Algorithm. *IEEE Trans. Intell. Transp. Syst.* **2017**, *18*, 92–102. [CrossRef]

21. Nadi, S.; Houshyaripour, A.H.; Nadi, S.; Houshyaripour, A.H. A new model for fuzzy personalized route planning using fuzzy linguistic preference relation. *Int. Arch. Photogramm. Remote Sens. Spatial Inf. Sci.* **2017**, *XLII-4/W4*, 417–421. [CrossRef]

22. Dai, J.; Yang, B.; Guo, C.; Ding, Z. Personalized route recommendation using big trajectory data. In Proceedings of the IEEE International Conference on Data Engineering, Seoul, Korea, 13–17 April 2015; pp. 543–554.

23. Letchner, J.; Krumm, J.; Horvitz, E. Trip router with individualized preferences (TRIP): Incorporating personalization into route planning. In Proceedings of the National Conference on Artificial Intelligence and the Eighteenth Innovative Applications of Artificial Intelligence Conference, Boston, MA, USA, 16–20 July 2006; pp. 1795–1800.

24. Zhu, X.; Hao, R.; Chi, H.; Du, X. FineRoute: Personalized and Time-Aware Route Recommendation Based on Check-ins. *IEEE Trans. Veh. Technol.* **2017**, *66*, 10461–10469. [CrossRef]

25. Long, Q.; Zeng, G.; Zhang, J.; Zhang, L. Dynamic route guidance method for driver's individualized demand. *J. Cent. South Univ. (Nat. Sci. Ed.)* **2013**, *44*, 2124–2129.

26. Ness, M.; Herbert, M. A prototype low cost in-vehicle navigation system. In Proceedings of the 1993 Vehicle Navigation and Information Systems Conference, Ottawa, ON, Canada, 12–15 October 1993; pp. 56–59.

27. Tang, T.Q.; Xu, K.W.; Yang, S.C.; Shang, H.Y. Analysis of the traditional vehicle's running cost and the electric vehicle's running cost under car-following model. *Mod. Phys. Lett. B* **2016**, *30*, 1650084. [CrossRef]

28. Kapoor, A.; Singhal, A. A comparative study of K-Means, K-Means++ and Fuzzy C-Means clustering algorithms. In Proceedings of the International Conference on Computational Intelligence & Communication Technology, Ghaziabad, India, 9–10 February 2017; pp. 1–6.

29. Balteanu, A.; Jossé, G.; Schubert, M. Mining Driving Preferences in Multi-cost Networks. In Proceedings of the International Conference on Advances in Spatial and Temporal Databases, Munich, Germany, 21–23 August 2013; Springer: New York, NY, USA, 2013; pp. 74–91.

30. Zhou, L.; Tong, L.; Chen, J.; Tang, J.; Zhou, X. Joint optimization of high-speed train timetables and speed profiles: A unified modeling approach using space-time-speed grid networks. *Transp. Res. Part B Methodol.* **2017**, *97*, 157–181. [CrossRef]

31. Yang, M.S.; Nataliani, Y. Robust-learning fuzzy c-means clustering algorithm with unknown number of clusters. *Pattern Recognit.* **2017**, *71*, 45–59. [CrossRef]

32. Wang, Q.; Zhang, Y.; Xiao, Y.; Li, J. Kernel-based fuzzy C-means clustering based on fruit fly optimization algorithm. In Proceedings of the 2017 International Conference on Grey Systems and Intelligent Services, Stockholm, Sweden, 8–11 August 2017; pp. 251–256.

33. Taha, A.A.; Hanbury, A. An Efficient Algorithm for Calculating the Exact Hausdorff Distance. *IEEE Trans. Pattern Anal. Mach. Intell.* **2015**, *37*, 2153–2163. [CrossRef] [PubMed]

34. Chen, Y.; He, F.; Wu, Y.; Hou, N. A local start search algorithm to compute exact Hausdorff Distance for arbitrary point sets. *Pattern Recogn.* **2017**, *67*, 139–148. [CrossRef]

35. Ji, P.; Zhang, H.Y. A subsethood measure with the hausdorff distance for interval neutrosophic sets and its relations with similarity and entropy measures. In Proceedings of the Control and Decision Conference, Yinchuan, China, 28–30 May 2016; pp. 4152–4157.

36. Li, T.; Pan, Q.; Gao, L.; Li, P. A novel simplification method of point cloud with directed Hausdorff distance. In Proceedings of the IEEE, International Conference on Computer Supported Cooperative Work in Design, Wellington, New Zealand, 26–28 April 2017; pp. 469–474.

37. Department of People's Republic of China Construction Department. *Urban Road Traffic Planning and Design Specification*; China Planning Press: Beijing, China, 1995.

38. Ren, F. *Road Capacity Manual: Special Report of the United States Traffic Research Committee No. 209*; China Construction Industry Press: Beijing, China, 1991.

39. Xiao, S.; Zhang, Y.; Zhang, B. Adaptive synchronization of delayed T-S type fuzzy neural networks. In Proceedings of the IEEE Control and Decision Conference, Qingdao, China, 23–25 May 2015; pp. 1726–1731.

40. Li, P.F.; Ning, Y.W.; Jing, J.F. Research on the detection of fabric color difference based on T-S fuzzy neural network. *Color Res. Appl.* **2017**, *42*, 609–618. [CrossRef]

41. Wei, X.; Chang, X.Q. The Optimization Design of Emergency Logistics Distribution Path Based on Ant Colony Algorithm. In *Proceedings of the 6th International Asia Conference on Industrial Engineering and Management Innovation*; Atlantis Press: Amsterdam, The Netherlands, 2016.

42. Yang, Q.; Chen, W.N.; Yu, Z.; Gu, T.; Li, Y.; Zhang, H.; Zhang, J. Adaptive Multimodal Continuous Ant Colony Optimization. *IEEE Trans. Evol. Comput.* **2017**, *21*, 191–205. [CrossRef]

43. Wang, R.; Zheng, D.; Zhang, H.; Mao, J.; Guo, N. A solution for simultaneous adaptive ant colony algorithm to memory demand vehicle routing problem with pickups. In Proceedings of the Control and Decision Conference, Yinchuan, China, 28–30 May 2016; pp. 2172–2176.

44. Liu, Z.; Jiang, J.; Yang, Y.; Wang, S. Adaptive ant colony algorithm based on cloud model. In Proceedings of the IEEE International Conference on Information and Automation, Lijiang, China, 8–10 August 2015; pp. 2654–2657.

![applied sciences logo] *applied*
sciences

MDPI

Article

Proposing Enhanced Feature Engineering and a Selection Model for Machine Learning Processes

Muhammad Fahim Uddin [1,*], Jeongkyu Lee [1], Syed Rizvi [2] and Samir Hamada [3]

[1] School of Computer Science and Engineering, University of Bridgeport, 126 Park Ave,
    Bridgeport, CT 06604, USA; jelee@bridgeport.edu
[2] Information Science and Technologies, Penn State University, 3000 Ivyside Park,
    Altoona, PA 16601, USA; srizvi@psu.edu
[3] Computer Systems, School of Business, Farmingdale State College, 2350 Broadhollow Rd,
    Farmingdale, NY 11735, USA; hamadas@farmingdale.edu
* Correspondence: muddin@bridgeport.edu; Tel.: +1-203-543-9688

Received: 6 March 2018; Accepted: 10 April 2018; Published: 20 April 2018

Featured Application: This module can be used independently in any Machine Learning project or can be used in a model that is engineered by boosting and blending of algorithms for better accuracy and fitness.

Abstract: Machine Learning (ML) requires a certain number of features (i.e., attributes) to train the model. One of the main challenges is to determine the right number and the type of such features out of the given dataset's attributes. It is not uncommon for the ML process to use dataset of available features without computing the predictive value of each. Such an approach makes the process vulnerable to overfit, predictive errors, bias, and poor generalization. Each feature in the dataset has either a unique predictive value, redundant, or irrelevant value. However, the key to better accuracy and fitting for ML is to identify the optimum set (i.e., grouping) of the right feature set with the finest matching of the feature's value. This paper proposes a novel approach to enhance the Feature Engineering and Selection (eFES) Optimization process in ML. eFES is built using a unique scheme to regulate error bounds and parallelize the addition and removal of a feature during training. eFES also invents local gain (LG) and global gain (GG) functions using 3D visualizing techniques to assist the feature grouping function (FGF). FGF scores and optimizes the participating feature, so the ML process can evolve into deciding which features to accept or reject for improved generalization of the model. To support the proposed model, this paper presents mathematical models, illustrations, algorithms, and experimental results. Miscellaneous datasets are used to validate the model building process in Python, C#, and R languages. Results show the promising state of eFES as compared to the traditional feature selection process.

Keywords: machine learning; enhanced feature engineering; parallel processing of model; feature optimization; eMLEE; eFES; overfitting; underfitting; optimum fitting

---

1. Introduction

One of the most important research directions of Machine Learning (ML) is Feature Optimization (FO) (collectively grouped as Feature Engineering (FE), Feature Selection (FS), and Filtering) [1]. For FS, a saying "Less is More" becomes the essence of this research. Dimensionality Reduction [2] has become a focus in the ML process to avoid unnecessary computing power/cost, overlearning, and predictive errors. In this regard, redundant features which may have similar predictive value to other feature(s), may be excluded without negatively affecting the learning process. Similarly, the irrelevant features should be excluded as well. FS and FE not only focuses on extracting a subset from the optimal feature set but also building new feature sets previously overlooked by ML techniques. This also includes reducing the higher dimensions into lower ones to extract the feature's value. Latest research has shown noteworthy progress in FE. In [3], the authors reviewed the latest progress in FS and associated algorithms. Out of a few, principal component analysis (PCA) [4] and Karhunen Loeve expansion [5] are widely used with eigen-values and eigen-vectors of the data covariance matrix for FO. The squared error is calculated as well in the mapping of orthonormal transformation to reduce general errors. Another approach is Bayes error probability [6] to evaluate a feature set. However, Bayes errors are generally unknown. Discriminant analysis are also used in FE. Hence, in the line with the latest progress and related study (See Section 2), the work proposed in this paper uses ML and mathematical techniques, such as statistical pattern classification [7], Orthonormalization [8], Probability theory [9], Jacobian [7], Laplacian [3], and Lagrangian distribution [10] to build the mathematical constructs and underlying algorithms (1 and 2). To advance such developments, a unique engineering of the features is proposed where the classifier learns to group an optimum set of features without consuming excessive computing power, regardless of the anatomy of the underlying datasets and predictive goals. This work also effectively addresses the known challenges of ML process such as overfitting, underfitting, predictive errors, poor generalization, and low accuracy.

1.1. Background and Motivation

Despite using the best models and algorithms, FO is crucial to the performance of the ML process and predictions. FS has been a focus in the fields of data mining [11], data discovery, text classification [12], and image processing [13]. Unfortunately, raw datasets pose no clear advice or insight into which variables must be focused on. Usually, datasets contain several variables/features but not all of them contribute towards predictive modeling. Another significance of such research is to determine the intra- and inter-relationships between the features. Their internal dependence and correlation/relevance greatly impact the way a model learns from the data. To make the process computationally inexpensive and keep the accuracy higher, features should be categorized by the algorithm itself. The existing literature proves that such work is rarely undertaken in ML research.

1.2. Parent Research

The proposed model eFES is a participating module of the enhanced machine learning engine engineering (eMLEE) model, which is based on parallel processing and learns from its mistakes (i.e., processing and storing the wrong predictions). Other than eFES, the rest of the four modules as shown in Figure 1 are beyond the scope of this paper. Specifically, eMLEE modules are: (i) enhanced algorithm blend and tuning (eABT) to optimize the classifier performance; (ii) enhanced feature engineering and selection (eFES) to optimize the features handling; (iii) enhanced weighted performance metric (eWPM) to validate the fitting of the model; and (iv) enhanced cross validation and split (eCVS) to tune the validation process. Out of these, eCVS is in its infancy in the research work. Existing research, as discussed in Section 2, has shown the limitations of general purpose algorithms in Supervised Learning (SL) for predictive analytics, decision making, and data mining. Thus, eFES (i.e., the part of eMLEE) fills the gaps that Section 2 discusses.

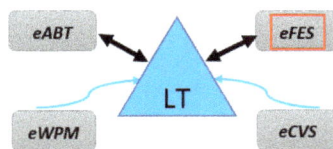

Figure 1. This illustration shows the elevated system externals of eMLEE. Logical Table (LT) interacts primarily with eFES and eABT as compared to the other two modules. It coordinates and regulates the metrics of the learning process in the parallel mode.

1.3. Our Contributions

Our contributions are the following.

a. Improved feature search and quantification for unknown or previously unlabeled features in the datasets for new insights and the relevance of predictive modeling.
b. Outlier identification to minimize the effects on classifier learning.
c. Constructing a feature grouping function (FGF) to add or remove a feature once we have scored them in their correlation, relevance, and non-redundancy nature of predictive value. Identifying the true nature of the feature vs attribute so bias can be reduced. Features tend to gain or lose their significance (predictive value) from one dataset to another. A perfect example would be an attribute "Gender" (e.g., Gender/Sex may not have any predictive value in a certain type of the dataset/prediction). However, it may have significant value in the different dataset.
d. Constructing a logical 3D space where each feature is observed for its fitness value. Each feature can be quantified based on a logical point in 3D space. Its contribution towards overfitting (x), underfitting (y), and optimum-fitting (z) can be scored, recorded, and then weighted for adding or removing in FGF.
e. Developing a unique approach of utilizing an important metric in ML (i.e., error). We have trained our model to be governed by maximum and minimum bounds of the error, so we can maintain acceptable bias and fitness including overlearning. Our maximum and minimum bounds for errors are 80% and 20% respectively. These error bounds can be considered one of our novel ideas in the proposed work. The logic goes thus: models are prone to overfitting, bias, high errors, and low accuracy. We tried to envision if the proposed model can be governed by some limits of the errors. Errors above 80% or below 20% are considered red flags. Such may indicate, bias, overlearning or under-learning of the model. Picking 80% and 20% was our rule of thumb to validate our theory with experiments on a diverse dataset (discussed in the appendix).
f. Finally, engineering local gain (LG) and global gain (GG) functions to improve the feature tuning and optimization.

Figure 2 shows the elevated level block diagram of the eFES Unit.

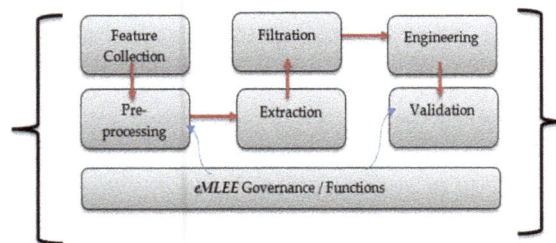

Figure 2. eFES elevated Level.

2. Related Study

To identify the gaps in the latest state of the art in the field of FO, we considered area of ML where FO was of high relevance. In general, every ML problem is affected by feature selection and feature processing. Predictive modeling, as our focus for the consumption of the proposed model, is a great candidate to be looked at, for FO opportunities. One of the challenges in FO is to mine the hidden features that are previously unknown and may hide a great predictive value. If such knowledge can be extracted and quantified, ML process can dramatically be improved. On the other hand, new features can also be created by aggregating existing features. Also, two irrelevant features can be combined, and their weighted function can become a productive feature with higher predictive value.

Clearly, in-depth comprehensive review of the FO and related state of the art are outside the scope of this paper. However, in this section we provide the related study of noteworthy references and then list the gaps we identified. We also present the comparisons of some of the techniques in Section 5.

Li et al. [3] presented a detailed review of the latest development in the feature selection segment of machine learning. They provided various frameworks, methods, and comparisons in both Supervised Learning (SL) and Unsupervised Learning (UL). However, their comparative study did not reveal any development where each feature can achieve a run-time predictive scoring and can be added or removed algorithmically as the learning process continues. Vergara and Estevez [14] reviewed feature selection methods. They presented updates on results in a unifying framework to retrofit successful heuristic criteria. The goal was to justify the need of the feature selection problem in depth concepts of relevance and redundancy. However, their work was only focused on unifying frameworks and placed the generalization on broader scale. Mohsenzadeh et al. [15] utilized a sparse Bayesian learning approach for feature sample selection. Their proposed relevance sample feature machine (RSFM) is an extension of RVM algorithm. Their results showed the improvement in removing irrelevant features and producing better accuracy in classification. Additionally, their results also demonstrated better generalization, less system complexity, reduced overfitting, and computational cost. However, their work needs to be extended to more SL algorithms. Ma et al. [16] utilized Particle Swarm Optimization (PSO) algorithm to develop their proposed approach for detection of falling elderly people. Their proposed research enhances the selection of variables (such as hidden neurons, input weights, etc.) The experiments showed higher sensitivity, specificity, and accuracy readings. Their work though in the domain of healthcare industry does not address the application of approach to a different industry with an entirely different dataset. Lam et al. [17] proposed a unsupervised feature-learning process to improve the speed and accuracy, using the Unsupervised Feature Learning (UFL) algorithm, and fast radial basis function (RBF) for further feature training. However, the UFL may not fit when applied. SL. Han et al. [18] used circle convolutional restricted Boltzmann machine method for 3D feature learning in unsupervised process of ML. The goal was to learn from raw 3D shapes and to overcome the challenges of irregular vertex topology, orientation ambiguity on the surface, and rigid transformation invariances in shapes. Their work using 3D modeling needs to be extended to SL domains and feature learning. Zeng et al. [19] used the deep perceptual features for traffic sign recognition in the kernel extreme learning machines. Their proposed DP-KELM algorithm showed high efficiency and generalization. However, the proposed algorithm needs to be tested across different traffic systems in the world for more distinctive features than those they have considered. Wang et al. [20] discussed the process of purchase decision in subject minds using MRI scanning images through ML methods. Using the recursive cluster elimination-based SVM method, they obtained higher accuracy (71%) as compared to previous findings. They utilized Filter (GML) and wrapping methods (RCE) for feature selection. Their work also needs to be extended to other image techniques in healthcare. Lara et al. [21] provided a survey on ML application for wearable sensors, based on human activity recognition. They provided a taxonomy of learning approach and their related response time on their experiments. Their work also supported feature extraction as an important phase of ML process. ML has also shown a promising role in engineering, mechanical, and thermo-dynamic systems. Zhang et al. [22] worked on ML techniques to do the prediction in the thermal systems for systems components. Besides many different units and technique adoptions, they also utilized FS methods based on correlation feature selection algorithm. They used Weka data-mining tools and came up with the reduced feature set of 16 for improved accuracy. However, their study did not reveal how exactly they came up with this number and whether different number of the features would have helped any further. Wang et al. [23] used the supervised feature method to remove redundant features and considered the important ones for their gender classification. However, they used the neural network method as a feature extraction method, which is mostly common in unsupervised learning. Their work is yet to be tested for more computer vision tasks including image recognition tasks in which bimodal vein modeling becomes significant. Liu et al. [24] utilized the concept of

F-measure optimization for FS. They developed a cost-sensitive feature approach to determine the best F-measure-based feature for the selection by ML process. They argued F-measure to be better than accuracy, for purposes of performance measurement. However, accuracy is not sufficient to be considered a baseline for performance reflection of any model or process. Abbas et al. [25] proposed solutions for IoT-based feature models using the multi-objective optimum approach. They enhanced the binary pattern for nested cardinality constraints using three paths. The second path was observed to increase the time complexity due to the increasing group of features. Though their work was not directly in ML methodologies, their work showed performance improvement in the 3rd path when the optional features were removed.

Here are the gaps that we identified based on a comprehensive literature review and comparisons made, evaluated, and presented in Section 5 later in this paper.

a.   Parallel processing of the features, in which features can be evaluated one by one, has not been done, while the model metrics are being measured and recorded simultaneously to see the real-time effect.

b.   Improved grouping of features is needed across diverse feature types in datasets for improved performance and generalization.

c.   3D modeling is rarely done. The 3D approach can help for side-by-side metric evaluation.

d.   Accuracy is taken as granted to measure the performance of the model. However, we argue on the relevance of this metric and support other metrics in conjunction with it. We engineer our model to incline towards the metrics that are found relevant for a given problem based on the classifier learning.

e.   Feature quantification and function building governed by algorithms the way we presented is not found in the literature, and the dynamic ability of such a design, as our work indicated, can be a good filler of this gap in the state of the art.

f.   Finally, FO has not been directly addressed. FO helps to regulate the error biasing, outlier detection, and poor generalization.

3. Groundwork of eFES

This section presents background on the underlying theory, mathematical constructs, definitions, algorithms, and the framework.

The elevated-level framework shown in Figure 3 elaborates on the importance of each definition, incorporated with the other units of eMLEE and the ability to implement parallel processing by design. In general computing, parallel processing is done by dividing program instructions to be run by multiple processors, so the time efficiency can be improved. This also ensures the maximum utilization of otherwise idle processors. Similar concepts can be implemented on the algorithms and ML models. ML algorithms depend on the problem and data types and require sequential training of each of the data models. However, the parallel processing can dramatically improve the learning process, especially for the blended model, such as eMLEE. In light of the latest work of parallel processing in ML, such as in [26], the authors introduced the parallel framework on ML algorithms for large graphs. They experimented with aggregation and sequential steps in their model to allow researchers to improve the usage of various algorithms. Another study was done in [27], where authors used induction to improve the parallelism in the decision trees. Python and R libraries have come a long way to help provide useful libraries and classes to develop various ML techniques. Authors in [28] introduced a python library Qjan to parallelize the ML algorithms in compliance by MapReduce. A PhD thesis [29] work done by a student at the University of California at Berkeley used concurrency control method to parallelize the ML process. Another study done in [30] utilized parallel processing approaches in ML techniques for detection in big-data networks. Therefore, similar progresses have motivated us to incorporate parallel processing in the proposed model.

Our proposed model parallelism is done in two layers:

(i) Outer layer to eFES, where eFES unit communicates with other units of the eMLEE such as eABT, eWPM, eCVS and LT. Parallelism is done through real-time metric measurement with LT object and based on classifier learning, eFES reacts to the inner layer (defined next). Other units such as eABT and eCVS enhance the algorithm blend and test-training split in parallel, while eFES is being trained. In other words, all four units including eFES regulated by LT unit, are run in parallel to improve the speed of the learning process and validation for every feature as processed in the eFES unit and every algorithm processed in the eABT unit. However, eFES can also work without being related to the other units, if researchers and industrialists may however choose so.

(ii) Inner layer to eFES, where adding and removing of the feature are done in parallel. When the qualifying feature is added, the metrics are measured by the model to see if fitness improves, and then features are added and removed one by one to see the effect on the fitness function. This may be done sequentially, but parallelism improves the insurance that each feature is evaluated at the same time; the classifier is incorporating metrics reading from LT object and speed of the process especially when the huge dataset is being processed.

Figure 3 illustrates the inner layer parallel processing of each construct that constitutes the eFES unit. It shows the high-level block diagram of eFES unit modeling and related functions. Each definition is explained in plain English next.

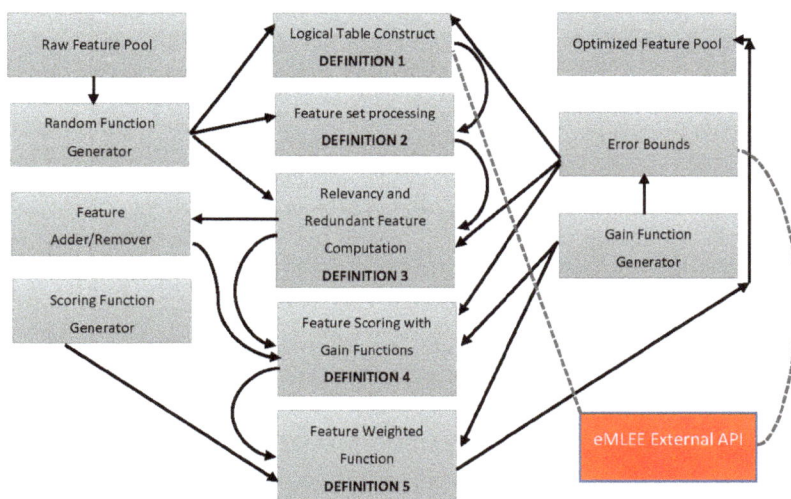

Figure 3. Theoretical Foundation Illustration on the elevated-level.

Definition 1 covers the theory of LT unit, which works as a main coordinator assisting and regulating all the sub-units of eMLEE engine such as eFES. It inherently is based on parallel processing at a low level. While the classifier is in the learning process, LT object (in parallel) performs the measurements, records it, updates it as needed, and then feeds the learning process back. During classifier learning, LT object (governed by LT algorithm, outside of scope of this paper, but will be available as an API) creates a logical table (with rows and columns) where it adds or removes the entry of each feature as a weighted function, while constantly measuring the outcome of the classifier learning.

Definition 2 covers the creation of the feature set from raw input via random process, as shown above. As discussed earlier, it uses 3D modeling using x, y, and z coordinates. Each feature is quantized based on the scoring of these coordinates (x being representative of overfit, y being underfit and z being an optimum fit).

Definition 3 covers the core functions of this unit to quantify the scoring of the features, based on their redundancy and irrelevancy. It does this in a unique manner. It should be noted that not every irrelevant feature with high score will be removed by the algorithm. That is the beauty of it. To increase the generalization of such model with a diverse dataset that it has not seen during test (i.e., prediction), features are quantified, and either kept, removed, or put on the waiting list for re-processing of addition or removal evaluation. The benefit of this approach is that it will not do injustice to any feature without giving a second chance later in the learning process. This is because features, once aggregated with a new feature or previously unknown feature, can dramatically improve scores to participate in the learning process. However, the deep work of "unknown feature extraction" is kept for future work, as discussed in the future works section.

Definition 4 utilizes definition 1 to 3 and introduces a global and local gain functions to evaluate the optimum feature-set. Therefore, the predictor features, accepted features, and rejected features can be scored and processed.

Definition 5 finally covers the weight function to observe the 3D weighted approach for each feature that passes through all the layers, before each feature is finally added to the list of the final participating features.

The rest of the section is dedicated to the theoretical foundation of mathematical constructs and underlying algorithms.

3.1. Theory

eFES model manifests itself into specialized optimization goals of the features in the datasets. The most crucial step of all is the Extended Feature Engineering (EFE) that we refer when we build upon existing EF techniques. These five definitions help build the technical mechanics of the proposed model of eFES unit.

Definition 1. *Let there be a Logical Table (LT) module that regulates the ML process during eFES constructions. Let LT have 3D coordinates as x, y, and z to track, parallelize, and update the $x \leftarrow overfit(0:1)$, $y \leftarrow underfit(0:1)$, $z \leftarrow optimumfit(-1:+1)$. Let there be two functions, Feature Adder as $+\mathbb{F}$, and Feature Remover as $-\mathbb{F}$, based on linearity of the classifier for each feature under test for which the RoOpF (Rule. 1) is valid. Let Lt. RoOpF > 0.5 to be considered of acceptable predictive value.*

eFES LT module builds very important functions at initial layers for adding a good fit feature and removing a bad fit feature from the set of features available to it, especially when algorithm blend is being engineered. Clearly, not all features will have an optimum predictive value and thus identifying them will count towards optimization. The feature adder function is built as:

$$+ \mathbb{F}(x,y,z) = +\mathbb{F}_{F_n} = (F_n \cup F_{n+1}) \sum_{i=1}^{Z} (LT.score\ (i)) + \sum_{j,k=1}^{x,y} (LT.score\ (j,k)) \tag{1}$$

The feature remover function is built as:

$$- \mathbb{F}(x,y,z) = -\mathbb{F}_{F_n} = (F_n \cap F_{n+1}) \sum_{j,k=1}^{x,y} (LT.score\ (j,k)) - \sum_{i=1}^{z} (LT.score(i)) \tag{2}$$

Very similar to *k*-means clustering [12] concept, that is highly used in unsupervised learning, LT implements feature weights mechanism (FWM) so it can report a feature with high relevancy score and non-redundant in a quantized form. Thus, we define:

$$FWM(X,Y,Z) = \sum_{x=1}^{X} \sum_{y=1}^{Y} \sum_{z=1}^{Z} (u_x w_x . \ u_y w_y . u_z w_z) (\Delta(x,y,z)) \tag{3}$$

$$\Delta(x,y,z) = \begin{cases} \prod_{l=1}^{L}(u_{lx}w_{lx}), & \text{if } z \neq 0, \; AND \; z > (0.5, y) \\ u_i \in \{0,1\}, & -1 \leq i \leq L \\ \prod_{l=1}^{L}(u_{ly}w_{ly}), & \text{if } z \neq 0, \; AND \; z > (0.5, x) \end{cases} \quad (4)$$

Illustration in Figure 4 shows the concept of LT modular elements in 3D space as discussed earlier. Figure 5 shows the variance of the LT. Figure 6 shows that it is based on binary weighted classification scheme to identify the algorithm for blending and then assign a binary weight accordingly in LT logical blocks. The diamond shape shows the err distribution that is observed and recorded by LT module as new algorithm is added or existing is removed. The complete mathematical model for eFES LT is beyond the scope of this paper. We finally provide the eFES LT functions as:

$$eFES^{\boxplus} = [\mathbb{R}_{eFES} = \frac{1}{N_e}\left(\sqrt{\frac{err}{err + Err}}\right)^2] \times \sum_{n=1}^{N} F_n(f(x,y,z)\Big|exp\left(\frac{+\mathbb{F}_{F_n}}{+\mathbb{F}_{F_n} + (-\mathbb{F}_{F_n})}\right)\Big| \quad (5)$$

where *err* = local error (LE), *Err* = global error (GE). $f(x,y,z)$ is the main feature set in 'F' for 3D.

RULE 1
If (LTObject.ScoFunc (A (i) > 0.5)
        Assign "1"
Else Assign "0"

PROCEDURE 1
Execute *LT.ScoreMetrics (Un.F, Ov.F)*
Compute *LT.Quantify (*LT)*
Execute *LT.Bias (Bias.F, *)*
_Shows the pointer to the LT object.

Definition 2. *$F_n = \{F_1, F_2, F_3, \ldots \ldots, F_n\}$ indicates all the features appears in the dataset, where each feature $F_i \in F_n | f_w \geq 0$. f_w indicates the weighted feature value in the set. Let $F_{ran}(x,y,z)$ indicates the randomized feature set.*

We estimate the cost function based on randomized functions. Las Vegas and Monte Carlo algorithms are popular randomized algorithms. The key feature of the Las Vegas algorithm is that it will eventually have to make the right solution. The process involved is stochastic (i.e., not deterministic) and thus guarantee the outcome. In case of selecting a function, this means the algorithm must produce the smallest subset of optimized functions based on some criteria, such as the accuracy of the classification. Las Vegas Filter (LVS) is widely used to achieve this step. Here we set a criterion in which we expect each feature at random gets a random maximum predictive value in each run. ∅ shows the maximum inconsistency allowed per experiment. Figure 5 shows the cost function variation in LT object for each coordinate.

PROCEDURE 2
$Score_{best} \leftarrow$ *Import all attributes as 'n'*
$Cost_{best} \leftarrow n$
For $j \leftarrow$ *1~to Iteration$_{max}$* **Do**
        Cost $\leftarrow$ *Generate random number between 0~and $Cost_{best}$*
        Score $\leftarrow$ *Randomly select item from Cost feature*
        If *LT.InConsistance ($Score_{best}$, Training Set)* $\leq \emptyset$ **Then**
            $Score_{best} \leftarrow Score$
            $Cost_{best} \leftarrow C$
    Return ($Score_{best}$)

Definition 3. *Let lt.IrrF and lt.RedF be two functions to store the irrelevancy and redundancy score of each feature for a given dataset.*

Let us define a Continuous Random Vector $CRV \in Q^N$, and Discrete Random Variable $DRV \in H = \{h_1, h_2, h_3, \ldots \ldots, h_n\}$. The density function of the random vector based on cumulative probability is $P(CRV) = \sum_{i=1}^{N} P_H(h_i)p\ CRV \mid DRV$, $P_H(h_i)$ being a priori probability of class.

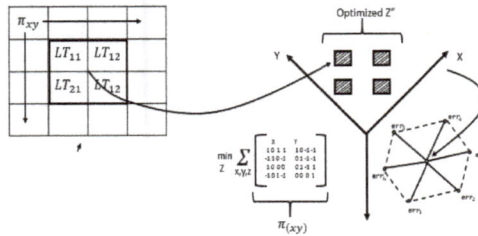

Figure 4. Illustration of the conceptual view of LT Modules in 3D space.

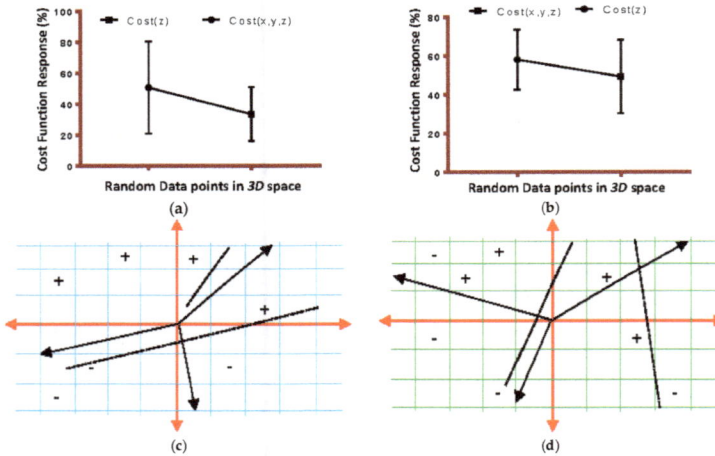

Figure 5. (**a**) This test shows the variance of the LT module for the cost function for all three co-ordinates and then z (optimum-fitness). This is the ideal behavior; (**b**) This test shows the real (experimental) behavior; (**c**) This shows the ideal shift of all 3 coordinates in space while they are tested by the model in parallel. Each coordinate (x, y, z) lies on the black lines in each direction. Then, based on the scoring reported by LT object (and cost function), they either sit on positive point or negative as shown; (**d**) This shows the ideal spread of each point, when z is optimized with the lowest cost function.

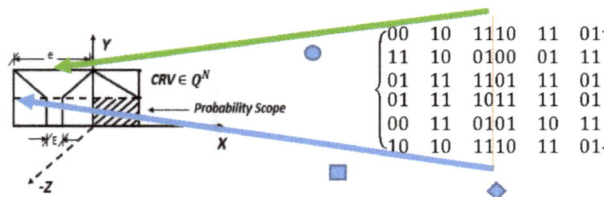

Figure 6. Illustration of probability-based feature binary classification. Overlapped matrices for *Red.F* and *Irr.F*, for which the probability scope resulted in acceptable errors by stepping into vector space, for which 0.8 > err > 0.2.

As we observe that the higher error limit (e) (err, green line, round symbol) and lower error limit (E), (Err, blue line, square symbol) bound the feature correlation in this process. Our aim is to spread the distribution in z-dimension for optimum fitting as features are added. The red line (diamond symbol) that separates the binary distribution of Redundant Feature (Red.F) and Irrelevant Features (Irr.F) based on error bounds. The green and red lines define the upper and lower limit of the error, in which all features correlate. Here, we build a mutual information (MI) function [14] so we can quantify the relevance of a feature upon other in the random set and this information is used to build the construct for *Irr.F*, since once our classifier learns, it will mature the *Irr.F* learning module as defined in the Algorithm 1.

$$MI(Irr.F\,(x,y,z)|f_i,f_{i+1} = \sum_{a=1}^{N}\sum_{b=1}^{N} p\,(f_i(a),f_{i+1}(b).\log\left(\frac{f_i(a),f_{i+1}(b)}{p\,(f_i(a)\cdot f_{i+1}(b)}\right) \tag{6}$$

We expect $MI \leftarrow 0$, for features to be statistically independent, so we build the construct in which the MI will be linearly related to the entropies of the features under test for *Irr.F* and *Red.F*, thus:

$$M.I(f_i, f_{i+1}) = \begin{cases} H(f_i) - H\,(f_i|f_{i+1}) \\ H\,(f_{i+1} - H\,(f_{i+1}|f_i) \\ H(f_i) + H(f_{i+1}) - H\,(f_i, f_{i+1}) \end{cases} \tag{7}$$

We use the following construct to develop the relation of '*Irr.F*' and '*Red.F*' to show the irrelevancy factor and Redundant factor based on binary correlation and conflict mechanism as illustrated in above table.

$$Irr.F = \sum_{i,j}^{K}\begin{Bmatrix} f_{ii} & f_{ij} \\ f_{ji} & f_{ji} \end{Bmatrix} Red.F = \begin{cases} MI(f_i;\,Irr.F) > 0.5 & \textit{Strong Relevant Feature} \\ MI(f_i;\,Irr.F) < 0.5 & \textit{Weak Relevant Feature} \\ MI(f_i;\,Irr.F) = 0.5 & \textit{Neutral Relevant Feature} \end{cases} \tag{8}$$

Definition 4. *Globally in 3-D space, there exist three types of features types (variables), as predictor features: $PF = \{pf_1, pf_2, pf_3, \ldots\ldots pf_n\}$, and accepted features to be $AF = \{af_1, af_2, af_3, \ldots\ldots, af_n\}$ and rejected features to be $RF = \{rf_1, rf_2, rf_3, \ldots..rf_n\}$, in which $\mathbb{G} \geq (g+1)$, global gain for all experimental occurrence of data samples. ' $\mathbb{G}$ ' being the global gain (GG). ' g ' being the local gain (LG). Let PV be the predictive value. Accepted features are $af_n \in PV$, strongly relevant to the sample data set ΔS, if there exist at-least one x and z or y and z plane with score ≥ 0.8, AND a single feature $f \in F$ is strongly relevant to the objective Function 'ObF' in distribution 'd' if there exist at-least a pair of example in data set $\{\Delta S_1, \Delta S_2, \Delta S_3, \ldots..., \Delta S_n \in I\}$, such that $d\,(\Delta S_i) \neq 0$ and $d\,(\Delta S_{i+1}) \neq 0$. Let $\nabla\,(\varphi, \rho, \omega)$ correspond to the acceptable maximum 3-axis function for possible optimum values of x, y, and z respectively.*

We need to build an ideal classifier that learns from data during training and estimate the predictive accuracy, so it generalizes well on the testing data. We can use probabilistic theory of Bayesian [31] to develop a construct similar to direct table lookup. We assume a random variable to be 'rV' that will appear with many values in set of $\{rV_1, rV_2, rV_3, \ldots, rV_n\}$ that appear as a class. We will use prior probability $P\,(rV_i)$. Thus, we represent a class or set of classes as rV_i, and the greatest $P\,(rV_i)$, for given pattern of evidence (pE) that classifier learns on $P\,(rV_i\,|\,pE) > P\,(rV_j|\,pE)$ *valid for all* $i \neq j$.

Because we know that

$$P\,(rV_i\,|\,pE) = \frac{P\,(pE\,|\,rV_i)\,P\,(rV_i)}{(P(pE))} \tag{9}$$

Therefore, we can write the conditional equation where $P\,(pE)$ is considered with regard to probability of (pE) is $P\,(pE\,|\,rV_i)P\,(rV_i) > P\,(pE\,|\,rV_j)P\,(rV_j)$ *valid for all* $i \neq j$. Finally, we can write the probability of the error for the above given pattern, as $P\,(pE)|error$, assuming the cost function for all correct classification is 0, and for all incorrect is 1, then as stated earlier, the Bayesian classification will put the instance in the class labelling the highest posterior probability as $P\,(pE) = \sum_{i=1}^{k} P\,(rV_i)\,P\,(pE|rV_i)$. Therefore, the construct can thus be determined as

$P(pE)|error = Error [1 - max\{P(rV_1) | pE, \ldots\ldots, P(rV_k|pE)\}]$. Let us construct the matrix function of all features, accepted and rejected features, based on GG and LG, as

$$\mathbb{G}(x,y,z) = \frac{1}{N} \sum_{i=1}^{n} \{(g_i) \times MH\} \qquad (10)$$

$$MH = \left\{ \begin{matrix} pf_{x1y1} & \cdots & pf_{x1yn} \\ \vdots & \ddots & \vdots \\ pf_{xny1} & \cdots & pf_{xnyn} \end{matrix} \right\} = \left\{ \begin{matrix} af_{11} & af_{12} & \cdots & af_{1n} \\ af_{n1} & af_{n2} & \cdots & af_{nm} \end{matrix} \right\} \times \left\{ \begin{matrix} rf_{11} & rf_{1n} \\ rf_{21} & rf_{2n} \\ rf_{2n} & rf_{mn} \end{matrix} \right\} \pm \nabla(\varphi, \rho, \omega) \qquad (11)$$

Table 1 shows the various ranges for Minimum, Maximum and Middle points of the all three functions as discussed earlier.

Table 1. Typical observations of the functions.

Function	Min	Mid	Max
$\nabla(\varphi, \rho, \omega)$	(0.21, 0.71, 0.44)	(0.43, 55, 49)	(0.81, 0.76, 58)
$g(x, y, z)$	(0.34, 0.51, −0.11)	(0.55, 0.51, 0.68)	(0.67, 71, 89)
$\mathbb{G}(x, y, z)$	(0.44, 0.55, 0.45)	(0.49, 0.59, 0.58)	(0.52, 0.63, 0.94)

Using Naïve Bayes multicategory equality as:

$$P_{1,2,3,\ldots,N} \left[\sum_j x_j \right] + \left[\sum_j y_j \right] + \left[\sum_j z_j \right] = \sum_k Var(x, y, z) [z^{*i}] \qquad (12)$$

where $z^*(n) \, argmaxP(z) \prod_{k=1}^{n} p([z]).z_k$, and Fisher score algorithm [3] can be used in FS to measure the relevance of each feature based on Laplacian score, such that $B(i,j) = \begin{cases} \frac{1}{N_l} \; if \; u_i = u_i = 1 \\ 0 \; otherwise, \end{cases}$.

N_l shows the no. of data samples in test class shown subscript '*l*'. Generally, we know that based on specific affinity matrix, $FISHER_{score}(f_i) = 1 - \frac{1}{LAPCLACIAN_{score}(f_i)}$.

To group the features based on relevancy score, we must ensure that each group member of the features exhibit low variance, medium stability and their score is based on optimum-fitness, thus each member follows k ($k \in K$, *where* $K \leq f(0:1)$). This also ensure that we address the high dimensionality issue, as when feature appears in high dimension, they tend to change their value for training mode, thus, we determine the information gain using entropy function as:

$$Entropy(F_n) = \sum_{t=1}^{V1} -p_t \log p_t \qquad (13)$$

V_1 indicates the number of various value of the target features in set of F. and p_t is the probability of the type of value of t in a complete subset of the feature tested. Similarly, we can calculate the entropy for each feature in x, y, z dimension as:

$$Entropy(F_{n \in x,y,z}) = \sum_{t \in T(x,y,z)} \frac{|F_{(t:x,y,z)}|}{|F_t|} Entropy(F_n) \qquad (14)$$

Consequently, gain function in probability of entropy in 3D is determined as:

$$Gain\left(I, F_{(t:x,y,z)}\right) = Entropy(F_n) - Entropy(F_{n \in x,y,z}) \qquad (15)$$

We develop a ratio of gain for each feature in z-dimension as this ensure the maximum fitness of the feature set for the given predictive modeling in the given dataset for which ML algorithm needs to be trained. Thus, gR indicates the ratio between:

$$gR(z) = \frac{Gain\left(I, F_{(t:x,y,z)}\right)}{\mathbb{G}(x,y,z)} |P(pE) > P(pE)|error \qquad (16)$$

Figure 7 shows the displacement of the local gain and global gain functions based on probability distributions. As discussed earlier, LG and GG functions are developed to regulate the accuracy and thus validity of the classifier is measured initially based on accuracy metric.

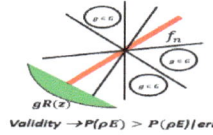

Figure 7. Illustration on probability of LG and GG function.

Table 2 shows the probability of local and global error limits based on probability function (Figure 7) in terms of local (g) and global gain function (G).

Table 2. Typical observations.

	g (max)	G (min)	G (max)	g (min)
P (err)	0.25	0.45	0.31	0.56
P (Err)	0.32	0.41	0.49	0.59

RULE 2

If (g (err) < 0.2) *Then*

Flag 'O.F'

Elseif (g (err) > 0.8) Flag 'U.F'

If we assume the fact of $\{\Delta S_1, \Delta S_2, \Delta S_3, \ldots\ldots, \Delta S_n \in I\}$, such that $d\ (\Delta S_i) \neq 0$ and $d\ (\Delta S_{i+1}) \neq 0$, where '$I$' is the global input of testing data. We also confirm the relevance of the feature in the set using objective Function construct in distribution 'd', thus:

$$ObF\ (d,\ I)\ =\ \frac{\log\ \left(Gain\ \left(I,\ F_{(t:x,y,z)}\right)\right)}{(err[max:1],\ err[min:0])}\ \ |\ d\ (\Delta S_i)\ \neq\ 0\ |\ for\ every\ F_i\ in\ group \tag{17}$$

Then, Using Equations (14)–(17), we can finally get

$$F.Eng(x,y,z)\ =\ \frac{1}{(k\times M)}\ \sum_{t=1}^{K}\prod_{t=k}^{M}\ ObF\ (d,\ I)\ \times MH_t) \tag{18}$$

$$F.Grp(x,y,z)\ =\ F.Eng(x,0,0)\ +\ F.Eng(0,y,0)\ -\ F.Eng(0,0,z) \tag{19}$$

Figure 8 Illustration of Feature Engineering and Feature Group as constructed in the mathematical model and governed by the Algorithms 1 and 2, defined later. Metrics API is available from eMLEE package. The white, yellow, and red orbital shapes indicate the local gain progression through 3D space. The little 3D shapes (x, y, and z) in the accepted feature space in grouping indicates several (theoretically unlimited) instances of the optimized values as the quantization progresses.

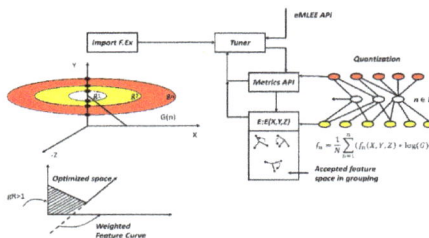

Figure 8. Illustration of Feature Engineering and Feature Group as constructed in the mathematical model.

Definition 5. *Feature selection is governed by satisfying the scoring function (score) in 3D space (x: Over-Fitness, y: Under-Fitness, z: Optimum-Fitness) for which evaluation criterion needs to be maximized, such that Evaluation Criterion : f'. There exist a weighted, $W(\varnothing)\{\nabla(\varphi, \rho, \omega), 1\}$ function that quantifies the score for each feature, based on response from eMLEE engine with function eMLEE$_{return}$, such that each feature in $\{f_1, f_2, f_3, \ldots\ldots, f_n,\}$, has associated score for $(\varphi : x, y, z, \rho : x, y, z, \omega : x, y, z)$.*

Two or more features may have the same predictive value and will be considered redundant. The non-linear relationship exists between two or more features (variables) that affects the stability and linearity of the learning process. If the incremental accuracy is improved, then non-linearity of a variable is ignored. As the number of the features are added or removed in the given set, the OF, UF, and B changes. Thus, we need to quantify their convergence, relevance, and covariance distribution across the space in 3D. We implement weighted function for each metric using LVQ technique [1], in which, we measure each metric over several experimental runs for enhanced feature set, as reported back from the function explained in Theorems 1 and 2, such that we optimize the z-dimension for optimum fitness and reduce x and y dimension for over-fitness and under-fitness. Let us define:

$$W(\varnothing) = \frac{1}{\int_{S_t} p(x)dx} \sum_{\gamma=1}^{\sigma} N_\gamma^T \cdot N_\gamma \int_{S_\gamma} p(x)dx \tag{20}$$

where the piecewise effective decision border is $S_t = \sum_{\gamma=1}^{\sigma} S_\gamma$, In addition, the unit normal vector, (N_γ) for border S_γ, $\gamma = 1, 2, 3, 4, \ldots\ldots\sigma$ is valid for all cases in space. Let us define the probability distribution of data on $S_\gamma : \mathbb{Q}_\gamma = \int_{S_\gamma} p(x)dx$. Here, we can use the Parzen method [32], to restore the nonparametric density estimation method, to estimate the $\mathbb{Q}_\gamma$.

$$\widehat{\mathbb{Q}_\gamma}(\Delta) = \sum_{j=1}^{K} \delta \left(d\left(x_i, S_\gamma\right) \leq \frac{\Delta}{2}\right) \tag{21}$$

where $d\left(x_i, S_\gamma\right)$ shows the Euclidean distance function. Figure 9 shows the Euclidean distance function based on binary weights for $W(\varnothing)$ function.

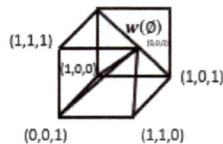

Figure 9. Illustration of function based on Euclidean Distance function.

Table 3 lists the quick comparison of the values of two functions as developed to observe the minimum and maximum bounds.

Table 3. Average Observations.

Function	Min	Max
$W(\varnothing)$	28.31%	78.34%
$\mathbb{Q}_\gamma$	+0.2834	−0.1893

We used the Las Vegas Algorithm approach that helps to get correct solution at the end. We used it to validate the correctness of our gain function. This algorithm guarantees correct outcome if the solution is returned or created. It uses the probability approximate functions to implement runnable time-based instances. For our feature selection problem, we will have a set of features that will guarantee the optimum minimum set of features for acceptable classification accuracy. We use linear regression to compute the value of features to detect the non-linearity relationship between features, we thus implement a function, $Func(U(t)) = a + b * t$. Where a, and b are two test features and values can be determined by using linear regression techniques, so $b = \frac{\sum_{t=1}^{T}(t - \bar{t})(U(t) - \bar{u})}{\sum_{t=1}^{T}(t - \bar{t})^2}$.

Where, $a = \bar{u} - b * \bar{t}, \bar{u} = \frac{1}{T}\sum_{t=1}^{T} U(t), \bar{t} = \frac{1}{T}\sum_{t=1}^{T} t$. These equations also minimize the squared error. To compute weighted function, we use feature ranking technique [12]. In this method, we will score each feature, based on quality measure such as information gain. Eventually, the large feature set will be reduced to a small feature set that is usable. The Feature Selection can be enhanced in several ways such as pre-processing, calculating information gain, error estimation, redundant feature or terms removal, and determining outlier's quantification, etc. The information gain can be determined as:

$$Gain_I(w) = -\sum_{j=1}^{M} P(M_j) . \log P(M_j) + P(w)\sum_{j=1}^{M} P(M_j \mid w).\log P(M_j \mid w) + P(\overline{w})\sum_{j=1}^{M} P(M_j \mid \overline{w}) . \log P(M_j \mid \overline{w}) \quad (22)$$

'M' shows the number of classes and 'P' is the probability. 'W' is the term that it contains as a feature. $P(M_j \mid w)$ is the conditional probability. In practice, the gain is normalized using Entropy, such as

$$Norm.Gain_I(w) = \frac{\{Gain_I(w)\}}{\left\{-\frac{n(w)}{n}\log\frac{n(w)}{n}\right\}} \quad (23)$$

Here we apply conventional variance-mean techniques. We can assume, $\max \nabla \sum_{i=1}^{n} \varphi_i \rho_i \omega_i - \sum_{i=1}^{n} \log \varphi_i \rho_i \omega_i$. The algorithm will ensure that 'EC {F.Sco (x,y,z), F.Opt (x,y,z) $\geq$ 0.5}' stays in optimum bounds. Linear combination of Shannon information terms [7] and conditional mutual information maximization (*CMIM*) [3] for $U_{MAX}(Z_k) = \max_{Z_k \in \Delta s}[Inf(Z_k : X, Y \mid (XY)_k)]$ builds the functions as

$$Score(X \mid Y) = \sum_{y_k \in Y} G(y_k) . \sum_{x_{k\prime} \in X} G(x_{k\prime}) \times \log(g(z)) \quad (24)$$

$$J_{MIN}(Z)^d = -\beta\left(\prod_{k,k\prime}^{K(0)} S(X:Y)_k + \gamma\left(\prod_{k,k\prime}^{K(0)} S(Y:X)_{k\prime}\right)\right) \quad (25)$$

By using Equations (23)–(27), we get

$$F.Sco(x,y,z) = Score(X \mid Y) + \sum_{i=1}^{n} W(\varnothing)_i - \sum_{j=1}^{n} Gain_j(w) \quad (26)$$

$$F.Opt(x,y,z) = J_{MIN}(Z)^d . \prod_{F.Soc(x,y,z)}^{N} \left\{\frac{F.Soc(x,y,z)}{1 + TNorm.Gain_j(w)}\right\} - \sum_{j=1}^{n} \Delta Err(j) \quad (27)$$

3.2. eFES Algorithms

The following algorithms aim: (i) to compute functions as raw feature extraction, related features identification, redundancy, and irrelevancy to prepare the layer for feature pre-processing; (ii) to compute and quantify the selection and grouping factor for the acceptance as model incorporates them; and (iii) to compute the optimization function of the model, based on weights and scoring functions. *objeMLEE* is the API call for accessing public functions

Following are the pre-requisites for the algorithms.

Initialization: Set the algorithm libraries, create subset of the dataset for random testing and then correlating (overlapping) tests.
Create: *LTObject* for eFES$^{\boxplus}$
Create: ObjeMLEE (h)/*create an object reference of eMLEE API */
Set: ObjeMLEE.PublicFunctions (h.eABT,h.eFES,h.eWPM,h.eCVS)/* Handles for all four constructs*/
Global Input: $A_n = \{A_1, A_2, A_3, \ldots\ldots, A_n\}, F_n = \{F_1, F_2, F_3, \ldots\ldots, F_n\}, DataSet(signal, noise)$

Dataset Selection: These algorithms require the dataset to be formatted and labelled with supervised learning in mind. These algorithms have been tested for miscellaneous datasets selected from different domains as listed in the appendix. Some preliminary clean-up may be needed depending upon the sources and raw format of the data. For our model building, we developed a Microsoft SQL Server-based data warehouse. However raw data files such as TXT and CSV are valid input files.

Overall Goal (Big Picture): The foremost goal of these two algorithms is to govern the mathematical model built-in eFES unit. These algorithms are essential to work in a chronological mode, as the output of Algorithm 1 is required for Algorithm 2. The core idea that algorithms utilize is to quantify each feature either in original, revealed or an aggregated state. Based on such scoring, which is very dynamic and parallelized while classifier learning is being governed by these algorithms, the feature is removed, added, or put on the waiting list, for the second round of screening. This is the beauty of it. For example, Feature-X may be scored low in the first round and because Feature-Y is now added, that impacts the scoring of the Feature-X, and thus Feature-X is upgraded by scoring function and included accordingly. Finally, algorithms accomplish the optimum grouping of the features from the dataset. This scheme maximizes the relevance, reduces the redundancy, improves the fitness, accuracy, and generalization of the model for improved predictive modeling in any datasets.

Algorithm 1 aims to compute the low-level function as *F.Prep* (x, y, z), based on final Equations (26) and (27) as developed in the model earlier. It uses the conditions of Irrelevant feature and Redundant feature functions and run the logic if the values are below 50% as a check criterion. This algorithm splits the training data based on popular approach as cross validation. However, it must be noted in line 6, that we use our model API for improving the value of k in the process, that we call enhanced cross validation. LT object regulates it and optimizes the value of k based on the classifier performance in the real time. It then follows the error rule (80%, 20%) and keeps track of each corresponding feature, as they are added or removed. Finally, it gets to the start using the gain function in 3D space for each fitting factor since our model is based on 3D scoring of each feature in the space where point is moved in x, y, and z values in space (logical tracking during classifier learning).

Algorithm 2 aims to use the output of algorithm 1 in conjunction with computing many other crucial functions to compute a final function of feature grouping function (FGF). It uses the weighted function to analyze each participating feature including the ones that were rejected. It also utilizes the LT object and its internal functions using the API. This algorithm slices the data into various non-overlapping segments. It uses one segment at a time, then randomly mixed them for more slices to improve the classifier generalization ability during the training phase. It uses eFES$^{\boxplus}$ as a LT object from the library of eMLEE and records the coordinates for each feature. This way, entry is made in LT class, corresponding to the gain function as shown in lines 6 to 19. From line 29 to 35, it also uses probability distribution function, as explained earlier. It computes two crucial functions of $\nabla\ (\varphi, \rho, \omega)$ and $\mathbb{G}\ (x, y, z)$. For this global gain (GG) function, each distribution of local gain $g\ (x, y, z)$ must be considered as features come in for each test. All the low probability-based readings are discarded for active computation but kept in waiting list in the LT object for the second run. This way, algorithm does justice to each feature and give it a second chance before finally discarding it. The rest of the features that qualify in first or second run, are then added to the FGF.

Example 1. *In one of the experiments (such as Figure 13) on dataset with features including 'RELIGION', we discovered something very interesting and intuitive. The data was based on survey from students, as listed in the appendix. We then formatted some of the sets from different regions and ran our Good Fit Student (GFS) and Good Fit job Candidate (GFjC) algorithms (as we briefly discuss in future works). GFS and GFjC are based on eMLEE model and utilize eFES. We noticed as a pleasant surprise that in some cases, it rejected the RELIGION feature for GFS prediction and this made sense as normally religion will not influence success in the studies of the student, but then we discovered that it gave some acceptable scoring to the same feature, because it was coming from a different GEOGRAPHICAL region of the world. It made sense as well, because religion's influence on the individual may be diverse depending on his or her background. We noticed that it successfully correlated with Correlation Factor (CF) > 0.83 on other features in the set and considered the associated feature to be given high* *score due to being appeared with the other features of the collateral importance (i.e., Geographical information). CF is one of the crucial factors in GFS and GFjC algorithms. GFS and GFjC are out of the scope of this paper.*

Algorithm 1. Feature Preparation Function—*F.Prep* (x, y, z)

Input: Sample Dataset (ΔS_n)
Output: *F.Prep* (x, y, z)

1:	*While* (GG (x,y,z) < 0.5) *Do*
2:	**Compute**: $P(CRV) \leftarrow \sum_{i=1}^{N} P_H(h_i)p\ CRV \mid DRV$
3:	Set: $x \leftarrow 0,\ y \leftarrow 0,\ z \leftarrow 0$
4:	Compute: err and Err (x,y,z) using ()
5:	*For* $(F = \{F_1, F_2, F_3, \ldots\ldots, F_n,\}$ *Do*
6:	Apply: Cross Validation on $DS(sig, noi)$
7:	Update: The Split Function using h.eCVS (k, F)
8:	Set: $Q^N \rightarrow DRV$/* Based on Mapping function */
9:	Compute: $h.MI(Irr.F(x,y,z) \mid f_i, f_{i+1}), L.Func(z), L.Cost\ (z)$
10:	Update: h.MI (F_i)
11:	*If* (*err is in bounds as per rule*) *Then*
12:	Mark: the feature F_i and Flag.
13:	Update: each $f \in F^{(n)}$, for which $f_n \geq F\{0.85, 0:1\}$ is valid
14:	Select: F (n) based on random function, and distribution in space: $D[f\ (T) \mid F^{(n)} \in \partial F\ (F^{(n)})]$
15:	*While* (Irr.F ≥ 0.5 *AND Red.F* ≥ 0.5) *Do*
16:	Compute:LTObject.Weighted (**eFES**⊞, **h.MI** (F_i))
17:	Extract: **MI − i** ← **I MI Index**
18:	Set: $Irr.F \leftarrow \sum_{i,j}^{K} \begin{Bmatrix} f_{ii} & f_{ij} \\ f_{ji} & f_{ji} \end{Bmatrix}$
19:	
20:	Compute: MI for Entropy, CF as correlating factor
21:	Re-compute: MI and Red.F (MI)
22:	*End While*
23:	*End if*
24:	Define: the categorical or numerical values, and set $F^{(n)} = Constant\ value$
25:	Compute: *F.Prep* $(x, y, z) \leftarrow$ h.blend (MI,z)
26:	*End for*
27:	Slice: Data Samples $\{\Delta S_n \in S\}$
28:	Compute: and Create Matrices
29:	Set: $\mathbb{G}\ (x, y, z) \leftarrow \frac{1}{N} \sum_{i=1}^{n}\{(g_i) \times MH$
30:	$Entropy\ (F_{n \in x,y,z}) \leftarrow \sum_{t \in T\ (x,y,z)} \frac{\mid F_{(t:x,y,z)} \mid}{\mid F_t \mid} Entropy\ (F_n)$
31:	Compute: $Gain\ (I, F_{(t:x,y,z)})$/* Using Equation (15) */
32:	Compute: **gR (z)** /* Using Equation (16) */
33:	*End While*
34:	Reset: x,y,z
35:	Update: h.Update (gR (z), h.Prep (x,y,z), CF)
36:	Compute: *F.Prep* $(x, y, z) \leftarrow$ h.Model (h*)
	Return: *F.Prep* (x, y, z)

Another example was encountered where this feature played significant role in the job industry where a candidate's progress can be impacted based on their religious background. Another example is of GENDER feature/attribute that we discuss in Section 6.1. This also explains our motivation towards creating FGF (Algorithm 2).

FGF function determines the right number and type of the features from a given data set during classifier learning and reports accordingly if satisfactory accuracy and generalization have not been reached. eFES unit, as explained in the model earlier, uses 3D array to store the scoring via LT object in the inner layer of the model. Therefore, eFES algorithms can tell the model if more features are needed to finally train the classifier for acceptable prediction in the real-world test.

Some other examples are data from healthcare, where a health condition (a feature) may become of high relevance if a certain disease is being predicted. For example, to predict the likelihood of cancer in a patient, DIABETES can have higher predictive score, because an imbalanced sugar can feed the cancerous cells. During learning, the classifier function starts identification of the features and then start adding or removing them based on effectiveness of the cost and time. Compared to other approaches where such proactive quantification is not done the eFES scheme dominates.

Algorithm 2. Feature Grouping Function (FGF)

Input: Sample Dataset (ΔS_n), *F.Prep* (x, y, z)
Output: **h. FGF**

1:    *While* $((W(\emptyset)\{ \, \nabla(\varphi, \rho, \omega), 1\}))\neq 0, \{\in 0, 1\}))$ *Do*
2:       Slice: Data Samples $\{\Delta S_n \in S\}$
3:       Compute: $W(\emptyset)$/* As per equation (20) */
4:       $Y \leftarrow (x, 0, z), X \leftarrow (0, y, z), Y \leftarrow (x, 0, z)$
5:       Execute: h.Train $(\{\Delta S_n\})$/* Sample training begins on data set */
6:       *If* (h.Train $\leq$ *ABS* $(\nabla(\varphi, \rho, \omega))$ *Then*
7:          Compute: h.Biasness $(W(\emptyset), \textbf{Prep}\,(X, Y, Z))$
8:          Compute: $\widehat{\mathbb{Q}_Y}(\Delta)$/* Using Equation (21) */
9:          Update: h.Record (LTObject (eFES$^{\boxplus}$, $\widehat{\mathbb{Q}_Y}(\Delta)$)
10:      *Else*
11:          Update: h.Record (h.Train)
12:          Set: $Y \leftarrow (x, 0, z), X \leftarrow (0, y, z), Y \leftarrow (x, 0, z)$
13:      *End If*
14:      Compute: h.localgain and h.globalgain
15:      *For* (g $\in$ $(g + 1, \Delta Gain_l(w))$ *Do*
16:          Compute: $Gain_l(w)$/* Using Equation (22) */
17:          Set: $Norm. Gain_l(w)$ to local minima
18:      *End For*
19:      Compute: $(F.Sco\,(x, y, z))$ as h.execute (*FF* as fitness factor),
20:      $(Gain_l(w), err, Err, F.Eng\,(x, y, z))$
21:      Compute: $(F.Opt\,(x, y, z))$ as h.concatenate $(F.Eng\,(x, y, z)$, Y,X,Z, $F(n))$
22:      *If* $(gR\,(z) < 0.5)$ *Then*
23:          Compute: err (z) and Err (x,y,z)
24:          Update: H.RecordErrors (err,Err, gR (z))
25:      *End If*
26:      Execute: h.Update (gR (z), $\mathbb{G}\,(x, y, z)$)
27:      Update: *LT* function, LT.Gain (h*)
28:      *For* (all tests in $P\,(rV_i \,|\, pE) > P\,(rV_j |\, pE))$ *Do*
29:          Re-compute: the LG and GG
30:          Update: the h.LT (P)
31:          *If* (P (pE | rV$_i$)P (rV$_i$) > 0.5) *Then*
32:             Compute: $\nabla\,(\varphi, \rho, \omega)$
33:             Compute: $\mathbb{G}\,(x, y, z)$ for all distributions of g (x, y, z)
34:          *End If*
35:          Compute: $(F.Sco\,(x, y, z))$/* Using Equation (26) */
36:          Compute: $(F.Opt\,(x, y, z))$/* Using Equation (27) */
37:          Compute: h.FGF (F.Sco,F.opt)
38:      *End For*
39:    *End While*
      Return (h. FGF))

Appl. Sci. **2018**, *8*, 646

Figure 10 simulations demonstrate the global gain and local gain functions coherences with respect to Loss and Cost functions. Figure 10a shows that gain function is unstable when eFES randomly created the feature set. Figure 10b shows that gain functions stabilize when eFES uses weighted function, as derived in the model.

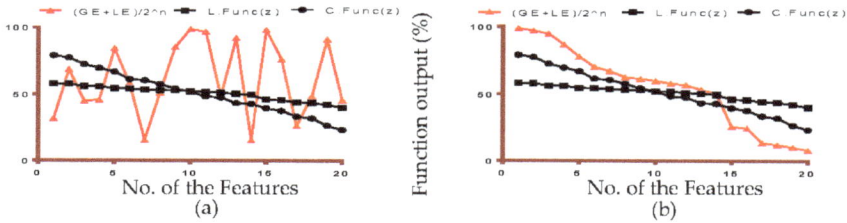

Figure 10. Gain Function correlation with Loss and Cost Function.

3.3. eFES Framework

Figure 11 illustrates the internal functions of the proposed module on a granular level. eMLEE API refers to the available functions that eMLEE model provides to each module such as eFES. Each grey box is a function. The diamond shapes represent a decision-making point in the flow.

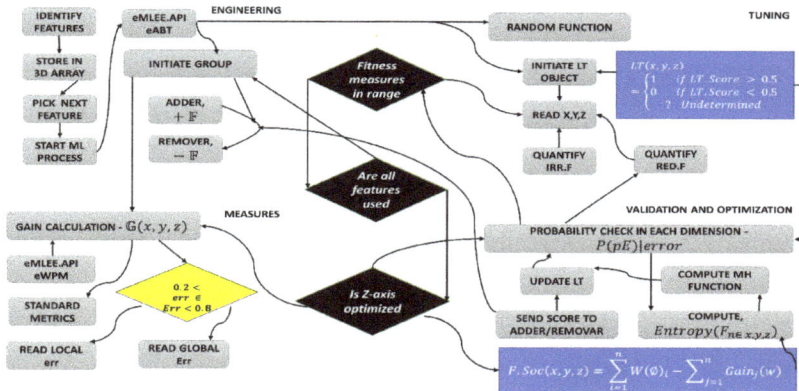

Figure 11. Framework to illustrate the internal processes of eFES module.

4. Results and Discussions

This section provides simulated results in 3D and 2D view to provide in-depth analysis of the outcome of the proposed model for various functions and metrics. Significant samples of the entire experimental results are provided at the latest state of this eFES model development stage. These simulations elaborate on processing features to observe the optimum fitness (i.e., z dimension). 3D visuals are selected for better analysis of how the curve moves in space when the learner is optimized in the dimensions. The equation below drives the experimental run for monitoring the z-dimension in correspondence to each of x, y, and z. It should be noted that the results shown are a snapshot of 100+ experimental runs for several data samples of the datasets. The equation shown for each indicates the sampling construct for the analysis being envisioned. Features were included in the experiments from the raw datasets. To improve the generalization of the model, various experiments were performed on standard numbers such as 5, 10, 15, 20 and 40. Clearly, less is more, as we stated earlier, but we leave it up to the model to finally group (FGF) the features that have the highest predictive value for learning and ensuring the maximum fitness and generalization. For each experiment, a miscellaneous dataset was used to improve the generalization ability of the model and underlying algorithms.

Figure 12 shows the 3D variance simulations of the functions. Figure 13 shows the comparison between features that were engineered (Enhanced Feature Engineering (EFE)) and that were not engineered (in blue). It is observed that EFE outperformed the FE. No FE indicates that the experiment took features set as per standard pick and ran the process. EFE indicates the enhanced feature engineering while incorporating mathematical constructs and algorithms, where features were added and removed based on metrics reading and eventually creating an optimum feature set, as engineered by eMLEE.

Figure 14a–d shows the tests on 20-experimental run. It should be noted that as the number of experiments were increased, the classifier learning was improved as per proposed model. The selection of 20 features were based on optimum number of the grouping function (FGF). Clearly, each dataset brings in different number of features. Out of these features, some features are irrelevant, redundant, and outliers. Some features are not known at the beginning of classifier learning. However, we standardized around number 20 for experimental purposes. However, it is up to the algorithm to tell the model how many features need to be qualified and then included in the learning process.

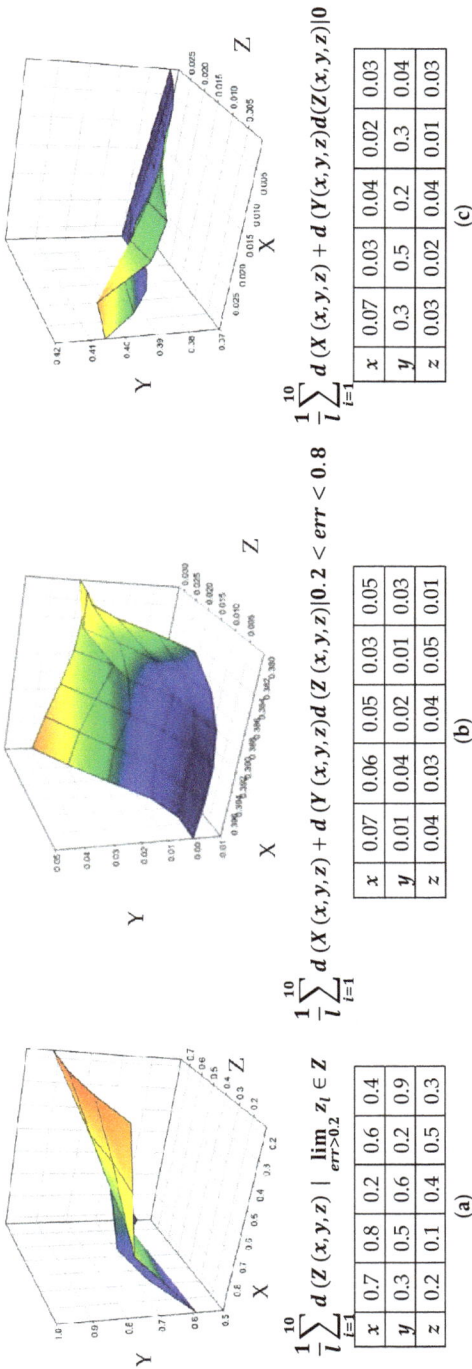

$$\frac{1}{l}\sum_{i=1}^{10} d(Z(x,y,z)) \mid \lim_{err>0.2} z_i \in Z$$

x	0.7	0.8	0.2	0.6	0.4
y	0.3	0.5	0.6	0.2	0.9
z	0.2	0.1	0.4	0.5	0.3

(a)

$$\frac{1}{l}\sum_{i=1}^{10} d(X(x,y,z)) + d(Y(x,y,z))d(Z(x,y,z))|0.2 < err < 0.8$$

x	0.07	0.06	0.05	0.03	0.05
y	0.01	0.04	0.02	0.01	0.03
z	0.04	0.03	0.04	0.05	0.01

(b)

$$\frac{1}{l}\sum_{i=1}^{10} d(X(x,y,z)) + d(Y(x,y,z))d(Z(x,y,z))|0$$

x	0.07	0.03	0.04	0.02	0.03
y	0.3	0.5	0.2	0.3	0.04
z	0.03	0.02	0.04	0.01	0.03

(c)

Figure 12. (a) It shows that variance in z is minimum on random datasets; (b) It shows the variance in all of axis as ideal, as what we wanted to observe; (c) It shows the variance in all axis to be real (practical), as what we observed.

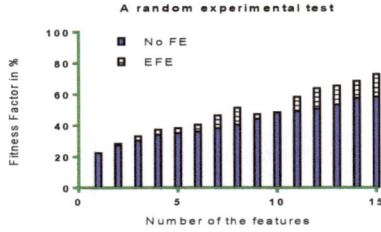

Figure 13. A random experiment on 15 features for FE vs. EFE Correlation study for the observed Fitness Factor.

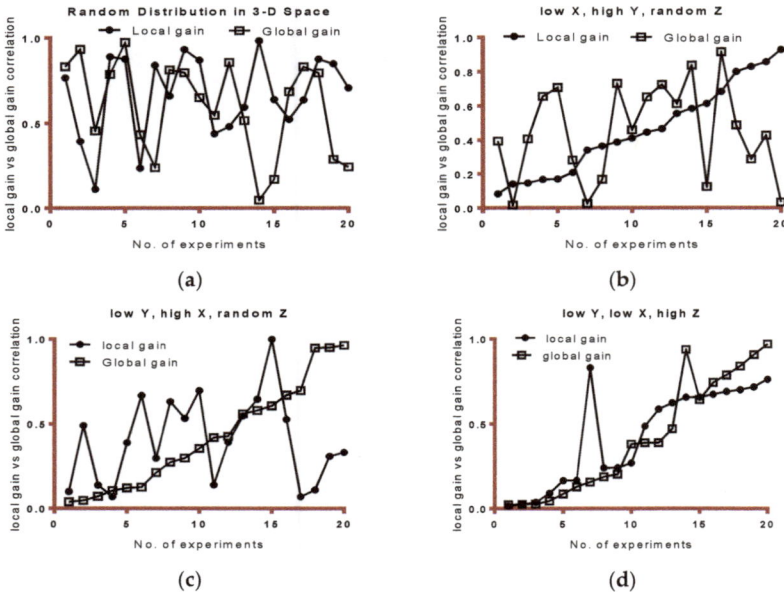

Figure 14. (**a**) We observe that LG and GG were very random throughout the tests; (**b**) We observe that LG showed linear correlation (regression) when x (overfitting) was found to be low, and z was kept random in 3D space. GG, as observed was random; (**c**) Observations on low y, where we found GG to be close to the linear response; (**d**) Finally, as the model is optimized (high z), we saw expected and desired linear regression. We observed some unexpected as shown by the peaks, which were suspected to be riding outliers.

Figure 15a–e shows the test on (5, 10, 15, 20 and 50) features set. It compares the EFE and FE correlation for Fitness Factor (FF). FF is computed by the eFES algorithms explained earlier.

Figure 15 shows the set of experiments for observation of diverse set of features to study the model's fitness factor. As it is observed that EFE keeps the linearity (stability) of the model. (e) was as special test of various metrics. "Engineered" refers to all the metrics of eFES model incorporated.

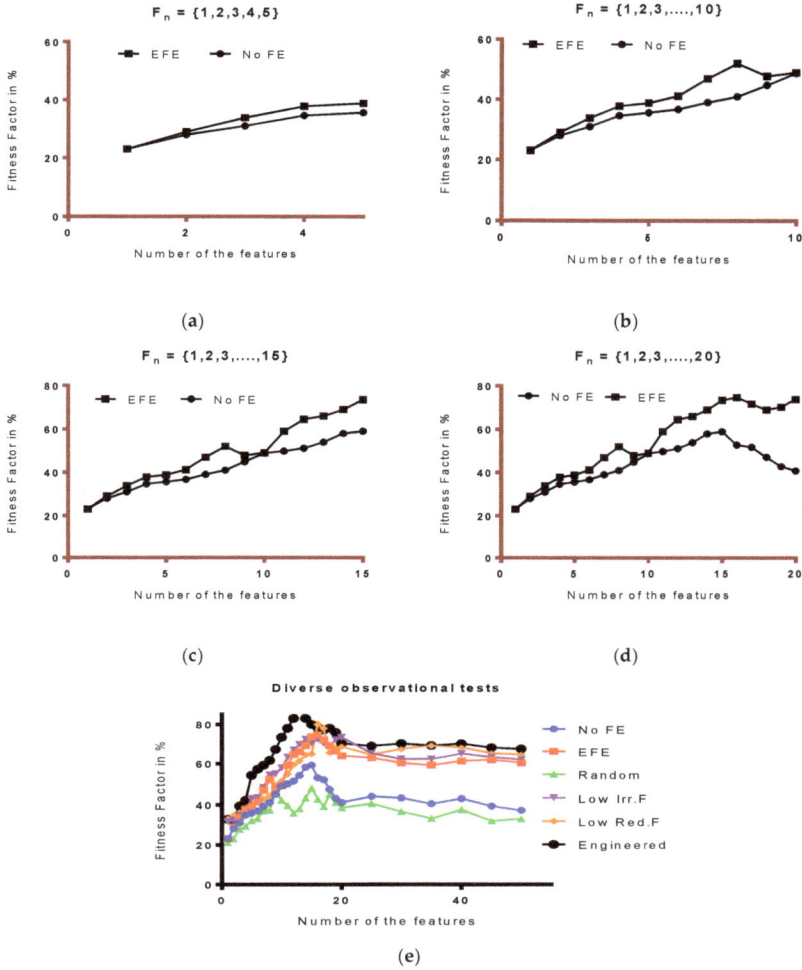

Figure 15. Set of experiments for observation of diverse set of features to study the model's fitness factor: (**a**) With features in higher range of 15, we observe the consistent stability; (**b**) This considers 10 features to evaluate the fitness function for both EFE and FE. Clearly, we observe the improvement in Fitness Function; (**c**) With features in higher range of 15, we observe the consistent stability; (**d**) However, as expected, we noticed that features up to 20, the maximum range of fitness function is around 80%; (**e**) This shows the comparison of the various metrics and read the relevant value of the fitness factor for each study of the metrics as shown by distinct colors.

Figure 16 shows the three sets of 20-grouped feature sets. The goal of these experiments was to study the model ability to improve the accuracy for the features (Accepted, Rejected and Mixed) from the given data set.

Figure 17 shows the candlestick (commonly used for stock analysis) analysis for LE and GE bounds. It is observed that the model learned to stay in the bounds of 20% and 80% for LE and GE. Negative 50 range is shown to elaborate on potential of error swings (i.e., for invalid values). The green and purple sticks are for reference only.

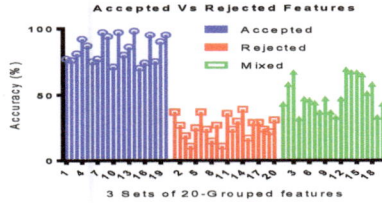

Figure 16. Accuracy Validation for Feature Optimization.

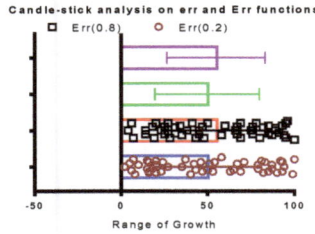

Figure 17. Observation of the candle-stick analysis for Global (Err) error bounds.

Figures 18 and 19 shows the observation of the bias of the model for 20-experimental analysis. Correlation Factor (CF) is computed by the eFES algorithms. Figure 18 shows the error and accuracy increases during higher end of quantification range as shown. Figure 19 on the other hand shows that the model has achieved the desired correlation of Err and Accuracy function.

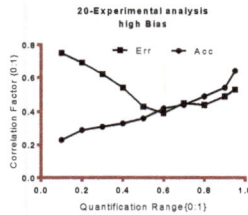

Figure 18. Poor Correlation of Err and Accuracy during high bias.

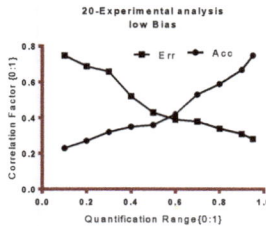

Figure 19. Expected and desired correlation of Err and Accuracy function.

Table 4 shows the outcome of our 10-experimental analysis test, where we tuned the model to discover the maximum possible practical measures as shown. We used 500+ iterations on several different datasets to validate the model stability with real world data with the functions we built in our proposed model. Red values are found to be in error. Further work is needed to investigate it.

Table 4. 10th experimental values for functions shown.

Internal Functions	1	2	3	4	5	6	7	8	9	10
MI	0.009	0.012	0.023	0.034	−0.931	0.563	0.611	0.678	0.712	0.731
$Irr.F(x,y,z)$	0.119	0.217	0.241	0.298	0.381	0.383	0.512	0.629	0.672	0.681
$Red.F$	0.191	0.200	−0.001	0.289	0.321	0.341	0.440	0.512	0.525	0.591
err	0.821	0.781	0.732	0.612	0.529	0.489	0.410	0.371	0.330	0.319
Err	0.901	0.900	0.871	0.844	0.731	−0.321	0.620	0.521	0.420	0.381
$F.Sco(x,y,z)$	0.390	0.421	0.498	0.534	0.634	0.721	0.770	0.812	0.856	0.891
$F.Opt(x,y,z)$	0.110	0.230	0.398	0.491	0.540	0.559	−0.210	0.639	0.776	0.791

5. Comparative Analysis

This section provides a brief comparison of the latest techniques with the proposed model of eFES. The data sources are detailed in the Appendix A. Table 5 shows the listing of the dataset used. Table 6 lists the methods considered for the comparisons. Table 7 lists the table structure of the division for results in Tables 8–19. The Python and R packages used are detailed in the tools sub-section of the Appendix A. All the values are normalized to range between 0 and 1 for our functions' standard outcome measures. It should be noted that the data shown in Tables 8–19 are a subset of our total experimental analysis. The rest of the results are left out for our literature review and model comparison paper for future work/writings.

Table 5. Data Sources (DS).

1	Breast Cancer Wisconsin Data Set
2	Car Evaluation
3	Iris species
4	Twitter User Gender Classification
5	College Scoreboard
6	Pima Indians Diabetes Database
7	Student Alcohol Consumption
8	Education Statistics
9	Storm Prediction center
10	Fatal Police Shootings
11	2015 Flight Delays and Cancellations
12	Credit Card Fraud Detection
13	Heart disease data set
14	Japan Census data
15	US Mass Shootings
16	Adult Census income
17	1.88 Million US Wildfires
18	S & P 500 stock Data
19	Zika Virus epidemic
20	Retail Data Analytics

Table 6. Methods.

1	Information Gain	IG
2	Chi-squared	CS
3	Pearson Correlation	PC
4	Analysis of Variance	ANOVA
5	Weight of Evidence	WOE
6	Recursive Feature Elimination	RFE
7	Sequential Feature Selector	SFS
8	Univariant Selection	US
9	Principal Component Analysis	PCA
10	Random Forest	RF
11	Least Absolute Shrinkage and Selection Operator	LASSO
12	RIDGE Regression	RR
13	Elastic Net	EN
14	Gradient Boosted Machines	GBM
15	Linear discriminant analysis	LDA
	Multiple Discriminant Analysis	MDA
16	Joint Mutual Information	JMI
17	Non-negative matrix factorization	NNMF

Table 7. Tables structures for 8 to 19.

Table Number	Measure	Number of the Datasets	Number of the Methods
8	Accuracy	1–10	1–10
9	Accuracy	11–20	11–17
10	Error	1–10	1–10
11	Error	11–20	11–17
12	Precision Score	1–10	1–10
13	Precision Score	11–20	11–17
14	C-Index	1–10	1–10
15	C-Index	11–20	11–17
16	AUC	1–10	1–10
17	AUC	11–20	11–17
18	Cohen's Kappa	1–10	1–10
19	Cohen's Kappa	11–20	11–17

Table 8. Performance Evaluation (PE) = Accuracy, ideally higher values are desired.

DS	IG	CS	PC	ANOVA	WOE	RFE	SFS	US	PCA	RF	eFES	Winner
1	0.489	0.581	0.493	0.821	0.432	0.718	0.562	0.321	0.612	0.937	0.827	RF
2	0.391	0.501	0.452	0.732	0.479	0.700	0.590	0.476	0.842	0.842	0.863	**eFES**
3	0.429	0.543	0.572	0.683	0.481	0.681	0.623	0.419	0.693	0.534	0.633	PCA
4	0.444	0.492	0.365	0.601	0.538	0.710	0.512	0.478	0.793	0.792	0.825	**eFES**
5	0.492	0.572	0.392	0.621	0.593	0.661	0.652	0.563	0.824	0.312	0.881	**eFES**
6	0.482	0.563	0.592	0.910	0.582	0.633	0.692	0.673	0.847	0.118	0.792	ANOVA
7	0.523	0.557	0.673	0.666	0.523	0.502	0.619	0.693	0.734	0.492	0.700	PCA
8	0.542	0.599	0.962	0.732	0.710	0.682	0.638	0.478	0.892	0.692	0.983	**eFES**
9	0.423	0.630	0.921	0.802	0.504	0.623	0.742	0.732	0.872	0.631	0.902	PC
10	0.491	0.612	0.683	0.678	0.864	0.644	0.699	0.535	0.418	0.683	0.789	WOE

Table 9. Performance Evaluation (PE) = Accuracy.

DS	LASSO	RR	EN	GBM	LDA	MDA	JMI	NNMF	eFES	Winner
11	0.662	0.572	0.723	0.882	0.704	0.772	0.663	0.823	0.801	GBM
12	0.606	0.623	0.221	0.803	0.772	0.828	0.920	0.613	0.934	**eFES**
13	0.512	0.691	0.378	0.723	0.834	0.583	0.612	0.593	0.860	**eFES**
14	0.731	0.612	0.143	0.703	0.234	0.524	0.683	0.444	0.792	**eFES**
15	0.771	0.745	0.123	0.856	0.803	0.890	0.583	0.798	0.812	MDA
16	0.924	0.703	0.426	0.323	0.866	0.597	0.421	0.231	0.690	LASSO
17	0.832	0.791	0.484	0.428	0.792	0.890	0.792	0.166	0.942	**eFES**
18	0.883	0.723	0.923	0.573	0.723	0.748	0.842	0.772	0.793	EN
19	0.811	0.596	0.573	0.803	0.436	0.800	0.723	0.724	0.942	**eFES**
20	0.698	0.582	0.590	0.777	0.494	0.784	0.683	0.682	0.825	**eFES**

Table 10. Performance Evaluation (PE) = Error, ideally lower values are desired.

DS	IG	CS	PC	ANOVA	WOE	RFE	SFS	US	PCA	RF	eFES	Winner
1	0.254	0.443	0.623	0.234	0.112	0.213	0.487	0.126	0.111	0.173	0.223	PCA
2	0.231	0.193	0.423	0.278	0.321	0.183	0.215	0.193	0.213	0.213	0.201	RFE
3	0.593	0.318	0.283	0.318	0.294	0.143	0.368	0.216	0.172	0.229	0.045	**eFES**
4	0.443	0.244	0.342	0.087	0.221	0.193	0.125	0.259	0.193	0.281	0.112	ANOVA
5	0.183	0.289	0.124	0.213	0.084	0.426	0.258	0.442	0.145	0.342	0.014	**eFES**
6	0.392	0.173	0.192	0.282	0.103	0.083	0.159	0.044	0.039	0.293	0.023	**eFES**
7	0.361	0.111	0.009	0.045	0.115	0.063	0.193	0.310	0.135	0.183	0.216	PC
8	0.183	0.325	0.289	0.183	0.183	0.222	0.329	0.331	0.173	0.312	0.128	**eFES**
9	0.498	0.310	0.423	0.192	0.435	0.215	0.229	0.283	0.132	0.073	0.024	**eFES**
10	0.389	0.300	0.528	0.216	0.392	0.376	0.402	0.194	0.081	0.082	0.006	**eFES**

Table 11. Performance Evaluation (PE) = Error.

ΔΣ	ΛΑΣΣΟ	PP	EN	ΓΒΜ	ΛΔΑ	ΜΔΑ	ϑΜΙ	ΝΝΜΦ	εΦΕΣ	Ωιννεϱ
11	0.092	0.175	0.134	0.097	0.281	0.145	0.113	0.130	0.073	εΦΕΣ
12	0.125	0.179	0.111	0.173	0.173	0.100	0.193	0.193	0.034	εΦΕΣ
13	0.214	0.231	0.190	0.239	0.151	0.182	0.231	0.210	0.111	εΦΕΣ
14	0.163	0.200	0.138	0.166	0.088	0.163	0.009	0.003	0.183	ΝΝΜΦ
15	0.193	0.193	0.083	0.129	0.210	0.219	0.122	0.122	0.178	EN
16	0.236	0.437	0.238	0.321	0.134	0.110	0.177	0.191	0.088	εΦΕΣ
17	0.173	0.432	0.110	0.146	0.443	0.325	0.212	0.253	0.134	EN
18	0.113	0.267	0.193	0.191	0.392	0.283	0.154	0.099	0.004	εΦΕΣ
19	0.172	0.167	0.183	0.219	0.108	0.200	0.121	0.111	0.312	ΛΔΑ
20	0.283	0.045	0.128	0.183	0.214	0.204	0.231	0.131	0.021	εΦΕΣ

Table 12. Performance Evaluation (PE) = Precision Score, ideally higher values are desired.

DS	IG	CS	PC	ANOVA	WOE	RFE	SFS	US	PCA	eFES	Winner
1	0.677	0.899	0.961	0.520	0.755	0.816	0.820	0.639	0.792	0.723	CS
2	0.936	0.755	0.553	0.600	0.522	0.690	0.776	0.764	0.841	0.802	IG
3	0.861	0.874	0.779	0.834	0.647	0.844	0.677	0.907	0.744	0.689	CS
4	0.545	0.603	0.882	0.850	0.725	0.637	0.887	0.554	0.754	0.956	**eFES**
5	0.767	0.584	0.894	0.861	0.761	0.915	0.753	0.513	0.909	0.932	**eFES**
6	0.753	0.557	0.664	0.707	0.706	0.732	0.622	0.714	0.804	0.762	PCA
7	0.814	0.585	0.865	0.667	0.790	0.620	0.781	0.773	0.933	0.903	PCA
8	0.546	0.706	0.852	0.902	0.619	0.710	0.732	0.738	0.638	0.967	**eFES**
9	0.859	0.760	0.610	0.627	0.617	0.673	0.591	0.803	0.575	0.992	**eFES**
10	0.698	0.710	0.702	0.674	0.821	0.691	0.503	0.781	0.746	0.710	US

Table 13. Performance Evaluation (PE) = Precision Score.

DS	RF	LASSO	RR	EN	GBM	LDA	MDA	JMI	NNMF	eFES	Winner
11	0.733	0.763	0.824	0.473	0.610	0.786	0.522	0.731	0.862	0.893	**eFES**
12	0.838	0.864	0.549	0.772	0.584	0.910	0.760	0.706	0.631	0.845	LDA
13	0.611	0.964	0.928	0.781	0.565	0.703	0.550	0.827	0.908	0.923	LASSO
14	0.905	0.923	0.754	0.807	0.643	0.670	0.605	0.531	0.650	0.982	**eFES**
15	0.640	0.601	0.820	0.950	0.512	0.948	0.827	0.786	0.662	0.734	EN
16	0.530	0.690	0.621	0.622	0.808	0.934	0.630	0.537	0.931	0.956	**eFES**
17	0.547	0.825	0.512	0.711	0.740	0.877	0.766	0.697	0.561	0.893	**eFES**
18	0.632	0.642	0.891	0.670	0.864	0.665	0.774	0.902	0.702	0.962	**eFES**
19	0.728	0.671	0.720	0.726	0.743	0.582	0.550	0.781	0.631	0.704	JMI
20	0.836	0.746	0.574	0.585	0.979	0.872	0.758	0.941	0.952	0.942	GBM

Table 14. Performance Evaluation (PE) = C-Index, ideally lower values are desired.

DS	IG	CS	PC	ANOVA	WOE	RFE	SFS	US	PCA	eFES	Winner
1	0.412	0.333	0.236	0.294	0.226	0.062	0.421	0.427	0.215	0.118	**eFES**
2	0.319	0.256	0.393	0.106	0.433	0.078	0.361	0.128	0.235	0.056	**eFES**
3	0.207	0.251	0.271	0.118	0.134	0.307	0.222	0.338	0.211	0.312	ANOVA
4	0.523	0.220	0.058	0.052	0.203	0.325	0.061	0.439	0.040	0.189	PCA
5	0.134	0.534	0.476	0.137	0.144	0.387	0.199	0.114	0.105	0.210	PCA
6	0.627	0.425	0.285	0.243	0.448	0.274	0.488	0.186	0.181	0.034	**eFES**
7	0.113	0.193	0.152	0.498	0.200	0.036	0.025	0.149	0.071	0.092	SFS
8	0.167	0.273	0.410	0.128	0.105	0.435	0.139	0.193	0.148	0.192	WOE
9	0.291	0.221	0.096	0.291	0.326	0.448	0.161	0.235	0.211	0.073	**eFES**
10	0.093	0.293	0.407	0.488	0.200	0.179	0.341	0.472	0.040	0.002	**eFES**

Table 15. Performance Evaluation (PE) = C-Index.

DS	RF	LASSO	RR	EN	GBM	LDA	MDA	JMI	NNMF	eFES	Winner
11	0.054	0.314	0.314	0.036	0.150	0.219	0.201	0.190	0.019	0.112	NNMF
12	0.268	0.217	0.293	0.252	0.552	0.209	0.318	0.394	0.219	0.243	LDA
13	0.542	0.325	0.320	0.327	0.346	0.245	0.249	0.322	0.403	0.296	LDA
14	0.282	0.141	0.251	0.132	0.136	0.360	0.217	0.224	0.232	0.106	**eFES**
15	0.043	0.060	0.210	0.021	0.093	0.264	0.091	0.247	0.136	0.129	RF
16	0.060	0.053	0.200	0.100	0.055	0.252	0.155	0.056	0.078	0.031	**eFES**

Table 15. *Cont.*

DS	RF	LASSO	RR	EN	GBM	LDA	MDA	JMI	NNMF	eFES	Winner
17	0.200	0.053	0.331	0.140	0.040	0.107	0.216	0.335	0.247	0.013	**eFES**
18	0.327	0.233	0.258	0.295	0.290	0.346	0.334	0.378	0.329	0.297	LASSO
19	0.094	0.196	0.312	0.309	0.066	0.216	0.128	0.164	0.258	0.032	**eFES**
20	0.073	0.263	0.204	0.064	0.053	0.206	0.010	0.239	0.047	0.024	MDA

Table 16. Performance Evaluation (PE) = AUC, ideally higher values are desired.

DS	IG	CS	PC	ANOVA	WOE	RFE	SFS	US	PCA	eFES	Winner
1	0.569	0.808	0.739	0.633	0.848	0.563	0.518	0.540	0.874	0.810	PCA
2	0.513	0.796	0.800	0.643	0.610	0.659	0.618	0.664	0.589	0.762	CS
3	0.784	0.636	0.781	0.589	0.499	0.585	0.539	0.858	0.717	0.96	**eFES**
4	0.592	0.834	0.498	0.788	0.789	0.713	0.911	0.830	0.645	0.976	**eFES**
5	0.655	0.698	0.805	0.504	0.880	0.574	0.638	0.885	0.742	0.699	WOE
6	0.590	0.741	0.791	0.825	0.654	0.826	0.698	0.679	0.962	0.892	PCA
7	0.802	0.626	0.680	0.510	0.896	0.745	0.646	0.735	0.974	0.740	PCA
8	0.805	0.560	0.550	0.826	0.609	0.812	0.659	0.704	0.814	0.894	**eFES**
9	0.642	0.802	0.769	0.891	0.504	0.482	0.629	0.830	0.734	0.836	ANOVA
10	0.872	0.898	0.858	0.785	0.921	0.573	0.831	0.754	0.868	0.971	**eFES**

Table 17. Performance Evaluation (PE) = AUC.

DS	RF	LASSO	RR	EN	GBM	LDA	MDA	JMI	NNMF	eFES	Winner
11	0.725	0.835	0.889	0.751	0.545	0.706	0.676	0.562	0.518	0.774	RR
12	0.889	0.819	0.532	0.555	0.890	0.751	0.946	0.688	0.778	0.903	MDA
13	0.568	0.835	0.520	0.525	0.502	0.764	0.605	0.651	0.487	0.952	**eFES**
14	0.780	0.728	0.606	0.870	0.792	0.545	0.553	0.855	0.990	0.962	NNMF
15	0.602	0.615	0.833	0.700	0.804	0.493	0.645	0.616	0.899	0.867	NNMF
16	0.736	0.649	0.589	0.665	0.848	0.847	0.905	0.621	0.897	0.952	**eFES**
17	0.541	0.711	0.777	0.511	0.868	0.884	0.691	0.904	0.665	0.962	**eFES**
18	0.796	0.525	0.768	0.762	0.755	0.513	0.759	0.910	0.599	0.852	**eFES**
19	0.873	0.481	0.606	0.639	0.558	0.575	0.783	0.842	0.675	0.820	RF
20	0.860	0.365	0.893	0.603	0.893	0.840	0.829	0.646	0.496	0.824	GBM

Table 18. Performance Evaluation (PE) = Cohen's Kappa, ideally higher values are desired.

DS	IG	CS	PC	ANOVA	WOE	RFE	SFS	US	PCA	eFES	Winner
1	0.666	0.816	0.913	0.621	0.206	0.656	0.930	0.978	0.586	0.912	US
2	0.762	0.754	0.502	0.926	0.959	0.774	0.915	0.566	0.875	0.925	WOE
3	0.921	0.207	0.691	0.757	0.920	0.520	0.846	0.932	0.758	0.623	US
4	0.693	0.542	0.673	0.500	0.765	0.924	0.647	0.501	0.824	0.957	**eFES**
5	0.773	0.533	0.775	0.615	0.814	0.535	0.682	0.536	0.878	0.856	PCA
6	0.685	0.910	0.568	0.606	0.698	0.831	0.646	0.902	0.851	0.945	**eFES**
7	0.635	0.716	0.676	0.793	0.593	0.802	0.843	0.671	0.930	0.991	**eFES**
8	0.667	0.877	0.918	0.751	0.854	0.930	0.794	0.527	0.936	0.875	PCA
9	0.897	0.644	0.454	0.517	0.762	0.802	0.685	0.865	0.650	0.834	IG
10	0.599	0.824	0.803	0.802	0.827	0.875	0.933	0.851	0.724	0.925	**eFES**

Table 19. Performance Evaluation (PE) = Cohen's Kappa.

DS	RF	LASSO	RR	EN	GBM	LDA	MDA	JMI	NNMF	eFES	Winner
11	0.892	0.929	0.819	0.874	0.537	0.662	0.833	0.581	0.857	0.983	**eFES**
12	0.954	0.660	0.944	0.489	0.582	0.869	0.753	0.786	0.771	0.973	**eFES**
13	0.576	0.952	0.686	0.588	0.744	0.712	0.658	0.927	0.671	0.910	LASSO
14	0.519	0.780	0.505	0.850	0.603	0.731	0.942	0.975	0.958	0.846	JMI
15	0.985	0.846	0.903	0.591	0.584	0.750	0.617	0.945	0.892	0.904	RF
16	0.786	0.804	0.605	0.673	0.814	0.635	0.909	0.573	0.732	0.973	**eFES**
17	0.827	0.567	0.814	0.772	0.867	0.890	0.670	0.771	0.763	0.734	LDA
18	0.751	0.733	0.820	0.813	0.760	0.637	0.871	0.739	0.867	0.923	**eFES**
19	0.698	0.645	0.636	0.801	0.727	0.886	0.969	0.954	0.781	0.845	MDA
20	0.583	0.646	0.795	0.930	0.953	0.523	0.681	0.565	0.524	0.578	EN

6. Final Remarks

6.1. Conclusions

This paper reports the latest progress of the proposed model for enhanced Feature Engineering and Selection (eFES) including mathematical constructs, framework, and the algorithms. eFES is a module of enhanced Machine Learning Engine Engineering (eMLEE) parent model. eFES is based on the following building blocks: (a) a features set is processed through standard methods and records the measured metrics; (b) features are weighted based on learning process where accepted features and rejected features are separated using 3D-based training through building Local Gain (LG) and Global Gain (GG) functions; (c) features are then scored and optimized so the ML process can evolve into deciding which features need to be accepted or rejected for improved generalization of the model; (d) finally features are evaluated, tested, and the model is completed with feature grouping function (FGF). This paper reports observation on several hundreds of experiments and then implements 10 experimental approaches to tune the model. The 10th experimental rule was adopted to narrow down (i.e., slice) the result extraction from several hundred runs. The LG and GG functions were built and optimized in 3D space. The included results show promising outcomes of the proposed scheme of the eFES model. It supports the use of feature sets to further optimize the learning process of ML models for supervised learning. Using the novel approach of Local Error and Global Error bounds of 20% to 80%, we could tune our model more realistically. If the errors were above 80% or below 20%, we flag it to be an invalid fit. This unique approach of engineering a model turns out to be very effective in our experiments and observations, as reported and discussed in this paper. This model though is based on parallel processing but using high-speed hardware or a Hadoop-based system will help further.

Features (i.e., attributes) in the datasets are often irrelevant and redundant and may have less predictive value. Therefore, we constructed these two functions. A) Irrelevant *Irr.F* (Equation (8)) and B) *Red.F* (Algorithm 1). The real-world data may have more features and based on this exact fact, we realized the gap to fill with our work. For ML model classifier learning, features play a crucial role when it comes to speed, performance, predictive accuracy, and reliability of the model. Too many features or too few features may overfit or underfit the model. Then the question becomes, what is the optimum (i.e., the right number) feature set that should be filtered for a ML process, that is where our work comes in. We wanted to have the model decides for itself as it continues to learn with more data. Certain features such as" Gender" or "Sex" may have extreme predictive value (i.e., weight) for building predictive modeling for an academic data from a part of the world where gender bias is high. However, the same feature may not play a significant role when it is included in a set from a domain, where gender bias may not exist. Moreover, we also do not anticipate that based on our thoughts, but we let our model tell us which feature should be included or removed, thus we have two functions, Adder ($+\mathbb{F}(x, y, z)$) Equation (1) and Remover ($-\mathbb{F}(x, y, z)$) Equation (2). Number "20" for features was selected to optimize around this number. Figure 15a–e shows the tests for 5, 10, 15 and 20 to observe the fitness factor. Tables 7–12 in Section 5 show the promising state of eFES as compared to the existing techniques. It performed very well, generally. Parallel processing and 3D engineering of the features functions greatly improved the FO as we intended to investigate and improved with our work. Future work will further enhance the internals of it.

In some of the experimental tests, we came across some invalid outcomes, where we had to re-tune our model. Clearly, every model build-up process contains such issues where more work/investigation is always needed. We have found that such issues are not reflective of any huge inaccuracy in the results or instability of the model. Specially, in our diverse and stress testing, the errors and unexpected behavior and readings were very little as compared to stable and expected results. It should be watched closely with future enhancements and results, so it does not grow and become a real bug. This model is based on supervised learning algorithms.

Appl. Sci. **2018**, *8*, 646

6.2. Future Works

To further improve the current state of the eMLEE and its components (such as reported in this paper), we will be testing more data specifically from http://www.kaggle.com, www.data.gov, and www.mypersonality.org. We will be developing/testing more algorithms, especially in the domains of unsupervised learning for new insights into feature engineering and selection. Also, eFES needs further extensions towards exploring and engineering unknown features that are normally not encountered by the learning process but may have great predictive value. We are improving/developing a model known as the "Predicting Educational Relevance for an Efficient Classification of Talent (PERFECT)" Algorithm Engine (PAE). PAE is based on eMLEE and incorporates three algorithms known as Noise Removal and Structured Data Detection (NR-SDD), Good Fit Student (GFS), and Good Fit job Candidate (GFjC). We have published the preliminary results [33] and are working to apply the eFES (i.e., eMLEE) model in its latest form to study/explore/validate further enhancements.

Author Contributions: All authors contributed to the research and related literature. F.U. wrote the manuscript as a part of his Ph.D. dissertation work. J.L. advised the F.U. throughout his research work. All the authors discussed the research with F.U. during production of this paper. J.L., S.R. and S.H. reviewed the work presented in this paper and advised changes and improvement throughout the paper preparation and model experimental process. F.U. incorporated the changes. F.U. conducted the experiments and simulations. All authors reviewed and approved the work.

Conflicts of Interest: The authors declare no conflict of interest.

Key Notations

x	Point of overfitting (OF)
y	Point of underfitting (UF)
z	Point of optimum-fitting (OpF)
F_n	Complete raw feature set
$+\mathbb{F}$	Feature Remover Function
$-\mathbb{F}$	Feature Adder Function
LT	Logical Table Function
$A\ (i)$	ith ML algorithm such as SVM
$\mathbb{R}_{eFES}$	Ratio of normalized error between local and global errors
$F_{ran}\ (x,y,z)$	Randomized feature set
f_w	Weighted feature value
$\Delta(x,y,z)$	Regulating Function in LT object to obey the reference of 50% for training
$err\ (e),\ LE$	Local error
$Err\ (e),\ GE$	Global error
$\varnothing$	maximum inconsistency
Q^N	nth random generator
f_{ii}	Position of a feature in 2D space
$g\ (LG)$	Local gain
$\mathbb{G}\ (GG)$	Global gain
af_i	ith accepted feature
rf_i	ith rejected feature
pf_i	ith predictive feature
ΔS_i	ith dataset item
$\nabla\ (\varphi,\rho,\omega)$	Acceptable parameter function for x,y,z
ObF	Objective function
$k \in K$	Predictor ID in the group of K
EC	Evaluation Criterion
$W(\varnothing)$	Weighted Function
$N_\gamma,\ S_\gamma$	Border unit normal vectors
$\mathbb{Q}_\gamma\ (\Delta)$	Probability distribution based on nonparametric density estimation
$Gain_I(w)$	Information gain
$J_{MIN}(Z)^d$	Jacobian minimization

Appendix A.

Appendix A.1. Dataset Sources

We have utilized the data from the following domains listed below. Some datasets were raw, CSV, and SQL lite format with parameters and field definitions. We transformed all our input data into the SQL Server data warehouse. Some of datasets are found to be ideal for doing healthcare preventive medicine, stock market, epidemic, and crime control prediction.

1. http://www.Kaggle.com—Credit Card Fraud Detection, Iris species, Human Resource Analytics, 2015 Flight Delays and Cancellations, Daily news for Stock Market Prediction, 1.88 Million US Wildfires, SMS Spam Collection Dataset, Twitter User Gender Classification, Brest Cancer Wisconsin Data Set, Retail Data Analytics, US Dept. of Education: College Scoreboard, Death in the United States, US Mass Shootings, Adult Census income, Fatal Police Shootings, Exercise Pattern Prediction, Netflix Prize Data, Pima Indians Diabetes Database, Job Posts, Student Survey, FiveThirtyEight, S & P 500 stock Data, Zika Virus epidemic, Student Alcohol Consumption, Education Statistics, Storm Prediction center.
2. http://snap.standford.edu—Facebook, Twitter, Wiki and bitcoin data set, Social networking APIs
3. https://docs.google.com/forms/d/1l57Un32YH6SkltntirUeLVpgfn33BfJuFLcYupg43oE/viewform?edit_requested=true—online questionnaire from students across 12 campuses in the world
4. http://archive.ics.uci.edu/ml/index.php—Iris, Car Evaluation, Heart disease data set, Bank Marketing Data set,
5. https://aws.amazon.com/datasets/—Enron Email Data, Japan Census data, 1000 Genomics Project,
6. https://cloud.google.com/bigquery/public-data/—We are experimenting it using BigQuery in our Sandbox environment and will publish results in future.
7. https://www.reddit.com/r/bigquery/wiki/datasets
8. https://docs.microsoft.com/en-us/azure/sql-database/sql-database-public-data-sets.

Appendix A.2. Tools

Due to the years of background in databases and data architecture, we selected the Microsoft SQL Server [34] (Business Intelligence, SQL Server Analysis Services, and Data mining) as our data warehouse. Preliminary work is being conducted in Microsoft Azure machine learning tools. We used Microsoft Excel data mining tools [35,36]. Due to our programing background, we used Microsoft C# (mostly for learning in the beginning) and Python and R language for main building of this model, and algorithms. There are various popular and useful Python data analysis and scientific libraries (https://wiki.python.org/moin/NumericAndScientific) such as Pandas, Numpy, SciPy (https://www.scipy.org/), Matplotlib, scikit-learn, Statsmodels, ScientificPython, Fuel, SKdata, Fuel, MILK, etc. For R language (https://cran.r-project.org/), there are various libraries such as gbm, KlaR, tree, RWeka, ipred, CORELearn, MICE Package, rpart, PARTY, CARET, randomForest. We used some of them as they were relevant to our work and we are in the process of learning, experimenting and using more of them for future work. We also used GraphPad Prism (https://www.graphpad.com/scientific-software/prism/) to produce simulated results. Some of the python and R packages used are the following: FSelector-package, sklearn.feature_extraction, sklearn.decomposition, from sklearn.ensemble, nsprcomp R package, R (RFE), R (varSelRF), R (Boruta package), calc.relaimpo (Relative important) (r), earth package, Step-wise Regression, Weight of Evidence (WOE).

References

1. Guyon, I.; Elisseeff, A. An Introduction to Variable and Feature Selection. *J. Mach. Learn. Res.* **2003**, *3*, 1157–1182.
2. Globerson, A.; Tishby, N. Sufficient Dimensionality Reduction. *J. Mach. Learn. Res.* **2003**, *3*, 1307–1331.
3. Li, J.; Cheng, K.; Wang, S.; Morstatter, F.; Trevino, R.P.; Tang, J.; Liu, H. Feature Selection: A Data Perspective. *ACM Comput. Surv.* **2017**, *50*, 94. [CrossRef]
4. Dayan, P. Unsupervised learning. In *The Elements of Statistical Learning*; Springer: New York, NY, USA, 2009; pp. 1–7.
5. Tuia, D.; Volpi, M.; Copa, L.; Kanevski, M.; Munoz-Mari, J. A Survey of Active Learning Algorithms for Supervised Remote Sensing Image Classification. *IEEE J. Sel. Top. Signal Process.* **2011**, *5*, 606–617. [CrossRef]
6. Chai, K.; Hn, H.T.; Cheiu, H.L. Naive-Bayes Classification Algorithm. Bayesian online classifiers for text classification and Filterin. In Proceedings of the 25th Annual International ACM SI GIR Conference on Research and Development in Information Retrieval, Tampere, Finland, 11–15 August 2002; pp. 97–104.
7. Kaggle. *Feature Engineering*; Kaggle: San Francisco, CA, USA, 2010; pp. 1–11.
8. Lin, C. *Optimization and Machine Learning*; MIT Press: Cambridge, MA, USA, 2013.
9. Armstrong, H. *Machines that Learn in the Wild*; NESTA: London, UK, 2015; pp. 1–18.

10. Shalev-Shwartz, S.; Ben-David, S. *Understanding Machine Learning*; Cambridge University Press: Cambridge, UK, 2014.
11. Liu, H.; Motoda, H. *Feature Selection for Knowledge Discovery and Data Mining*; Springer Science & Business Media: Berlin, Germany, 1998.
12. Forman, G. Feature Selection for Text Classification. *Comput. Methods Feature Sel.* **2007**, *16*, 257–274.
13. Nixon, M.S.; Aguado, A.S. *Feature Extraction & Image Processing for Computer Vision*; Academic Press: Cambridge, MA, USA, 2012.
14. Vergara, J.R.; Estévez, P.A. A Review of Feature Selection Methods Based on Mutual Information. *Neural Comput. Appl.* **2015**, *24*, 175–186. [CrossRef]
15. Mohsenzadeh, Y.; Sheikhzadeh, H.; Reza, A.M.; Bathaee, N.; Kalayeh, M.M. The relevance sample-feature machine: A sparse bayesian learning approach to joint feature-sample selection. *IEEE Trans. Cybern.* **2013**, *43*, 2241–2254. [CrossRef] [PubMed]
16. Ma, X.; Wang, H.; Xue, B.; Zhou, M.; Ji, B.; Li, Y. Depth-based human fall detection via shape features and improved extreme learning machine. *IEEE J. Biomed. Health Inform.* **2014**, *18*, 1915–1922. [CrossRef] [PubMed]
17. Lam, D.; Wunsch, D. Unsupervised Feature Learning Classification with Radial Basis Function Extreme Learning Machine Using Graphic Processors. *IEEE Trans. Cybern.* **2017**, *47*, 224–231. [CrossRef] [PubMed]
18. Han, Z.; Liu, Z.; Han, J.; Vong, C.M.; Bu, S.; Li, X. Unsupervised 3D Local Feature Learning by Circle Convolutional Restricted Boltzmann Machine. *IEEE Trans. Image Process.* **2016**, *25*, 5331–5344. [CrossRef] [PubMed]
19. Zeng, Y.; Xu, X.; Shen, D.; Fang, Y.; Xiao, Z. Traffic Sign Recognition Using Kernel Extreme Learning Machines With Deep Perceptual Features. *IEEE Trans. Intell. Transp. Syst.* **2016**, *18*, 1–7. [CrossRef]
20. Wang, Y.; Chattaraman, V.; Kim, H.; Deshpande, G. Predicting Purchase Decisions Based on Spatio-Temporal Functional MRI Features Using Machine Learning. *IEEE Trans. Auton. Ment. Dev.* **2015**, *7*, 248–255. [CrossRef]
21. Lara, O.D.; Labrador, M.A. A Survey on Human Activity Recognition using Wearable Sensors. *IEEE Commun. Surv. Tutor.* **2013**, *15*, 1192–1209. [CrossRef]
22. Zhang, K.; Guliani, A.; Ogrenci-Memik, S.; Memik, G.; Yoshii, K.; Sankaran, R.; Beckman, P. Machine Learning-Based Temperature Prediction for Runtime Thermal Management Across System Components. *IEEE Trans. Parallel Distrib. Syst.* **2018**, *29*, 405–419. [CrossRef]
23. Wang, J.; Wang, G.; Zhou, M. Bimodal Vein Data Mining via Cross-Selected-Domain Knowledge Transfer. *IEEE Trans. Inf. Forensics Secur.* **2018**, *13*, 733–744. [CrossRef]
24. Liu, M.; Xu, C.; Luo, Y.; Xua, C.; Wen, Y.; Tao, D. Cost-Sensitive Feature Selection by Optimizing F-measures. *IEEE Trans. Image Process.* **2017**, *27*, 1323–1335. [CrossRef]
25. Abbas, A.; Member, S.; Siddiqui, I.F.; Uk, S.; Lee, J.I.N. Multi-Objective Optimum Solutions for IoT-Based Feature Models of Software Product Line. *IEEE Access* **2018**, in press. [CrossRef]
26. Haller, P.; Miller, H. Parallelizing Machine Learning-Functionally. In Proceedings of the 2nd Annual Scala Workshop, Stanford, CA, USA, 2 June 2011.
27. Srivastava, A.; Han, E.-H.S.; Singh, V.; Kumar, V. Parallel formulations of decision-tree classification algorithms. In Proceedings of the 1998 International Conference Parallel Process, Las Vegas, NV, USA, 15–17 June 1998; pp. 1–24.
28. Batiz-Benet, J.; Slack, Q.; Sparks, M.; Yahya, A. Parallelizing Machine Learning Algorithms. In Proceedings of the 24th ACM Symposium on Parallelism in Algorithms and Architectures, Pittsburgh, PA, USA, 25–27 June 2012.
29. Pan, X.; Sciences, C. Parallel Machine Learning Using Concurrency Control. Ph.D. Thesis, University of California, Berkeley, CA, USA, 2017.
30. Siddique, K.; Akhtar, Z.; Lee, H.; Kim, W.; Kim, Y. Toward Bulk Synchronous Parallel-Based Machine Learning Techniques for Anomaly Detection in High-Speed Big Data Networks. *Symmetry* **2017**, *9*, 197. [CrossRef]
31. Kirk, M. *Thoughtful Machine Learning*; O'Reilly Media: Newton, MA, USA, 2015.
32. Kubat, M. *An Introduction to Machine Learning*; Springer: Berlin, Germany, 2015.
33. Uddin, M.F.; Lee, J. Proposing stochastic probability-based math model and algorithms utilizing social networking and academic data for good fit students prediction. *Soc. Netw. Anal. Min.* **2017**, *7*, 29. [CrossRef]
34. Tang, Z.; Maclennan, J. *Data Mining With SQL Server 2005*; John Wiley & Sons: Hoboken, NJ, USA, 2005.

35. Linoff, G.S. *Data Analysis Using SQL and Excel*; John Wiley & Sons: Hoboken, NJ, USA, 2008.
36. Fouché, G.; Langit, L. Data Mining with Excel. In *Foundations of SQL Server 2008 R2 Business Intelligence*; Apress: Berkeley, CA, USA, 2011; pp. 301–328.

![applied sciences logo] **applied sciences**

MDPI

Article

Uncertainty Flow Facilitates Zero-Shot Multi-Label Learning in Affective Facial Analysis

Wenjun Bai, Changqin Quan * and Zhiwei Luo

School of System Informatics, Kobe University, 1-1, Rokkodai-cho, Nada-ku, Kobe 657-8501, Japan;
zokbwj@gmail.com (W.B.); luo@gold.kobe-u.ac.jp (Z.L.)
* Correspondence: quanchqin@gold.kobe-u.ac.jp

Received: 21 December 2017; Accepted: 13 February 2018; Published: 19 February 2018

Featured Application: The proposed Uncertainty Flow framework may benefit the facial analysis with its promised elevation in discriminability in multi-label affective classification tasks. Moreover, this framework also allows the efficient model training and between tasks knowledge transfer. The applications that rely heavily on continuous prediction on emotional valance, e.g., to monitor prisoners' emotional stability in jail, can be directly benefited from our framework.

Abstract: To lower the single-label dependency on affective facial analysis, it urges the fruition of multi-label affective learning. The impediment to practical implementation of existing multi-label algorithms pertains to scarcity of scalable multi-label training datasets. To resolve this, an inductive transfer learning based framework, i.e., **Uncertainty Flow**, is put forward in this research to allow knowledge transfer from a single labelled emotion recognition task to a multi-label affective recognition task. I.e., the model uncertainty—which can be quantified in **Uncertainty Flow**—is distilled from a single-label learning task. The distilled model uncertainty ensures the later efficient zero-shot multi-label affective learning. On the theoretical perspective, within our proposed **Uncertainty Flow** framework, the feasibility of applying weakly informative priors, e.g., uniform and Cauchy prior, is fully explored in this research. More importantly, based on the derived weight uncertainty, three sets of prediction related uncertainty indexes, i.e., **soft-max uncertainty**, **pure uncertainty** and **uncertainty plus** are proposed to produce reliable and accurate multi-label predictions. Validated on our manual annotated evaluation dataset, i.e., the multi-label annotated FER2013, our proposed **Uncertainty Flow** in multi-label facial expression analysis exhibited superiority to conventional multi-label learning algorithms and multi-label compatible neural networks. The success of our proposed **Uncertainty Flow** provides a glimpse of future in continuous, uncertain, and multi-label affective computing.

Keywords: affective computing; Bayesian neural network; Multiple Label Learning; transfer learning

1. Introduction

1.1. Challenges in Affective Facial Analysis

Affective facial analysis, which is assessed as one of most primitive functions in vivo, has yet to be successfully implemented in machine. Previous attempts in accomplishing this goal focused on improving the accuracy of emotion classification tasks. Less attention was paid to reveal the uniqueness of affective classification. The uniqueness is on the intrinsic ambiguity of emotion per se. The same facial expression may be interpreted differently dependent upon its associated contexts, spatial and temporal cues [1,2]. Hence, affective classification, in a nutshell, is ambiguous [3]. The past effort in resolving this ambiguity has been reflected in lowering the single-label dependency in producing emotion categories [4].

To further lowering the single-label dependency, one stream of research aims in 'softening' the label space in production of soft labels, allowing affective prediction along the continuous axis. I.e., it allows the relaxation of a discrete label into a partial continuous one. Like in Bai et al. [5], the pseudo soft labels can be crafted by a continuous approximation to the original labels. However, this relaxation trick merely provides a provisional resolution in tackling the ambiguity (cf. [5]).

Instead of the foregoing proximal solution (hard label relaxation), here, we suggest a distal approach: extending the single-label discrimination to the multi-label domain. The research on multi-label affective discrimination is also in line with the finding that the decision boundaries among classes are less ostentatious in affective analysis compare to other categorisation problems, e.g., object classification [2,3]. Benefited from previous researches on multi-label classification in general, it appears straightforward to extend affective computing along this direction. However, there is one difficulty that hinders the success application of multi-label affective recognition: it is laborious and expensive to collect the multi-label training data [6].

1.2. Uncertainty Flow in Zero-Shot Multi-Label Learning

In combating with the scarcity of multi-label training data, unlike conventional approaches, we resort on inductive transfer learning [7] that allows the knowledge to be distilled from a source task, i.e., a single-label affective learning task, and to be applied on a similar but more complex target task, i.e., a multi-label affective discrimination task. But instead of transferring the mere knowledge, i.e., the model parameters between source and target tasks, we propose the **Uncertainty Flow** framework to transfer the model uncertainty between tasks. The crux of our proposed **Uncertainty Flow** is on the quality of uncertainty quantification. To measure this quantity, instead of non-Bayesian neural networks, Bayesian neural networks are employed in quantification of model uncertainty. Bayesian neural network—a recapitulation of a neural network under the direct probabilistic modelling—replaces the single point estimate of the model parameters with the distribution of the parameter. It allows the production of real probabilistic outputs, i.e., model uncertainty [8]. Contrast with conventional implementations on Bayesian neural networks, we further provide our suggestion on the usage of weakly informative priors, e.g., uniform and Cauchy prior, in perfecting the final production of model uncertainty.

The article is organised as following: we chiefly introduce the proposed **Uncertainty Flow** framework in sketch along with the description of four core components, e.g., Bayesian neural networks (More precisely, two hierarchical Bayesian neural networks); our suggested weakly informative priors; the quantification of model uncertainty; and three prediction related uncertainty indexes, e.g., **soft-max uncertainty**, **pure uncertainty** and **uncertainty plus**. To demonstrate the effectiveness of our proposed **Uncertainty Flow** framework, we then present the results from a large-scale comparative experiment. This large-scale experiment contains three levels of comparisons, i.e., the comparison among models, the comparison among different priors, and the comparison among three uncertainty indexes. The observed pronounced discriminability, i.e., 20 to 30 percent performance enhancement, proved the effectiveness of the proposed **Uncertainty Flow** framework.

This pioneer research should be credited under following contributions: (1) We develop a novel inductive transfer learning [3] based computational framework that allows multi-label affective prediction on single evoked source. (2) Unlike conventional inductive transfer learning, the proposed **Uncertainty Flow** focuses on model uncertainty rather than the mere model weights in knowledge distillation. (3) To obtain the model uncertainty, rather than the conventional used informative priors, the usage on weakly informative priors, e.g., uniform and Cauchy prior has also been proposed. (4) To further improve the discriminability of the **Uncertainty Flow**, two advanced prediction related uncertainty indexes, i.e., **pure uncertainty**, and **uncertainty plus** are also suggested in this research.

2. Related Works

2.1. Previous Works on Affective Learning

Past works on neural network based affective computing have focused on the segmentation of single facial expression into finer sub-components, which can be achieved via the added principal component analysis (PCA) [9] or the complex feature pre-processing engineering, e.g., the introduction of Sobel filters [10]. However, the complex in emotional representation demands affective analysis to move beyond the single label categorisation. The researches on multi-label learning have been divided into two streams: problem transformation and algorithm adaptation, respectively [11]. The former approach allows a multi-label learning problem to degrade to a single-label one. Two widely applied problem transformation algorithms are binary relevance [12] and hierarchical of multi-label classifier, AKA., ML-ARAM [13]. The latter approach directly tackles the multi-label learning via the reconstructed loss function. Within this scope, the representative models are ranging from k-nearest neighbour related ML-KNN [14], to label relevance based multi-label neural networks [15].

Despite of the bulk of researches on multi-label learning in general, their applications on affective computing are rarely documented. To fulfill this research gap, Mower et al. [16,17] proposed a feature-agglomerate extraction method to encompass all appeared distinctive emotions in single prediction. Their approach coincides with the foregoing ML-ARAM model in ensuring the structured multi-label predictions. However, their claimed confidence rating—the computed Euclidean distance between input space and feature hyperplane—is mere an metric to index the importance of a feature. Another study that aimed in applying multi-label learning in affective classification relied on a novel regularisation to further penalise the max margin loss [18]. In spite of their claimed effectiveness in extracting multi-label affective features, the success of their proposed Group LASSO regulariser depended heavily on their manual and recursive feature extraction process.

2.2. Previous Works on Bayesian Neural Networks

Previous efforts in developing Bayesian neural networks need to be mentioned here. Deep neural networks are suffered from their inability in outputting authentic probabilistic output [19]. In literature, the history of probabilistic neural networks can be dated back to the early proposal of using the 'soft-max' function to transform a real-value prediction to a probabilistic one [20,21]. The mathematical role of this added 'soft-max' function is to normalise all real-valued outputs into $[0, 1]$ range. However, this added 'soft-max' function is not sufficient to craft real probabilistic account for each classification prediction [8,22]. Therefore, the production of real probabilistic outputs demands the binding of a neural network with a direct probabilistic model. The resultant model is a Bayesian neural network. However, Bayesian neural networks have long been criticised for their imprecise prior-to-posterior inferences and unreliable posterior samplings in practice. Credits to the recent advance in variational inference, i.e., the achievement in deriving rapid and precise variational method to tackle the issue of intractable posterior inferences, it allows the scalable training of Bayesian neural networks [23].

Although the exhaustive review of Bayesian neural network is out of our scope, we focus on the priors in Bayesian neural networks. In general, a prior can be classified as either informative or non-informative. Despite of fruitful researches on informative priors, e.g., Gaussian and Laplace priors [24], the work on non-informative or weakly informative priors is still at early stage [25]. Early researches on non-informative priors, e.g., Jeffery prior and reference prior, emphasised on pursuing the invariance prosperities of non-informative priors (It is resistible to all types of differentiable transformation of the input variables) [26,27]. However, these non-informative priors are neither applicable in multiple parameter modelling nor asymptotically inconsistent in deriving the posterior [28]. To merge the gap between the informative and non-informative priors, a proposal of using weakly informative priors, i.e., semi-flat priors, has been put up in literature [29]. The practical advantages of weakly informative priors over informative ones have been witnessed in other Bayesian models, e.g., generalised linear model [29].

3. Uncertainty Flow Framework

Sketched in Figure 1, the proposed **Uncertainty Flow** framework is consisted of four components, i.e., a dual Bayesian neural networks, the weakly informative priors, the derived model uncertainty, and prediction related uncertainty indexes. The work pipeline of **Uncertainty Flow** initiates at standard supervised training of a source Bayesian neural network(BNN) with a weakly informative prior, e.g., uniform or Cauchy prior, follows the computation of the weight posterior in preparation of model uncertainty in the source BNN, then this distilled model uncertainty is transferred to a target BNN, which is specialised in outputting multi-label predictions. Finally, three distinctive prediction related uncertainty indexes are introduced in perfecting the final outputs from the target BNN.

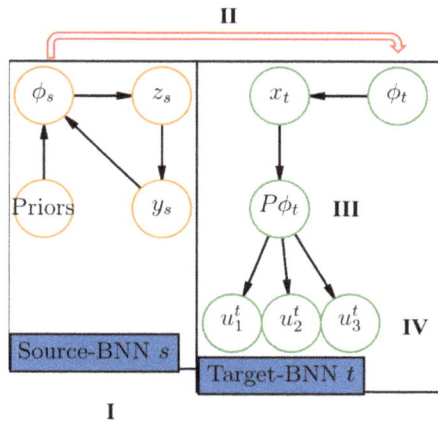

Figure 1. Uncertainty Flow Framework. The graphic model explanation of our proposed **Uncertainty Flow** framework shows four essential elements. I.e., the dual Bayesian neural networks in **I**, e.g., a source and a target BNN(separate by different colours in Figure 1); the weakly informative prior in **II**; the quantification of model uncertainty in **III**; and three proposed uncertainty indexes in perfecting the final multi-label categorisation in **IV**.

3.1. I. Bayesian Neural Network

The core part of proposed **Uncertainty Flow** is the dual BNNs. Under Bayesian learning, a deep neural network—a stacked multiple non-linearity transformations of affine computations—is perceived as sequential layer-wise prior-to-posterior inferences. To allow the model to be flexible enough, we resort on the hierarchical architecture. In a standard classification task set-up, where both input and output variables are observable, i.e., $\{X|x_nY|y_n\}$, and each input is comprised of D features, i.e., $x_n \in \Re^D$, the likelihood functions for our dual BNNs are specified in following Equations (1) to (3):

$$p(y_n|\theta, x_n, \sigma^2) = Categorical(y_n|NN_{train;test}(x_n; \theta), \sigma^2) \tag{1}$$

$$NN(x_n; w)_{source} = softmax \circ \left(tanh \circ \sum_{i=1}^{n}(x \cdot \theta)\right) \tag{2}$$

$$NN(x_n; w)_{target} = sigmoid \circ \left(tanh \circ \sum_{i=1}^{n}(x \cdot \theta)\right) \tag{3}$$

For simplicity, only *tanh* non-linearity is considered as the activation function in BNN. Notice here, we tailer the target BNN in concord with the differentiated task demand. I.e., to allow a Bayesian neural network to produce multiple outputs, the soft-max function in the source BNN is replaced with a real-value function, e.g., sigmoid function, in the target BNN.

Armed with foregoing likelihood functions, a full hierarchical Bayesian neural network is derived from following Equations (4) to (6):

$$\sigma \sim Normal(0, I) \tag{4}$$

$$\theta \sim Normal(w|0, \sigma^2) \tag{5}$$

$$y_n|\theta, \sigma^2 \sim Categorical(y_n|NN_{source;target}(x_n; \theta), \sigma^2) \tag{6}$$

Here, we narrow our discussion in the most simplified version of a hierarchical Bayesian neural network, which contains one hyper-parameter, i.e., σ. This hyper-parameter, e.g., σ directly controls the variance of a prior in production of weight posterior.

3.2. II. Weakly Informative Priors

The prior, which determines the first and second order statistics of model parameters, is de facto the driving force in bayesian learning. Hence, the proper specification of a model ties closely with the choice of an applicable prior for a given task. Unfortunately, the majority works on Bayesian learning pay overwhelmed attention towards the prior that are informative and conjugate for their analytical convenience, the family of uninformative and weakly informative priors had been largely ignored.

Argued in [30], differ than conventional implementation on Bayesian neural networks with the common used informative prior, the usage of weakly informative prior, i.e., a semi-flat prior, yielded superior predicative performance in single-label discrimination. Therefore, it is rational to extend this finding in multi-label learning. The formal definitions of informative and weakly informative priors are rendered below, and their corresponding probabilistic density curves are plotted in Figure 2. Their differentiated effects on a simple simulation is shown in Figure 3.

- Informative Prior

  - Normal Prior

$$\theta_N \sim N(\mu, \sigma^2) \tag{7}$$

$$p(\theta_N) = \frac{1}{\sqrt{2\pi\sigma^2}} exp(-\frac{1}{2\sigma^2}(\theta - \mu)^2) \tag{8}$$

- Weakly Informative Prior

  - Uniform Prior

$$\theta_U \sim U(\alpha, \beta) \tag{9}$$

$$p(\theta_U) = \frac{1}{\beta - \alpha} \tag{10}$$

  - Cauchy Prior

$$\theta_C \sim Cauchy(\alpha, \beta) \tag{11}$$

$$p(\theta_C) = \frac{1}{\pi\beta(1 + (\frac{x-\alpha}{\beta})^2)} \tag{12}$$

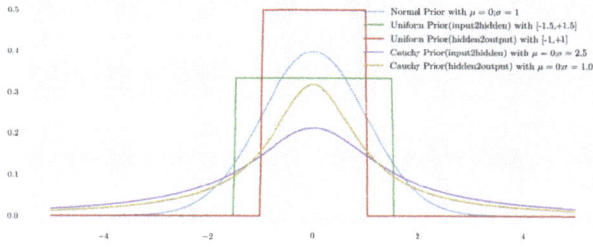

Figure 2. The Probabilistic Density Curves of Normal, Uniform, and Cauchy Priors. This graph demonstrates the probabilistic density curve for each of the prior applied in this research, i.e., Normal, Uniform and Cauchy priors. Here, the further specification of input-to-hidden, hidden-to-hidden priors are defined as the hierarchical shrinkage in variance of the corresponding priors.

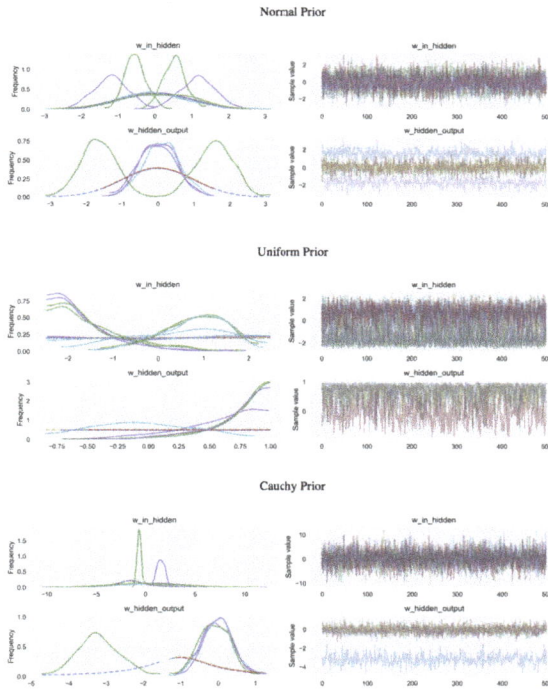

Figure 3. Effects of Different Priors on Posterior in a Simulated Binary Classification. Note, here, the simulated posterior inferences were based on a simple three layer Bayesian neural network with five units in the hidden layer on a binary classification problem. Left panel of Figure 3 reflects how prior, i.e., denoted as red and blue dotted lines, transformed to the weight posterior, i.e., identified as green and light blue lines. The right panel of Figure 3 demonstrates the sampling value of yielded weight posterior, based on 500 posterior samples. Reflected by the separateness of yielded weight posterior, it is clear that Cauchy prior achieved the most discriminability compare to other two priors, e.g., Normal and Uniform priors.

3.3. III. Model Uncertainty

In order to craft model uncertainty, we rely on the quantitative analysis of posterior predictive distribution under Bayesian neural networks. As the production of posterior predicative distribution entails the computation of the intractable parameter posterior, the common way is to approximate it via minimising the KL-divergence [23,31] ($KL(q||p) = \sum_\theta q(\theta)log\frac{q(\theta)}{p(x|\theta)}$) between approximated variational distribution, i.e., $q(\theta)$ and true posterior, i.e., $p(\theta|x)$, therefore solving for optimal posterior, i.e., $q^*(\theta)$, becomes:

$$q^*(\theta) = argmin_{q(\theta) \in Q} KL(q(\theta)||p(\theta|x)) \tag{13}$$

As we cannot compute the KL-divergence directly, the common approach is to resort on optimising an alternative objective, i.e., maximising the ELBO(evidence lower bound), derived as

$$ELBO(q^*) = \mathbb{E}[logp(x|\theta)] - KL(q(\theta)||p(\theta)) \tag{14}$$

Whereas *ELBO* can be seen as a sum of the expected log likelihood of the data with the negative divergence between the variational variance and the model priors. Then it is customary to use the mean-field variational family to complete the specification of the above-mentioned optimisation. The mean-field variational family for each latent model parameter, i.e., θ, can be defined:

$$q(\theta) = \Pi_{j=1}^m q_j(\theta_j) \tag{15}$$

Hence, finding the intractable posterior degrades to a coordinate ascent optimisation in obtaining the optimal θ^* in maximising *ELBO* (cf. Algorithm 1 in [23] for detailed review). The learned optimal parameter posterior, e.g., $(\theta_s|x)$ serves as a surrogate to be used in the parameter posterior in target BNN, i.e., $(\theta_t|x)(\theta_s|x)$. The model uncertainty, which is distilled from the source task, is now flowed to the target task.

3.4. IV. Prediction Related Uncertainty Indexes

With the flowed parameter posterior in the target task, i.e., $(\theta_t|x)$, it is feasible to form the predictive posterior distribution for each upcoming novel observation, i.e., x_{new}, where,

$$p(x_{new}|x) = \int_{\theta^*} p(x_{new}|\theta^*)p(\theta|x)d\theta \tag{16}$$

Armed with this predictive posterior distribution, it allows the production of prediction related uncertainty indexes. As lengthy discussion in previous literature [20,32], one overwhelming claim insists that the probabilistic outputs can be produced by the soft-max function in (It is often placed in the final layer of neural networks to allow the real-valued prediction to be 'pushed' in presenting the pseudo-probabilisitic output.) permitting the averaging over the repetitive forward propagations of new observation in either Bayesian neural network or non-Bayesian neural networks. This type of probabilistic output merely tells the most probable output given the input, not how uncertain is the prediction. For the comparative purpose, we refer this type of uncertainty index as **soft-max uncertainty**. The quantification of this **soft-max uncertainty** has been previously approximated via averaged T times of forward model(input) propagation [33], expressed in Equations (17) and (18):

$$Soft - Max = \mathbb{E}(y*) \approx \frac{1}{T}\sum_{t=1}^T (y*(x_{new}))|_{p(\theta|x_{new})} \tag{17}$$

$$= Class - Type|_{Non-Bayesian} \tag{18}$$

From Bayesian learning perspective, the above-mentioned **softmax uncertainty** reflects the belief of applying predictive mean in indexing the prediction uncertainty. Numerically, this type of uncertainty index captures the mere classification type in multiple-object discrimination, is equivalent with the class type probability in non-Bayesian neural networks. However, as the predictive mean does not capture the full picture of parameter posterior distribution, we draw our attention towards the predictive variance instead.

We argue that the yielded predictive variance reflects the degree of uncertainty that is associated with each prediction. As a result, based on the approximated weight posterior, i.e., $p(\theta|x)$, a better prediction related uncertainty index is expressed below in Equation (19):

$$Pure = Var[y*] \approx Var\Big[\sum_{t=1}^{T}\int_{\Theta} p(x_{new})p(\theta|x)d\theta\Big] \tag{19}$$

We denote this measure of prediction uncertainty index as **pure uncertainty**.

One step further, rather than the dichotomous uncertainty indexes, e.g., **pure uncertainty** and **soft-max uncertainty**, these two indexes can be fused together, which allows the uncertainty index to reflect both class-type probabilistic prediction and the model uncertainty associated with each prediction. In consistent with the previous naming tradition, this type of uncertainty index is marked as **uncertainty plus**. The production of this **uncertainty plus** shows that each class type probabilistic prediction should be proportionately adjusted according to its associated prediction uncertainty, expressed in Equation (20):

$$Uncert+ \approx \frac{\frac{1}{T}\sum_{t=1}^{T}(\hat{y}*(x_{new}))|_{p(\theta|x_{new})}}{Var\Big[\sum_{t=1}^{T}\int_{\Theta} p(x_{new})p(\theta|x)d\theta\Big]} \tag{20}$$

For illustrative purposes, how each of three above mentioned uncertainty indexes, e.g., **soft-max uncertainty**, **pure uncertainty**, and **uncertainty plus**, influences on a simple binary classifier, is demonstrated in Figure 4.

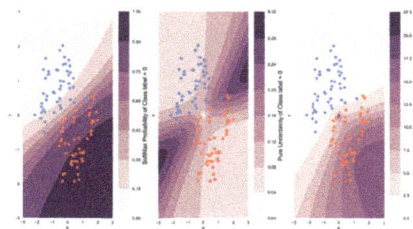

Figure 4. Comparison of Three Prediction Related Uncertainty Indexes on a Binary Classifier. In this figure, three different means of crafting uncertainty boundary for obtaining classification prediction on a simple binary classification problem, i.e., two classes are separated by blue and red, is delineated.

4. Experiment

Relying on the transferred model uncertainty, the proposed **Uncertainty Flow** framework allows a learner to output multi-label predictions under single-label training curriculum. Empirical validation of our proposed framework contains two enquiries that need to be addressed. I.e., one is to investigate that whether or not our proposed **Uncertainty Flow** is superior to conventional multi-label learning algorithms in facilitating the zero-shot multi-label learning; whereas the other is to see which uncertainty index leads to the most significant performance elevation. Also, in order to investigate the role of suggested weakly informative priors, we additional specify our **Uncertainty Flow** into three types of priors. Hence, in total, there are three-level comparisons in our experiment, i.e., *model comparison*; *prior comparison*; and *uncertainty comparison*. The entire experiment is written in Python, using Theano [34], and Pymc3 [35] libraries. The partial code to produce this study is available from: https://github.com/LeonBai/Uncertainty-Flow.

4.1. Dataset

4.1.1. Training Dataset

We selected the first 1500 images from FER2013 [36] as our training dataset. The reason for intentional lowered size of training dataset is to enforce the similar model complexity between a source BNN and a target BNN. (cf. Figure 1). We leave the relaxation of such restriction to future research. FER2013 is a well researched public dataset, which is comprised of facial expression images for pictorial sentiment discrimination. Prior to our implementation, all images in this truncated version of FER2013 had gone through the standard preprocessing process, e.g., fixate the faces at centre, standardise the image size to 48 by 48 pixels in resolution, and all faces are properly registered. We then normalised the pixel values of input images. The original FER2013 images are labelled as one of seven emotion categories, e.g., angry, disgust, fear, happy, sad, surprise, and neutral.

4.1.2. Testing Dataset

To allow the evaluation of outputted multi-label predictions, it is imperative to rely on some existing benchmark annotations. Unfortunately, there is no current reliable multi-label annotations for FER2013 facial expressions. For this, we conducted a small-scale, i.e., 200 images, experiment on manual annotating the multi-label version of FER2013. The descriptive statistics of this annotated multi-label testing dataset is summarised in Appendix A, and the raw data can be found on https://github.com/LeonBai/Uncertainty-Flow. Preliminary statistic test revealed the high similarity between the original single-label and yielded multi-label annotations. I.e., treating the original single-label FER2013 annotations as ground truth, the overlaps between multi-label annotations and ground truth reached 75%, suggesting high similarity between two annotations. Indicated by a acceptable Fleiss-Kappa coefficient value [37], i.e., 0.25 (between −1 to 1, higher is more reliable)—the measurement of inter-rater reliability—the annotated multi-label version of FER2013 can be served as our testing dataset.

4.2. Models

To conduct an experiment that contains above-mentioned three-level comparisons, i.e., *model comparison*, *prior comparison*, *uncertainty comparison*, it demands explicit specification of all models in current experiment. In *model comparison*, four widely used multi-label learning algorithms, ranging from adaption algorithms, e.g., Multi-Label K-means Nearest Neighbour (MLkNN), Multi-label Neurofuzzy Classifier(ML-ARAM), to problem transformation algorithms, e.g., Binary Relevance (BR) and Label Powerset (LP), are included. In addition, two multi-label compatible neural networks. i.e., a multi-label feedforward Neural network (ML-FNN) and a multi-label convolutional neural network (ML-CNN), are also included in *model comparison* comparison.

In *prior comparison* and *uncertainty comparison*, depending on the prior type, i.e., informative or weakly informative, and different prediction related uncertainty indexes, e.g., **soft-max uncertainty**, **pure uncertainty**, **uncertainty plus**, the **Uncertainty Flow** generates 9 variants, denoting as BNN-normal-soft; BNN-normal-pure; BNN-normal-plus; BNN-uniform-soft; BNN-uniform-pure; BNN-uniform-plus; BNN-cauchy-soft; BNN-cauchy-pure; BNN-cauchy-plus. To further elevate the discriminative performance in multi-label prediction, we additional frame a convolutional neural network under Bayesian learning, producing Bayesian convolutional neural network within the proposed **Uncertainty Flow** framework, with its associated 9 variants, i.e., BCNN-normal-soft; BCNN-normal-pure; BCNN-normal-plus; BCNN-uniform-soft; BCNN-uniform-pure; BCNN-uniform-plus; BCNN-cauchy-soft; BCNN-cauchy-pure; BCNN-cauchy-plus. The configurations of above-mentioned models are summarised in Appendix B.

4.3. Evaluation Metrics

Different than the uniformed metric that used in single-label classification, i.e., classification accuracy, diversified evaluation metrics have been proposed. In line with the rouge classification from [11], we adhere the dichotomy classification of the evaluation metrics as bipartition and ranking based. For illustrative purposes, assuming a multi-label evaluation dataset is consist of input, i.e., x_i, and the set of true labels, i.e., Y_i, where $i = 1, ..., m$ and $Y_i \subseteq L$, L is the set of all correct labels. Under this notation, the set of predicated labels are denoted as Z_i, where $i = 1, ..., m$, while the rank predicted by learning method for a label λ is denoted as $r_i(\lambda)$. The most relevant label, receives the highest rank (1), while the least relevant one, receives the lowest rank (q).

4.3.1. Bipartition Based

Delegated from the single-label metric, bipartition based metrics are proposed to capture the differences between actual and predicted sets of labels over all evaluation dataset. These differences can be computed in various means via either averaged over all samples or all label sets.

1. Hamming loss

$$Hamming - Loss = \frac{1}{m} \sum_{i=1}^{m_i} \frac{|Y_i \triangle Z_i|}{M} \tag{21}$$

Where $\triangle$ represents the symmetric difference of two sets, i.e., predicted and true label sets. Contrast with other over-strict measures of multi-label classification accuracy, i.e., low tolerance on partial label misclassification, e.g., $\frac{1}{m} \sum_{i_1}^{m} I|Y_i = Z_i|$, the hamming loss, which sums up to 1, offers a mild criteria for wider range of measurement application.

2. Micro-Averaged F-Score & Average Precision

Inherited from classic binary evaluation in information retrieval tasks, F-score and average precision, which both reflect their corresponded combinations of averaging over precision and recall, are two readily applicable metrics in multi-label learning. Among various averaging operations, e.g., macro, weighted, and micro, the preferred operation is micro-average as it offers each sample-class pair an equal contribution to the overall metric. Consider a binary evaluation measure t_p, t_n, f_p, f_n that is computed via the number of true positives t_p, true negatives t_n, false positives f_p, false negatives f_n,the *n*th threshold for precision and recall are P_n and R_n, the interested micro-averaged F-score and average precision score(AP) are derived as following:

$$P_n = \frac{t_p}{t_p + f_p} \tag{22}$$

$$R_n = \frac{t_p}{t_p + f_n} \tag{23}$$

$$Micro Averaged(F_\beta) = (1 + \beta^2 \frac{P_n \times R_n}{\beta^2 P_n + R_n}) \tag{24}$$

$$Average Preision(AP) = \sum_n (R_n - R_{n-1}) P_n \tag{25}$$

4.3.2. Ranking Based

1. Converge

To measure the needed distance to cover all true label sets, i.e., Y_i in the predicted label sets, we resort on the converge error metric. It can be defined as following:

$$Coverage - Error = \frac{1}{m} \sum_{i=1}^{m} max_{\lambda \in Y_i} r_i(\lambda) - 1 \tag{26}$$

2. Ranking Loss

The ranking loss targets at the incorrect ordering of the predicted label sets. Presume $\bar{Y}_i$ is expressed as the complementary set of Y_i, its computation can be defined as following:

$$R - Loss = \frac{1}{m} \sum_{i=1}^{m} \frac{1}{|Y_i||\bar{Y}_i|} \{ (\lambda_a, \lambda_b) : r_i(\lambda_a) > r_i(\lambda_b), (\lambda_a, \lambda_b) \in Y_i \times \bar{Y}_i \} \tag{27}$$

4.4. Results & Discussion

The overall result of our conducted large-scale comparative experiment is chiefly presented in Table 1. For illustrative purposes, we grouped the results to highlight the comparison among different models. As we used various of evaluation metrics to assess the performance of corresponding models, it is difficulty to obtain a clear judgement that is based on single metric. I.e., when we pitted our approach, i.e., **Uncertainty Flow** against the MLkNN approach in conventional multi-label models, our approach, including all nine variations, is inferior to the MLkNN approach on the metric of *Hamming-Loss*. However, when we accessed the model according to its performance on *Average Precision*, our approach largely outperformed the MLkNN approach. Moreover, as we incorporated nine variations in **Uncertainty Flow**, the in-depth analyses of the prior types and uncertainty indexes are demanded. We then divided our discussion of the overall result into three parts, e.g., the results on model comparison, the results on prior comparison, and the results on uncertainty comparison.

Table 1. Model Comparison in Various Multi-label Evaluation Metrics.

Candidate Models	Hamming-Loss	Converge-Loss	Ranking-Loss	F-Score	Average Precision	Source
Conventional Multi-Label Models						
MLkNN	0.286	7.000	0.950	0.090	0.346	[14]
ML-ARAM	0.374	6.670	0.801	0.264	0.369	[13]
Binary Relevance	**0.275**	7.00	1.000	0.263	0.340	[38]
Label Powerset	0.328	6.940	0.837	0.215	0.370	[39]
Multi-Label Compatible Neural Networks						
ML-FNN	0.402	6.700	0.761	0.282	0.376	[40]
ML-CNN	0.387	6.600	0.733	0.3108	0.384	[41]
Uncertainty Flow - Bayesian Neural Networks						
BNN-normal-soft	0.404	6.500	0.75	0.279	0.360	This research
BNN-normal-pure	0.382	6.500	**0.723**	0.318	0.389	This research
BNN-normal-plus	0.402	6.673	0.759	0.282	0.525	This research
BNN-uniform-soft	0.414	6.750	0.7816	0.302	0.353	This research
BNN-uniform-pure	0.385	6.450	0.723	0.330	0.389	This research
BNN-uniform-plus	0.404	6.450	0.765	0.310	0.530	This research
BNN-cauchy-soft	0.400	6.7	0.759	0.285	0.378	This research
BNN-cauchy-pure	0.382	6.525	0.727	0.312	0.402	This research
BNN-cauchy-plus	0.401	**6.250**	0.741	0.290	0.527	This research
Uncertainty Flow - Bayesian Convolutional Neural Networks						
BCNN-normal-soft	0.421	6.750	0.791	0.250	0.385	This research
BCNN-normal-pure	0.384	6.700	0.737	0.319	0.449	This research
BCNN-normal-plus	0.400	6.675	0.751	0.288	0.561	This research
BCNN-uniform-soft	0.403	6.675	0.762	0.282	0.421	This research
BCNN-uniform-pure	0.404	6.800	0.769	0.279	0.311	This research
BCNN-uniform-plus	0.387	6.725	0.736	0.310	0.527	This research
BCNN-cauchy-soft	0.401	6.675	0.760	0.285	0.416	This research
BCNN-cauchy-pure	0. 396	6.800	0.753	**0.322**	0.390	This research
BCNN-cauchy-plus	0.401	6.75	0.758	0.285	**0.576**	This research

4.4.1. Model Comparison

Ruling out the factors of prior types and prediction related uncertainty indexes, the empirical comparison between **Uncertainty Flow** framework and the alternatives demonstrated mixed results, illustrated in Figure 5.

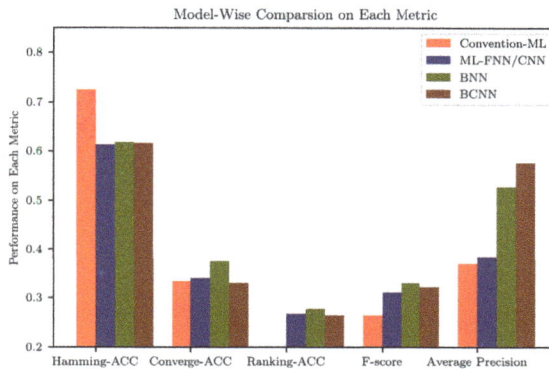

Figure 5. Model Wise Comparison on Multi-learning Metrics. Note here, for illustrative purposes, we used Hamming-ACC, Converge-ACC and Ranking-ACC instead of original loss based metrics. Rather than the averaging over the models in each category, e.g., Conventional-ML, we picked the most representative model in each category for different metric.

Previous findings from [11] and [6] stated that it is practical difficult to observe a single model or algorithm, which is competitive enough to beat others in every multi-label evaluation metric. Hence, it is imperative to investigate each loss metric independently. Focusing on *Hamming-Loss* (Hamming-ACC = 1 − Hamming loss), interestingly, the conventional multi-label learning models are particular good in minimising this type of loss. However, indicated in *Converge-Loss* and *Ranking-Loss*, both **Uncertainty Flow** and multi-label compatible neural networks, e.g., ML-FNN and ML-CNN are superior than the conventional multi-label learning alternatives.

Under two precision related metrics, e.g., *F-score* and *average precision*, with the help from a weakly informative prior, e.g., uniform or Cauchy prior, and an advanced prediction related uncertainty indexes, e.g., **pure uncertainty**, or **uncertainty plus**, both BNN and BCNN, exhibited clear performance advantage over their alternatives, e.g., ML-FNN, ML-CNN and conventional multi-label models. Especially on the metric of *average Precision*, the nontrivial performance enhancement, i.e., over 20% accuracy increase, demonstrated the superior discriminability that tags to our proposed **Uncertainty Flow**. Moreover, comparing the performance between BNN and BCNN, the convolutional architecture, e.g., BCNN, should be credited for overall performance improvement.

4.4.2. Prior Comparison

To verify the most applicable prior in our proposed **Uncertainty Flow**, the prior comparison among three candidate priors is worthy to be fully investigated. Shown in Figure 6b, despite some similarities in shapes, it is clear that each prior has its unique effect in shaping the corresponded posterior distribution of the weights. In specific, the effect of uniform prior on posterior weight is seen as the restriction on the approximated posterior weights, i.e., the posterior weights have to be higher than a fixed value, e.g., 1 in our implementation. This restriction effect may lower the discriminability of the uniform prior imposed model, shown in Figure 7. Interestingly, the posterior distribution of weights from imposed normal and Cauchy priors respectively rendered nearly identical distribution shape, shown in Figure 6a,c. The minute difference between these two is the enlarged variance for Cauchy prior induced posterior distribution of weights. Despite seemingly trivial, this difference in variance lead to the discrepancy in discriminative performance, shown in Figure 7. Overall, based on final induced discriminability, a Cauchy prior is considered as the most applicable prior in our proposed **Uncertainty Flow**.

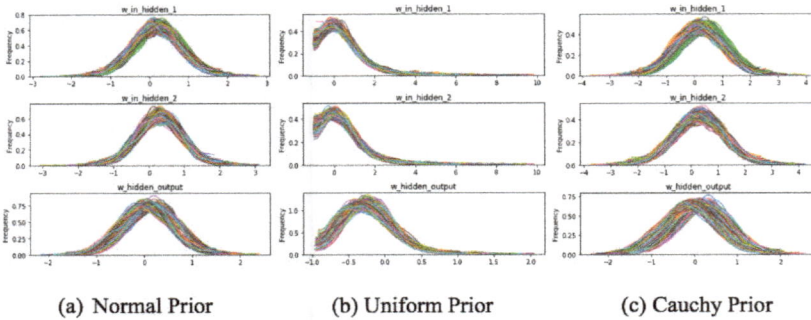

(a) Normal Prior   (b) Uniform Prior   (c) Cauchy Prior

Figure 6. Different Priors on Posterior Weights Under **Uncertainty Flow**. Note here, the notation $w_i n_h idden_1$ means posterior weights in first hidden layer in our implemented BNN or BCNN.

The performance enhancement that can be reflected by above-mentioned 'clustered' effect in weight posterior was observed in examination of the discriminability of three implemented prior. Plotted in Figure 7, focusing on the *average precision* evaluation metric, regardless of the variations in Bayesian neural networks, i.e., BNN or BCNN, the employment of Cauchy prior—as one kind of weakly informative prior—leaded competitive multi-label affective classification. However, as another implemented weakly informative prior, uniform prior was inferior to the used informative prior, e.g., normal prior. This observed attenuation in discriminability from uniform prior may due to the its above-mentioned spike-and-slab effect on weight posterior that requires extra training epochs to stabilise the pre-to-posterior inference.

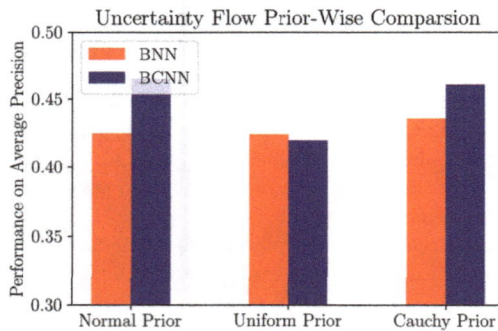

Figure 7. Prior Induced Discriminability Differences in **Uncertainty Flow**. Note here, we rule out the impacts of uncertainty indexes via averaging the performance of models with same implemented prior. Average precision is chosen as other metrics failed to render clear discriminability comparison.

4.4.3. Uncertainty Comparison

Undoubtedly, the most pronounced performance improvement is pertaining to the inclusion of advanced prediction related uncertainty indexes, e.g., **pure uncertainty** and **uncertainty plus**. To recall the foregoing definition of prediction related uncertainty indexes, the **soft-max uncertainty** is a mere indication of multi-class prediction type, which is equivalent with the predictions in non-Bayesian alternatives. The pure uncertainty, i.e., on the contrary—depends heavily on weight posterior—can be produced exclusively in our proposed **Uncertainty Flow** framework. Reflected in Figure 8, when the feedforward architecture was adopted, **soft-max uncertainty** is inferior to **pure**

uncertainty in producing multi-label prediction. Interestingly, when the convolutional architecture was chosen, it uncovered a different story, i.e., the discriminability from **pure uncertainty** became inferior to **soft-max uncertainty**.

Not surprisingly, the combination of **soft-max uncertainty** and **pure uncertainty**, i.e., the craft of **uncertainty plus**, allows a set of multi-label predictions to be tuned based on its uncertainty value. Shown in Figure 8, it is clear that the crafted predictions that are benefited from **Uncertainty plus** are superior to other two uncertainty indexes. I.e., its introduced improvement in *average precision* is over 20% compare to other two indexes. Combining the most applicable weakly informative prior and the advanced uncertainty indexe together, the two most efficient variants in feedforward and convolutional architectures are **BNN-cauchy-plus** and **BCNN-cauchy-plus**, respectively. We leave sensitivity analysis of our proposed advanced uncertainty indexes to future research.

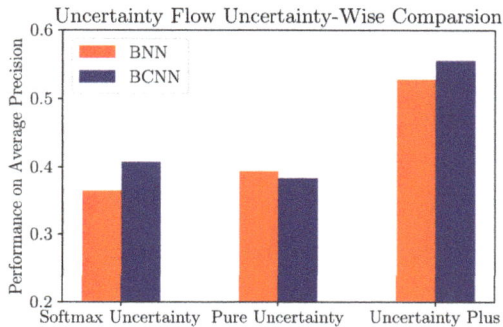

Figure 8. Discriminative Performance Across Different Prediction Related Uncertainty Indexes. Note here, the effects of priors were marginalised via prior-wise averaging.

5. Conclusions

Over reliance on single-label affective learning hinders the fruition of the automatic affective analysis. To free from this restriction, we resort on multi-label affective learning. However, current multi-learning algorithms are not scalable enough due to the scarcity of multi-label training samples. To tackle this issue, we propose a inductive transfer learning based framework, i.e., **Uncertainty Flow**. Under this pioneer framework, we argue that the model uncertainty can be distilled from a source single-label recognition task. The distilled knowledge is then fed to a to-be-learned multi-label affective recognition task. For predictions, three types of uncertainty indexes, i.e., **soft-max uncertainty**, **pure uncertainty**, and **uncertainty plus**, are further proposed. For empirical validation, the authors conducted a large-scale comparative experiment on the manual annotated multi-label FER2013 dataset across three levels of comparisons, i.e., *model comparison*, *prior comparison*, and *uncertainty comparison*. The observed performance superiority in **Uncertainty Flow** unequivocally renders the feasibility of applying this framework in zero-shot multi label affective learning.

However, even under the permitted computational resources, to run a full Bayesian posterior remains as a daunting task. How to speed up the posterior inference is an open research question. In terms of future researches, there are two streams of researches that are worthy to be further explored. One focuses on improving the discriminability of our novel proposed **Uncertainty Flow** framework. This entails the revision on the mainstream mean field based variational inference [23]. The other is to extend the current inductive transfer based framework to the transductive transfer domain [7], where has already been demonstrated in vivo [42].

Acknowledgments: This study is partially supported by the Okawa Foundation for Information and Telecommunications, and National Natural Science Foundation of China under Grant No. 61472117. We gracefully appreciate Sheng Cao, Dong Dong, Kawashima Koya, and Fujita Tomo for their helps in annotating the multi-label version of FER2013 testing images.

Author Contributions: W.B. and C.Q. conceived and designed the computational model; W.B. performed the experiments; W.B. and C.Q. analyzed the data; W.B. contributed analysis tools; W.B. wrote the paper; Z.-W.L. and C.Q. revised the paper.

Conflicts of Interest: The authors declare no conflict of interest. The founding sponsors had no role in the design of the study; in the collection, analyses, or interpretation of data; in the writing of the manuscript, and in the decision to publish the results.

Abbreviations

The following abbreviations are used in this manuscript:

BNN	Bayesian neural network
BCNN	Bayesian convolutional neural network
ML-kNN	Multi-Label adapted kNN(k Nearest Neighbour) classifier
ML-ARAM	Multi-Label fuzzy Adaptive Resonance Associative Map
ML-FNN	Multi-Label Compatible Feedforward Neural Network
ML-CNN	Multi-Label Compatible Convolutional Neural Network

Appendix A. Descriptive Statistics on Annotated FER2013 dataset

Table A1. Descriptive Statistics of Single-Label and Multi-Label FER2013 Datasets.

Name	# of Instances	# of Labels	Cardinality [11]	Source
Training	1500	7	1.0	[36]
Testing	200	7	1.89	This Research

Appendix B. Model Configurations

1. ML-kNN

   The number of k mixture components was set up to 4, and the default smoothing parameter was tuned at 0.

2. ML-ARAM

   The vigilance was set to 0.9 to reflect the high dataset dependence, the threshold was set to 0.02 in line with the original algorithm implementation [13].

3. Binary Relevance

   Base classifier: SVC(support vector classifier).

4. Label Powerset

   Base classifier: Naive Gaussian classifier.

5. ML-FNN

   Layer-wise Architecture:

   Dense (128) > Dropout ($p = 0.2$) > Dense (128) > Dropout ($p = 0.2$) > Dense (Output) (This notation indicates the information pathway from a dense connected layer with 128 units, to the final dense connected layer via intermediate dense connected and dropout layers).

   Epoch: 50 (1500 iterations)

6. ML-CNN

   Layer-Wise Architecture: Convolution (3×3) > Convolution (3×3) > Max Pooling (2×2) > Dropout ($p = 0.2$) > Dense (128) > Dense (Output). Epoch: 50 (1500 iterations)

7. BNN

   Layer-Wise Architecture: Same as *ML-FNN* Priors: Normal/Uniform/Cauchy Inference Method: Variational Mean Field Number of Posterior Sampling: 500.

8. BCNN

   Layer-Wise Architecture: Same as *ML-CNN* Priors: Normal/Uniform/Cauchy Inference Method: Variational Mean Field Number of Posterior Sampling: 500.

References

1. Hassin, R.R.; Aviezer, H.; Bentin, S. Inherently Ambiguous: Facial Expressions of Emotions, in Context. *Emot. Rev.* **2013**, *5*, 60–65.
2. Kumar, B.V. Face expression recognition and analysis: The state of the art. *arXiv* **2009**, arXiv:1203.6722.
3. Sariyanidi, E.; Gunes, H.; Cavallaro, A. Automatic Analysis of Facial Affect: A Survey of Registration, Representation, and Recognition. *IEEE Trans. Pattern Anal. Mach. Intell.* **2015**, *37*, 1113–1133.
4. Barsoum, E.; Zhang, C.; Ferrer, C.C.; Zhang, Z. Training Deep Networks for Facial Expression Recognition with Crowd-Sourced Label Distribution. In Proceedings of the 18th ACM International Conference on Multimodal Interaction, Tokyo, Japan, 12–16 November 2016.
5. Bai, W.; Luo, W. Hard Label Relaxation in Biased Pictorial Sentiment Discrimination. In Proceedings of the Natural Language Processing and Knowledge Engineering, Chengdu, China, 7–10 December 2017.
6. Madjarov, G.; Kocev, D.; Gjorgjevikj, D.; Džeroski, S. An extensive experimental comparison of methods for multi-label learning. *Pattern Recognit.* **2012**, *45*, 3084–3104.
7. Pan, S.J.; Yang, Q. A survey on transfer learning. *IEEE Trans. Knowl. Data Eng.* **2010**, *22*, 1345–1359.
8. Neal, R.M. *Bayesian Learning for Neural Networks*; Springer Science & Business Media: Berlin, Germany, 2012; Volume 118.
9. Perikos, I.; Ziakopoulos, E.; Hatzilygeroudis, I. Recognizing emotions from facial expressions using neural network. In *IFIP International Conference on Artificial Intelligence Applications and Innovations*; Springer: Berlin, Germany, 2014; pp. 236–245.
10. Filko, D.; Martinovic, G. Emotion Recognition System by a Neural Network Based Facial Expression Analysis. *Automatika* **2013**, *54*, 263–272.
11. Tsoumakas, G.; Katakis, I.; Vlahavas, I. Mining multi-label data. In *Data Mining and Knowledge Discovery Handbook*; Springer: Berlin, Germany, 2009; pp. 667–685.
12. Luaces, O.; Díez, J.; Barranquero, J.; del Coz, J.J.; Bahamonde, A. Binary relevance efficacy for multilabel classification. *Prog. Artif. Intell.* **2012**, *1*, 303–313.
13. Benites, F.; Sapozhnikova, E. HARAM: A Hierarchical ARAM neural network for large-scale text classification. In Proceedings of the 2015 IEEE International Conference on the Data Mining Workshop (ICDMW), Atlantic City, NJ, USA, 14–17 November 2015; pp. 847–854.
14. Zhang, M.L.; Zhou, Z.H. ML-KNN: A lazy learning approach to multi-label learning. *Pattern Recognit.* **2007**, *40*, 2038–2048.
15. Zhang, M.L.; Zhou, Z.H. Multilabel neural networks with applications to functional genomics and text categorization. *IEEE Trans. Knowl. Data Eng.* **2006**, *18*, 1338–1351.
16. Mower, E.; Mataric, M.J.; Narayanan, S. A Framework for Automatic Human Emotion Classification Using Emotion Profiles. *IEEE Trans. Audio Speech Lang. Process.* **2011**, *19*, 1057–1070.
17. Mower, E.; Metallinou, A.; Lee, C.C.; Kazemzadeh, A.; Busso, C.; Lee, S.; Narayanan, S. Interpreting ambiguous emotional expressions. In Proceedings of the 3rd International Conference on Affective Computing and Intelligent Interaction and Workshops, Amsterdam, The Netherlands, 10–12 September 2009; pp. 1–8.
18. Zhao, K.; Zhang, H.; Ma, Z.; Song, Y.Z.; Guo, J. Multi-label learning with prior knowledge for facial expression analysis. *Neurocomputing* **2015**, *157*, 280–289.
19. Ghahramani, Z. Probabilistic machine learning and artificial intelligence. *Nature* **2015**, *521*, 452–459.
20. Denker, J.S.; Lecun, Y. Transforming neural-net output levels to probability distributions. In Proceedings of the Advances in Neural Information Processing Systems, Denver, CO, USA, 2–5 December 1991; pp. 853–859.

21. Tishby, N.; Levin, E.; Solla, S.A. Consistent inference of probabilities in layered networks: Predictions and generalization. *Int. Jt. Conf. Neural Netw.* **1989**, *2*, 403–409.
22. Gal, Y.; Ghahramani, Z. Bayesian Convolutional Neural Networks with Bernoulli Approximate Variational Inference. *arXiv* **2015**, arXiv:1506.02158
23. Blei, D.M.; Kucukelbir, A.; McAuliffe, J.D. Variational Inference: A Review for Statisticians. *J. Am. Stat. Assoc.* **2017**, *112*, 859–877,
24. Williams, P.M. Bayesian Regularization and Pruning Using a Laplace Prior. *Neural Comput.* **1995**, *7*, 117–143.
25. Gülçehre, C.; Bengio, Y. Knowledge matters: Importance of prior information for optimization. *J. Mach. Learn. Res.* **2016**, *17*, 1–32.
26. Jeffreys, H. An invariant form for the prior probability in estimation problems. *Proc. R. Soc. Lond. A* **1946**, *186*, 453–461.
27. Bernardo, J.M. Reference posterior distributions for Bayesian inference. *J. R. Stat. Soc. Ser. B* **1979**, *41*, 113–147.
28. Gelman, A. Prior distributions for variance parameters in hierarchical models (comment on article by Browne and Draper). *Bayesian Anal.* **2006**, *1*, 515–534.
29. Gelman, A.; Jakulin, A.; Pittau, M.G.; Su, Y.S. A weakly informative default prior distribution for logistic and other regression models. *Ann. Appl. Stat.* **2008**, *2*, 1360–1383.
30. Bai, W.; Quan, C. Harness the Model Uncertainty via Hierarchical Weakly Informative Priors in Bayesian Neural Network. *Int. Rob. Auto. J.* **2017**, *3*, doi:10.15406/iratj.2017.03.00057.
31. Jordan, M.I.; Ghahramani, Z.; Jaakkola, T.S.; Saul, L.K. An introduction to variational methods for graphical models. *Mach. Learn.* **1999**, *37*, 183–233.
32. Lee, W.T.; Tenorio, M.F. *On Optimal Adaptive Classifier Design Criterion: How Many Hidden Units are Necessary for an Optimal Neural Network Classifier?*; Purdue University, School of Electrical Engineering: West Lafayette, IN, USA, 1991.
33. Srivastava, P.; Hopwood, N. A practical iterative framework for qualitative data analysis. *Int. J. Qual. Methods* **2009**, *8*, 76–84.
34. Bergstra, J.; Breuleux, O.; Bastien, F.; Lamblin, P.; Pascanu, R.; Desjardins, G.; Turian, J.; Warde-Farley, D.; Bengio, Y. Theano: A CPU and GPU math compiler in Python. In Proceedings of the 9th Python in Science Conference, Austin, TX, USA, 28 June–3 July 2010; pp. 1–7.
35. Salvatier, J.; Wiecki, T.V.; Fonnesbeck, C. Probabilistic programming in Python using PyMC3. *PeerJ Comput. Sci.* **2016**, *2*, e55.
36. Goodfellow, I.J.; Erhan, D.; Carrier, P.L.; Courville, A.; Mirza, M.; Hamner, B.; Cukierski, W.; Tang, Y.; Thaler, D.; Lee, D.H. Challenges in representation learning: A report on three machine learning contests. In *International Conference on Neural Information Processing*; Springer: Berlin, Germany, 2013; pp. 117–124.
37. Fleiss, J.L.; Cohen, J. The equivalence of weighted kappa and the intraclass correlation coefficient as measures of reliability. *Educ. Psychol. Meas.* **1973**, *33*, 613–619.
38. Read, J.; Pfahringer, B.; Holmes, G.; Frank, E. Classifier chains for multi-label classification. In *Machine Learning and Knowledge Discovery in Databases*; Springer: Berlin, Germany, 2009; pp. 254–269.
39. Tsoumakas, G.; Vlahavas, I. Random k-labelsets: An ensemble method for multilabel classification. In *Machine learning: ECML 2007*; Springer: Berlin, Germany, 2007; pp. 406–417.
40. Zhang, M.L. ML-RBF: RBF neural networks for multi-label learning. *Neural Process. Lett.* **2009**, *29*, 61–74.
41. Wei, Y.; Xia, W.; Lin, M.; Huang, J.; Ni, B.; Dong, J.; Zhao, Y.; Yan, S. HCP: A flexible CNN framework for multi-label image classification. *IEEE Trans. Pattern Anal. Mach. Intell.* **2016**, *38*, 1901–1907.
42. Nook, E.C.; Lindquist, K.A.; Zaki, J. A new look at emotion perception: Concepts speed and shape facial emotion recognition. *Emotion* **2015**, *15*, 569–578.

MDPI

St. Alban-Anlage 66

4052 Basel

Switzerland

Tel. +41 61 683 77 34

Fax +41 61 302 89 18

www.mdpi.com

Applied Sciences Editorial Office

E-mail: applsci@mdpi.com

www.mdpi.com/journal/applsci

www.ingramcontent.com/pod-product-compliance
Lightning Source LLC
Chambersburg PA
CBHW051727210326
41597CB00032B/5637